PARALLEL SCIENTIFIC COMPUTING AND OPTIMIZATION

Springer Optimization and Its Applications

VOLUME 27

Aims and Scope
Optimization has been expanding in all directions at an astonishing rate during the last few decades. New algorithmic and theoretical techniques have been developed, the diffusion into other disciplines has proceeded at a rapid pace, and our knowledge of all aspects of the field has grown even more profound. At the same time, one of the most striking trends in optimization is the constantly increasing emphasis on the interdisciplinary nature of the field. Optimization has been a basic tool in all areas of applied mathematics, engineering, medicine, economics and other sciences.

The series *Springer Optimization and Its Applications* publishes undergraduate and graduate textbooks, monographs and state-of-the-art expository works that focus on algorithms for solving optimization problems and also study applications involving such problems. Some of the topics covered include nonlinear optimization (convex and nonconvex), network flow problems, stochastic optimization, optimal control, discrete optimization, multi-objective programming, description of software packages, approximation techniques and heuristic approaches.

PARALLEL SCIENTIFIC COMPUTING AND OPTIMIZATION

Advances and Applications

By

RAIMONDAS ČIEGIS
Vilnius Gediminas Technical University, Lithuania

DAVID HENTY
University of Edinburgh, United Kingdom

BO KÅGSTRÖM
Umeå University, Sweden

JULIUS ŽILINSKAS
Vilnius Gediminas Technical University and Institute of Mathematics and Informatics, Lithuania

Springer

Raimondas Čiegis
Department of Mathematical Modelling
Vilnius Gediminas Technical University
Saulėtekio al. 11
LT-10223 Vilnius
Lithuania
rc@fm.vgtu.lt

David Henty
EPCC
The University of Edinburgh
James Clerk Maxwell Building
Mayfield Road
Edinburgh EH9 3JZ
United Kingdom
d.henty@epcc.ed.ac.uk

Bo Kågström
Department of Computing Science and
High Performance Computing
Center North (HPC2N)
Umeå University
SE-901 87 Umeå
Sweden
bokg@cs.umu.se

Julius Žilinskas
Vilnius Gediminas Technical University and
Institute of Mathematics and Informatics
Akademijos 4
LT-08663 Vilnius
Lithuania
julius.zilinskas@ktl.mii.lt

ISSN 1931-6828
ISBN 978-0-387-09706-0 ISBN 978-0-387-09707-7 (eBook)

Library of Congress Control Number: 2008937480

Mathematics Subject Classification (2000): 15-xx, 65-xx, 68Wxx, 90Cxx

Printed on acid-free paper

springer.com

Preface

The book is divided into four parts: Parallel Algorithms for Matrix Computations, Parallel Optimization, Management of Parallel Programming Models and Data, and Parallel Scientific Computing in Industrial Applications.

The first part of the book includes chapters on parallel matrix computations. The chapter by R. Granat, I. Jonsson, and B. Kågström, "RECSY and SCASY Library Software: Recursive Blocked and Parallel Algorithms for Sylvester-type Matrix Equations with Some Applications", gives an overview of state-of-the-art high-performance computing (HPC) algorithms and software for solving various standard and generalized Sylvester-type matrix equations. Computer-aided control system design (CACSD) is a great source of applications for matrix equations including different eigenvalue and subspace problems and for condition estimation. The parallelization is invoked at two levels: *globally* in a distributed memory paradigm, and *locally* on shared memory or multicore nodes as part of the distributed memory environment.

In the chapter by A. Jakušev, R. Čiegis, I. Laukaitytė, and V. Trofimov, "Parallelization of Linear Algebra Algorithms Using ParSol Library of Mathematical Objects", the mathematical objects library ParSol is described and evaluated. It is applied to implement the finite difference scheme to solve numerically a system of PDEs describing a nonlinear interaction of two counter-propagating laser waves.

The chapter by R.L. Muddle, J.W. Boyle, M.D. Mihajlović, and M. Heil, "The Development of an Object-Oriented Parallel Block Preconditioning Framework", is devoted to the analysis of block preconditioners that are applicable to problems that have different types of degree of freedom. The authors discuss the development of an object-oriented parallel block preconditioning framework within OOMPH-LIB, the object-oriented, multi-physics, finite-element library, available as open-source software. The performance of this framework is demonstrated for problems from non-linear elasticity, fluid mechanics, and fluid-structure interaction.

In the chapter by C. Denis, R. Couturier, and F. Jézéquel, "A Sparse Linear System Solver Used in a Distributed and Heterogenous Grid Computing Environment", the GREMLINS (GRid Efficient Linear Systems) solver of systems of linear equations is developed. The algorithm is based on multisplitting techniques, and a new balancing algorithm is presented.

The chapter by A.G. Sunderland, "Parallel Diagonalization Performance on High-Performance Computers", analyzes the performance of parallel eigensolvers from numerical libraries such as ScaLAPACK on the latest parallel architectures using data sets derived from large-scale scientific applications.

The book continues with the second part focused on parallel optimization. In the chapter by J. Žilinskas, "Parallel Global Optimization in Multidimensional Scaling", global optimization methods are outlined, and global optimization algorithms for multidimensional scaling are reviewed with particular emphasis on parallel computing. Global optimization algorithms are computationally intensive, and solution time crucially depends on the dimensionality of a problem. Parallel computing enables solution of larger problems.

The chapter by K. Woodsend and J. Gondzio, "High-Performance Parallel Support Vector Machine Training", shows how the training process of support vector machines can be reformulated to become suitable for high-performance parallel computing. Data is pre-processed in parallel to generate an approximate low-rank Cholesky decomposition. An optimization solver then exploits the problem's structure to perform many linear algebra operations in parallel, with relatively low data transfer between processors, resulting in excellent parallel efficiency for very-large-scale problems.

The chapter by R. Paulavičius and J. Žilinskas, "Parallel Branch and Bound Algorithm with Combination of Lipschitz Bounds over Multidimensional Simplices for Multicore Computers", presents parallelization of a branch and bound algorithm for global Lipschitz minimization with a combination of extreme (infinite and first) and Euclidean norms over a multidimensional simplex. OpenMP is used to implement the parallel version of the algorithm for multicore computers. The efficiency of the parallel algorithm is studied using an extensive set of multidimensional test functions for global optimization.

The chapter by S. Ivanikovas, E. Filatovas, and J. Žilinskas, "Experimental Investigation of Local Searches for Optimization of Grillage-Type Foundations", presents a multistart approach for optimal pile placement in grillage-type foundations. Various algorithms for local optimization are applied, and their performance is experimentally investigated and compared. Parallel computations is used to speed up experimental investigation.

The third part of the book covers management issues of parallel programs and data. In the chapter by D. Henty and A. Gray, "Comparison of the UK National Supercomputer Services: HPCx and HECToR", an overview of the two current UK national HPC services, HPCx and HECToR, are given. Such results are particularly interesting, as these two machines will now be operating together for some time and users have a choice as to which machine best suits their requirements. Results of extensive experiments are presented.

In the chapter by I.T. Todorov, I.J. Bush, and A.R. Porter, "DL_POLY_3 I/O: Analysis, Alternatives, and Future Strategies", it is noted that an important bottleneck in the scalability and efficiency of any molecular dynamics software is the I/O speed and reliability as data has to be dumped and stored for postmortem analysis. This becomes increasingly more important when simulations scale to many thousands of processors and system sizes increase to many millions of particles. This study outlines the problems associated with I/O when performing large classic MD runs and shows that it is necessary to use parallel I/O methods when studying large systems.

The chapter by M. Piotrowski, "Mixed Mode Programming on HPCx", presents several benchmark codes based on iterative Jacobi relaxation algorithms: a pure MPI version and three mixed mode (MPI + OpenMP) versions. Their performance is studied and analyzed on mixed architecture – cluster of shared memory nodes. The results show that none of the mixed mode versions managed to outperform the pure MPI, mainly due to longer MPI point-to-point communication times.

The chapter by A. Grothey, J. Hogg, K. Woodsend, M. Colombo, and J. Gondzio, "A Structure Conveying Parallelizable Modeling Language for Mathematical Programming", presents an idea of using a modeling language for the definition of mathematical programming problems with block constructs for description of structure that may make parallel model generation of large problems possible. The proposed structured modeling language is based on the popular modeling language AMPL and implemented as a pre-/postprocessor to AMPL. Solvers based on block linear algebra exploiting interior point methods and decomposition solvers can therefore directly exploit the structure of the problem.

The chapter by R. Smits, M. Kramer, B. Stappers, and A. Faulkner, "Computational Requirements for Pulsar Searches with the Square Kilometer Array", is devoted to the analysis of computational requirements for beam forming and data analysis, assuming the SKA Design Studies' design for the SKA, which consists of 15-meter dishes and an aperture array. It is shown that the maximum data rate from a pulsar survey using the 1-km core becomes about $2.7 \cdot 10^{13}$ bytes per second and requires a computation power of about $2.6 \cdot 10^{17}$ ops for a deep real-time analysis.

The final and largest part of the book covers applications of parallel computing. In the chapter by R. Čiegis, F. Gaspar, and C. Rodrigo, "Parallel Multiblock Multigrid Algorithms for Poroelastic Models", the application of a parallel multigrid method for the two-dimensional poroelastic model is investigated. A domain is partitioned into structured blocks, and this geometrical structure is used to develop a parallel version of the multigrid algorithm. The convergence for different smoothers is investigated, and it is shown that the box alternating line Vanka-type smoother is robust and efficient.

The chapter by V. Starikovičius, R. Čiegis, O. Iliev, and Z. Lakdawala, "A Parallel Solver for the 3D Simulation of Flows Through Oil Filters", presents a parallel solver for the 3D simulation of flows through oil filters. The Navier–Stokes–Brinkmann system of equations is used to describe the coupled laminar flow of incompressible isothermal oil through open cavities and cavities with filtering porous media. Two parallel algorithms are developed on the basis of the sequential numerical algorithm. The performance of implementations of both algorithms is studied on clusters of multicore computers.

The chapter by S. Eastwood, P. Tucker, and H. Xia, "High-Performance Computing in Jet Aerodynamics", is devoted to the analysis of methods for reduction of the noise generated by the propulsive jet of an aircraft engine. The use of high-performance computing facilities is essential, allowing detailed flow studies to be carried out that help to disentangle the effects of numerics from flow physics. The scalability and efficiency of the presented parallel algorithms are investigated.

The chapter by G. Jankevičiūtė and R. Čiegis, "Parallel Numerical Solver for the Simulation of the Heat Conduction in Electrical Cables", is devoted to the modeling of the heat conduction in electrical cables. Efficient parallel numerical algorithms are developed to simulate the heat transfer in cable bundles. They are implemented using MPI and targeted for distributed memory computers, including clusters of PCs.

The chapter by A. Deveikis, "Orthogonalization Procedure for Antisymmetrization of J-shell States", presents an efficient procedure for construction of the antisymmetric basis of j-shell states with isospin. The approach is based on an efficient algorithm for construction of the idempotent matrix eigenvectors, and it reduces to an orthogonalization procedure. The presented algorithm is much faster than the diagonalization routine `rs()` from the EISPACK library.

In the chapter by G.A. Siamas, X. Jiang, and L.C. Wrobel, "Parallel Direct Numerical Simulation of an Annular Gas–Liquid Two-Phase Jet with Swirl", the flow characteristics of an annular swirling liquid jet in a gas medium is examined by direct solution of the compressible Navier–Stokes equations. A mathematical formulation is developed that is capable of representing the two-phase flow system while the volume of fluid method has been adapted to account for the gas compressibility. Fully 3D parallel direct numerical simulation (DNS) is performed utilizing 512 processors, and parallelization of the code was based on domain decomposition.

In the chapter by I. Laukaitytė, R. Čiegis, M. Lichtner, and M. Radziunas, "Parallel Numerical Algorithm for the Traveling Wave Model", a parallel algorithm for the simulation of the dynamics of high-power semiconductor lasers is presented. The model equations describing the multisection broad-area semiconductor lasers are solved by the finite difference scheme constructed on staggered grids. The algorithm is implemented by using the ParSol tool of parallel linear algebra objects.

The chapter by X. Guo, M. Pinna, and A.V. Zvelindovsky, "Parallel Algorithm for Cell Dynamics Simulation of Soft Nano-Structured Matter", presents a parallel algorithm for large-scale cell dynamics simulation. With the efficient strategy of domain decomposition and the fast method of neighboring points location, simulations of large-scale systems have been successfully performed.

The chapter by Ž. Dapkūnas and J. Kulys, "Docking and Molecular Dynamics Simulation of Complexes of High and Low Reactive Substrates with Peroxidases", presents docking and parallel molecular dynamics simulations of two peroxidases (*ARP* and *HRP*) and two compounds (*LUM* and *IMP*). The study of docking simulations gives a clue to the reason of different reactivity of *LUM* and similar reactivity of *IMP* toward two peroxidases. In the case of *IMP*, −OH group is near FE=O in both peroxidases, and hydrogen bond formation between −OH and Fe=0 is possible. In the case of *LUM*, N-H is near Fe=0 in *ARP* and the hydrogen formation is possible, but it is farther in *HRP* and hydrogen bond with Fe=O is not formed.

The works were presented at the bilateral workshop of British and Lithuanian scientists "High Performance Scientific Computing" held in Druskininkai, Lithuania, 5–8 February 2008. The workshop was supported by the British Council through the INYS program.

The British Council's International Networking for Young Scientists program (INYS) brings together young researchers from the UK and other countries to make new contacts and promote the creative exchange of ideas through short conferences. Mobility for young researchers facilitates the extended laboratory in which all researchers now operate: it is a powerful source of new ideas and a strong force for creativity. Through the INYS program, the British Council helps to develop high-quality collaborations in science and technology between the UK and other countries and shows the UK as a leading partner for achievement in world science, now and in the future. The INYS program is unique in that it brings together scientists in any priority research area and helps develop working relationships. It aims to encourage young researchers to be mobile and expand their knowledge. The INYS supported workshop "High Performance Scientific Computing" was organized by the University of Edinburgh, UK, and Vilnius Gediminas Technical University, Lithuania. The meeting was coordinated by Professor R. Čiegis and Dr. J. Žilinskas from Vilnius Gediminas Technical University and Dr. D. Henty from The University of Edinburgh. The homepage of the workshop is available at http://techmat.vgtu.lt/~inga/inys2008/.

Twenty-four talks were selected from thirty-two submissions from young UK and Lithuanian researchers. Professor B. Kågström from Umeå University, Sweden, and Dr. I. Todorov from Daresbury Laboratory, UK, gave invited lectures. Review lectures were also given by the coordinators of the workshop.

This book contains review papers and revised contributed papers presented at the workshop. All twenty-three papers have been reviewed. We are very thankful to the reviewers for their recommendations and comments.

We hope that this book will serve as a valuable reference document for the scientific community and will contribute to the future cooperation between the participants of the workshop.

We would like to thank the British Council for financial support. We are very grateful to the managing editor of the series, Professor Panos Pardalos, for his encouragement.

Druskininkai, Lithuania
February 2008

Raimondas Čiegis
David Henty
Bo Kågström
Julius Žilinskas

Contents

Preface v

List of Contributors xix

Part I Parallel Algorithms for Matrix Computations

RECSY and SCASY Library Software: Recursive Blocked and Parallel Algorithms for Sylvester-Type Matrix Equations with Some Applications 3
Robert Granat, Isak Jonsson, and Bo Kågström
1 Motivation and Background 3
2 Variants of Bartels–Stewart's Schur Method 5
3 Blocking Strategies for Reduced Matrix Equations 6
3.1 Explicit Blocked Methods for Reduced Matrix Equations 6
3.2 Recursive Blocked Methods for Reduced Matrix Equations 7
4 Parallel Algorithms for Reduced Matrix Equations 9
4.1 Distributed Wavefront Algorithms 10
4.2 Parallelization of Recursive Blocked Algorithms 11
5 Condition Estimation 11
5.1 Condition Estimation in RECSY 13
5.2 Condition Estimation in SCASY 13
6 Library Software Highlights 13
6.1 The RECSY Library 13
6.2 The SCASY Library 14
7 Experimental Results 16
8 Some Control Applications and Extensions 18
8.1 Condition Estimation of Subspaces with Specified Eigenvalues 18
8.2 Periodic Matrix Equations in CACSD 19
References 22

Parallelization of Linear Algebra Algorithms Using ParSol Library of Mathematical Objects ... 25
Alexander Jakušev, Raimondas Čiegis, Inga Laukaitytė, and Vyacheslav Trofimov
1 Introduction ... 25
2 The Principles and Implementation Details of ParSol Library ... 27
2.1 Main Classes of ParSol ... 27
2.2 Implementation of ParSol ... 29
3 Parallel Algorithm for Simulation of Counter-propagating Laser Beams ... 29
3.1 Invariants of the Solution ... 30
3.2 Finite Difference Scheme ... 31
3.3 Parallel Algorithm ... 33
3.4 Results of Computational Experiments ... 34
4 Conclusions ... 34
References ... 35

The Development of an Object-Oriented Parallel Block Preconditioning Framework ... 37
Richard L. Muddle, Jonathan W. Boyle, Milan D. Mihajlović, and Matthias Heil
1 Introduction ... 37
2 Block Preconditioning ... 39
3 The Performance of the Block Preconditioning Framework ... 40
3.1 Reference Problem : 2D Poisson ... 40
3.2 Non-linear Elasticity ... 41
3.3 Fluid Mechanics ... 42
3.4 Fluid–Structure Interaction ... 44
4 Conclusions ... 45
References ... 45

A Sparse Linear System Solver Used in a Distributed and Heterogenous Grid Computing Environment ... 47
Christophe Denis, Raphael Couturier, and Fabienne Jézéquel
1 Introduction ... 47
2 The Parallel Linear Multisplitting Method Used in the GREMLINS Solver ... 48
3 Load Balancing of the Direct Multisplitting Method ... 50
4 Experimental Results ... 51
4.1 Experiments with a Matrix Issued from an Advection-Diffusion Model ... 51
4.2 Results of the Load Balancing ... 52
5 Conclusions and Future Work ... 55
References ... 56

Parallel Diagonalization Performance on High-Performance Computers . 57
Andrew G. Sunderland
1 Introduction 57
2 Parallel Diagonalization Methods 58
2.1 Equations for Matrix Diagonalizations in PRMAT 58
2.2 Equations for Matrix Diagonalizations in CRYSTAL 58
2.3 Symmetric Eigensolver Methods 59
2.4 Eigensolver Parallel Library Routines 60
3 Testing Environment 60
4 Results 61
5 Conclusions 65
References 66

Part II Parallel Optimization

Parallel Global Optimization in Multidimensional Scaling 69
Julius Žilinskas
1 Introduction 69
2 Global Optimization 70
3 Multidimensional Scaling 72
4 Multidimensional Scaling with City-Block Distances 75
5 Parallel Algorithms for Multidimensional Scaling 77
6 Conclusions 81
References 81

High-Performance Parallel Support Vector Machine Training 83
Kristian Woodsend and Jacek Gondzio
1 Introduction 83
2 Interior Point Methods 85
3 Support Vector Machines 86
3.1 Binary Classification 86
3.2 Linear SVM 86
3.3 Non-linear SVM 87
4 Parallel Partial Cholesky Decomposition 88
5 Implementing the QP for Parallel Computation 89
5.1 Linear Algebra Operations 90
5.2 Performance 90
6 Conclusions 92
References 92

Parallel Branch and Bound Algorithm with Combination of Lipschitz Bounds over Multidimensional Simplices for Multicore Computers 93
Remigijus Paulavičius and Julius Žilinskas
1 Introduction 93
2 Parallel Branch and Bound with Simplicial Partitions 94
3 Results of Experiments 96

4 Conclusions 100
References 102

Experimental Investigation of Local Searches for Optimization of Grillage-Type Foundations 103
Sergėjus Ivanikovas, Ernestas Filatovas, and Julius Žilinskas
1 Introduction 103
2 Optimization of Grillage-Type Foundations 104
3 Methods for Local Optimization of Grillage-Type Foundations 104
4 Experimental Research 105
5 Conclusions 111
References 112

Part III Management of Parallel Programming Models and Data

Comparison of the UK National Supercomputer Services: HPCx and HECToR 115
David Henty and Alan Gray
1 Introduction 115
2 System Overview 116
2.1 HPCx 116
2.2 HECToR 118
3 System Comparison 118
3.1 Processors 119
3.2 Interconnect 119
4 Applications Performance 120
4.1 PDNS3D 121
4.2 NAMD 122
5 Conclusions 123
References 123

DL_POLY_3 I/O: Analysis, Alternatives, and Future Strategies 125
Ilian T. Todorov, Ian J. Bush, and Andrew R. Porter
1 Introduction 125
2 I/O in DL_POLY_3 126
2.1 Serial Direct Access I/O 126
2.2 Parallel Direct Access I/O 127
2.3 MPI-I/O 127
2.4 Serial I/O Using NetCDF 128
3 Results and Discussion 128
4 Conclusions 131
References 132

Mixed Mode Programming on HPCx 133
Michał Piotrowski
1 Introduction 133
2 Benchmark Codes 134

3 Mixed Mode ... 135
4 Hardware ... 137
5 Experimental Results ... 139
6 Conclusions ... 142
References ... 143

A Structure Conveying Parallelizable Modeling Language for Mathematical Programming ... 145
Andreas Grothey, Jonathan Hogg, Kristian Woodsend, Marco Colombo, and Jacek Gondzio
1 Introduction ... 145
2 Background ... 146
2.1 Mathematical Programming ... 146
2.2 Modeling Languages ... 147
3 Solution Approaches to Structured Problems ... 149
3.1 Decomposition ... 149
3.2 Interior Point Methods ... 149
4 Structure Conveying Modeling Languages ... 150
4.1 Other Structured Modeling Approaches ... 151
4.2 Design ... 151
4.3 Implementation ... 153
5 Conclusions ... 155
References ... 155

Computational Requirements for Pulsar Searches with the Square Kilometer Array ... 157
Roy Smits, Michael Kramer, Ben Stappers, and Andrew Faulkner
1 Introduction ... 157
2 SKA Configuration ... 158
3 Computational Requirements ... 159
3.1 Beam Forming ... 159
3.2 Data Analysis ... 161
4 Conclusions ... 164
References ... 164

Part IV Parallel Scientific Computing in Industrial Applications

Parallel Multiblock Multigrid Algorithms for Poroelastic Models ... 169
Raimondas Čiegis, Francisco Gaspar, and Carmen Rodrigo
1 Introduction ... 169
2 Mathematical Model and Stabilized Difference Scheme ... 171
3 Multigrid Methods ... 172
3.1 Box Relaxation ... 173
4 Numerical Experiments ... 174
5 Parallel Multigrid ... 176

5.1 Code Implementation ... 176
5.2 Critical Issues Regarding Parallel MG ... 177
6 Conclusions ... 179
References ... 179

A Parallel Solver for the 3D Simulation of Flows Through Oil Filters ... 181
Vadimas Starikovičius, Raimondas Čiegis, Oleg Iliev, and Zhara Lakdawala
1 Introduction ... 181
2 Mathematical Model and Discretization ... 182
2.1 Time Discretization ... 183
2.2 Finite Volume Discretization in Space ... 184
2.3 Subgrid Approach ... 185
3 Parallel Algorithms ... 185
3.1 DD Parallel Algorithm ... 185
3.2 OpenMP Parallel Algorithm ... 189
4 Conclusions ... 190
References ... 190

High-Performance Computing in Jet Aerodynamics ... 193
Simon Eastwood, Paul Tucker, and Hao Xia
1 Introduction ... 193
2 Numerical Background ... 195
2.1 HYDRA and FLUXp ... 195
2.2 Boundary and Initial Conditions ... 196
2.3 Ffowcs Williams Hawkings Surface ... 196
3 Computing Facilities ... 197
4 Code Parallelization and Scalability ... 198
5 Axisymmetric Jet Results ... 199
5.1 Problem Set Up and Mesh ... 199
5.2 Results ... 200
6 Complex Geometries ... 200
6.1 Mesh and Initial Conditions ... 200
6.2 Results ... 203
7 Conclusions ... 205
References ... 205

Parallel Numerical Solver for the Simulation of the Heat Conduction in Electrical Cables ... 207
Gerda Jankevičiūtė and Raimondas Čiegis
1 Introduction ... 207
2 The Model of Heat Conduction in Electrical Cables and Discretization ... 208
3 Parallel Algorithm ... 211
4 Conclusions ... 212
References ... 212

Orthogonalization Procedure for Antisymmetrization of J-shell States . . . 213
Algirdas Deveikis
1 Introduction . . . 213
2 Antisymmetrization of Identical Fermions States . . . 214
3 Calculations and Results . . . 217
4 Conclusions . . . 220
References . . . 221

Parallel Direct Numerical Simulation of an Annular Gas–Liquid Two-Phase Jet with Swirl . . . 223
George A. Siamas, Xi Jiang, and Luiz C. Wrobel
1 Introduction . . . 223
2 Governing Equations . . . 224
3 Computational Methods . . . 226
3.1 Time Advancement, Discretization, and Parallelization . . 226
3.2 Boundary and Initial Conditions . . . 227
4 Results and Discussion . . . 229
4.1 Instantaneous Flow Data . . . 229
4.2 Time-Averaged Data, Velocity Histories, and Energy Spectra . . . 232
5 Conclusions . . . 234
References . . . 235

Parallel Numerical Algorithm for the Traveling Wave Model . . . 237
Inga Laukaitytė, Raimondas Čiegis, Mark Lichtner, and Mindaugas Radziunas
1 Introduction . . . 237
2 Mathematical Model . . . 240
3 Finite Difference Scheme . . . 242
3.1 Discrete Transport Equations for Optical Fields . . . 243
3.2 Discrete Equations for Polarization Functions . . . 244
3.3 Discrete Equations for the Carrier Density Function . . . 244
3.4 Linearized Numerical Algorithm . . . 244
4 Parallelization of the Algorithm . . . 246
4.1 Parallel Algorithm . . . 246
4.2 Scalability Analysis . . . 247
4.3 Computational Experiments . . . 248
5 Conclusions . . . 250
References . . . 250

Parallel Algorithm for Cell Dynamics Simulation of Soft Nano-Structured Matter . . . 253
Xiaohu Guo, Marco Pinna, and Andrei V. Zvelindovsky
1 Introduction . . . 253
2 The Cell Dynamics Simulation . . . 254
3 Parallel Algorithm of CDS Method . . . 256

3.1 The Spatial Decomposition Method 256
3.2 Parallel Platform and Performance Tuning 258
3.3 Performance Analysis and Results 259
4 Conclusions 262
References 262

Docking and Molecular Dynamics Simulation of Complexes of High and Low Reactive Substrates with Peroxidases 263
Žilvinas Dapkūnas and Juozas Kulys
1 Introduction 263
2 Experimental 265
2.1 *Ab Initio* Molecule Geometry Calculations 265
2.2 Substrates Docking in Active Site of Enzyme 265
2.3 Molecular Dynamics of Substrate–Enzyme Complexes . . 266
3 Results and Discussion 267
3.1 Substrate Docking Modeling 267
3.2 Molecular Dynamics Simulation 268
4 Conclusions 270
References 270

Index 273

List of Contributors

Jonathan W. Boyle
School of Mathematics, University of Manchester, Oxford Road, Manchester, M13 9PL, UK, e-mail: j.boyle@manchester.ac.uk

Ian J. Bush
STFC Daresbury Laboratory, Warrington WA4 4AD, UK

Marco Colombo
School of Mathematics, University of Edinburgh, Edinburgh, UK

Raphael Couturier
Laboratoire d Informatique del Université de Franche-Comté, BP 527, 90016 Belfort Cedex, France, e-mail: Raphael.Couturier@iut-bm.univ-fcomte.fr

Raimondas Čiegis
Vilnius Gediminas Technical University, Saulėtekio al. 11, LT-10223 Vilnius, Lithuania, e-mail: rc@fm.vgtu.lt

Žilvinas Dapkūnas
Vilnius Gediminas Technical University, Department of Chemistry and Bioengineering, Saulėtekio Avenue 11, LT-10223 Vilnius, Lithuania, e-mail: Zilvinas.Dapkunas@fm.vgtu.lt

Christophe Denis
School of Electronics, Electrical Engineering & Computer Science, The Queen's University of Belfast, Belfast BT7 1NN, UK, e-mail: C.Denis@qub.ac.uk
UPMC Univ Paris 06, Laboratoire d'Informatique LIP6, 4 place Jussieu, 75252 Paris Cedex 05, France, e-mail: Christophe.Denis@lip6.fr

Algirdas Deveikis
Department of Applied Informatics, Vytautas Magnus University, Kaunas, Lithuania, e-mail: a.deveikis@if.vdu.lt

Simon Eastwood
Whittle Laboratory, University of Cambridge, Cambridge, UK,
e-mail: se282@cam.ac.uk

Andrew Faulkner
Jodrell Bank Centre for Astrophysics, University of Manchester, Manchester, UK

Ernestas Filatovas
Institute of Mathematics and Informatics, Akademijos 4, LT-08663 Vilnius,
Lithuania, e-mail: ernest.filatov@gmail.com

Francisco Gaspar
Departamento de Matematica Aplicada, Universidad de Zaragoza, 50009 Zaragoza,
Spain, e-mail: fjgaspar@unizar.es

Jacek Gondzio
School of Mathematics, University of Edinburgh, The King's Buildings, Edinburgh,
EH9 3JZ, UK, e-mail: j.gondzio@ed.ac.uk

Robert Granat
Department of Computing Science and HPC2N, Umeå University, Umeå, Sweden,
e-mail: granat@cs.umu.se

Alan Gray
Edinburgh Parallel Computing Centre, The University of Edinburgh,
Edinburgh, UK, e-mail: a.gray@epcc.ed.ac.uk

Andreas Grothey
School of Mathematics, University of Edinburgh, Edinburgh, UK,
e-mail: A.Grothey@ed.ac.uk

Xiaohu Guo
School of Computing, Engineering and Physical Sciences, University of Central
Lancashire, Preston, Lancashire, PR1 2HE, UK, e-mail: XGuo@uclan.ac.uk

Matthias Heil
School of Mathematics, University of Manchester, Oxford Road, Manchester,
M13 9PL, UK, e-mail: m.heil@maths.man.ac.uk

David Henty
Edinburgh Parallel Computing Centre, The University of Edinburgh,
Edinburgh, UK, e-mail: d.henty@epcc.ed.ac.uk

Jonathan Hogg
School of Mathematics, University of Edinburgh, Edinburgh, UK

Oleg Iliev
Fraunhofer ITWM, Fraunhofer-Platz 1, D-67663 Kaiserslautern, Germany,
e-mail: iliev@itwm.fhg.de

Sergėjus Ivanikovas
Institute of Mathematics and Informatics, Akademijos 4, LT-08663 Vilnius,
Lithuania, e-mail: ivanikovas@gmail.com

Alexander Jakušev
Vilnius Gediminas Technical University, Saulėtekio al. 11, LT-10223, Vilnius, Lithuania, e-mail: alexj@fm.vgtu.lt

Gerda Jankevičiūtė
Vilnius Gediminas Technical University, Saulėtekio al. 11, LT-10223, Vilnius, Lithuania, e-mail: Gerda.Jankeviciute@fm.vgtu.lt

Fabienne Jézéquel
UPMC Univ Paris 06, Laboratoire d'Informatique LIP6, 4 place Jussieu, 75252 Paris Cedex 05, France, e-mail: Fabienne.Jezequel@lip6.fr

Xi Jiang
Brunel University, Mechanical Engineering, School of Engineering and Design, Uxbridge, UB8 3PH, UK, e-mail: Xi.Jiang@brunel.ac.uk

Isak Jonsson
Department of Computing Science and HPC2N, Umeå University, Umeå, Sweden, e-mail: isak@cs.umu.se

Bo Kågström
Department of Computing Science and HPC2N, Umeå University, Umeå, Sweden, e-mail: bokg@cs.umu.se

Michael Kramer
Jodrell Bank Centre for Astrophysics, University of Manchester, Manchester, UK

Juozas Kulys
Vilnius Gediminas Technical University, Department of Chemistry and Bioengineering, Saulėtekio Avenue 11, LT-10223 Vilnius, Lithuania, e-mail: Juozas.Kulys@fm.vgtu.lt

Zhara Lakdawala
Fraunhofer ITWM, Fraunhofer-Platz 1, D-67663 Kaiserslautern, Germany, e-mail: lakdawala@itwm.fhg.de

Inga Laukaitytė
Vilnius Gediminas Technical University, Saulėtekio al. 11, LT-10223, Vilnius, Lithuania, e-mail: Inga.Laukaityte@fm.vgtu.lt

Mark Lichtner
Weierstrass Institute for Applied Analysis and Stochastics, Mohrenstarsse 39, 10117 Berlin, Germany, e-mail: lichtner@wias-berlin.de

Milan D. Mihajlović
School of Computer Science, University of Manchester, Oxford Road, Manchester, M13 9PL, UK, e-mail: milan@cs.man.ac.uk

Richard L. Muddle
School of Computer Science, University of Manchester, Oxford Road, Manchester, M13 9PL, UK, e-mail: rlm@cs.man.ac.uk

Remigijus Paulavičius
Vilnius Pedagogical University, Studentu 39, LT-08106 Vilnius, Lithuania,
e-mail: r.paulavicius@vpu.lt
Institute of Mathematics and Informatics, Akademijos 4, LT-08663 Vilnius, Lithuania

Marco Pinna
School of Computing, Engineering and Physical Sciences, University of Central Lancashire, Preston, Lancashire, PR1 2HE, UK, e-mail: MPinna@uclan.ac.uk

Michał Piotrowski
Edinburgh Parallel Computing Centre, University of Edinburgh, Edinburgh, UK,
e-mail: mpiotrow@epcc.ed.ac.uk

Andrew R. Porter
STFC Daresbury Laboratory, Warrington WA4 4AD, UK

Mindaugas Radziunas
Weierstrass Institute for Applied Analysis and Stochastics, Mohrenstarsse 39, 10117 Berlin, Germany, e-mail: radziunas@wias-berlin.de

Carmen Rodrigo
Departamento de Matematica Aplicada, Universidad de Zaragoza, 50009 Zaragoza, Spain, e-mail: carmen.rodrigo.cardiel@gmail.com

George A. Siamas
Brunel University, Mechanical Engineering, School of Engineering and Design, Uxbridge, UB8 3PH, UK, e-mail: George.Siamas@brunel.ac.uk

Roy Smits
Jodrell Bank Centre for Astrophysics, University of Manchester, Manchester, UK,
e-mail: Roy.Smits@manchester.ac.uk

Ben Stappers
Jodrell Bank Centre for Astrophysics, University of Manchester, Manchester, UK

Vadimas Starikovičius
Vilnius Gediminas Technical University, Saulėtekio al. 11, LT-10223 Vilnius, Lithuania, e-mail: vs@sc.vgtu.lt

Andrew G. Sunderland
STFC Daresbury Laboratory, Warrington, UK, e-mail: a.g.sunderland@dl.ac.uk

Ilian T. Todorov
STFC Daresbury Laboratory, Warrington WA4 4AD, UK,
e-mail: i.t.todorov@dl.ac.uk

Vyacheslav Trofimov
M. V. Lomonosov Moscow State University, Vorob'evy gory, 119992, Russia,
e-mail: vatro@cs.msu.su

Paul Tucker
Whittle Laboratory, University of Cambridge, Cambridge, UK

Kristian Woodsend
School of Mathematics, University of Edinburgh, The King's Buildings, Edinburgh, EH9 3JZ, UK, e-mail: k.j.woodsend@sms.ed.ac.uk

Luiz C. Wrobel
Brunel University, Mechanical Engineering, School of Engineering and Design, Uxbridge, UB8 3PH, UK, e-mail: Luiz.Wrobel@brunel.ac.uk

Hao Xia
Whittle Laboratory, University of Cambridge, Cambridge, UK

Andrei V. Zvelindovsky
School of Computing, Engineering and Physical Sciences, University of Central Lancashire, Preston, Lancashire, PR1 2HE, UK, e-mail: AVZvelindovsky@uclan.ac.uk

Julius Žilinskas
Institute of Mathematics and Informatics, Akademijos 4, LT-08663 Vilnius, Lithuania, e-mail: julius.zilinskas@ktl.mii.lt
Vilnius Gediminas Technical University, Saulėtekio 11, LT-10223 Vilnius, Lithuania, e-mail: zilinskasjulius@gmail.com

Part I
Parallel Algorithms for Matrix Computations

RECSY and SCASY Library Software: Recursive Blocked and Parallel Algorithms for Sylvester-Type Matrix Equations with Some Applications

Robert Granat, Isak Jonsson, and Bo Kågström

Abstract In this contribution, we review state-of-the-art high-performance computing software for solving common standard and generalized continuous-time and discrete-time Sylvester-type matrix equations. The analysis is based on RECSY and SCASY software libraries. Our algorithms and software rely on the standard Schur method. Two ways of introducing blocking for solving matrix equations in reduced (quasi-triangular) form are reviewed. Most common is to perform a fix block partitioning of the matrices involved and rearrange the loop nests of a single-element algorithm so that the computations are performed on submatrices (matrix blocks). Another successful approach is to combine recursion and blocking.

We consider parallelization of algorithms for reduced matrix equations at two levels: globally in a distributed memory paradigm, and locally on shared memory or multicore nodes as part of the distributed memory environment. Distributed wavefront algorithms are considered to compute the solution to the reduced triangular systems. Parallelization of recursive blocked algorithms is done in two ways. The simplest way is so-called implicit data parallelization, which is obtained by using SMP-aware implementations of level 3 BLAS. Complementary to this, there is also the possibility of invoking task parallelism. This is done by explicit parallelization of independent tasks in a recursion tree using OpenMP. A brief account of some software issues for the RECSY and SCASY libraries is given. Theoretical results are confirmed by experimental results.

1 Motivation and Background

Matrix computations are fundamental and ubiquitous in computational science and its vast application areas. Along with the computer evolution, there is a continuing

Robert Granat · Isak Jonsson · Bo Kågström
Department of Computing Science and HPC2N, Umeå University, Sweden
e-mail: {granat · isak · bokg}@cs.umu.se

demand for new and improved algorithms and library software that is portable, robust, and efficient [13]. In this contribution, we review state-of-the-art high-performance computing (HPC) software for solving common standard and generalized continuous-time (CT) and discrete-time (DT) Sylvester-type matrix equations, see Table 1. Computer-aided control system design (CACSD) is a great source of applications for matrix equations including different eigenvalue and subspace problems and for condition estimation.

Both the RECSY and SCASY software libraries distinguish between *one-sided* and *two-sided* matrix equations. For one-sided matrix equations, the solution is only involved in matrix products of two matrices, e.g., $\mathrm{op}(A)X$ or $X\mathrm{op}(A)$, where $\mathrm{op}(A)$ can be A or A^T. In two-sided matrix equations, the solution is involved in matrix products of three matrices, both to the left and to the right, e.g., $\mathrm{op}(A)X\mathrm{op}(B)$. The more complicated data dependency of the two-sided equations is normally addressed in blocked methods from complexity reasons.

Solvability conditions for the matrix equations in Table 1 are formulated in terms of non-intersecting spectra of standard or generalized eigenvalues of the involved coefficient matrices and matrix pairs, respectively, or equivalently by nonzero associated sep-functions (see Sect. 5 and, e.g., [28, 24, 30] and the references therein).

The rest of this chapter is structured as follows. First, in Sect. 2, variants of the standard Schur method for solving Sylvester-type matrix equations are briefly described. In Sect. 3, two blocking strategies for dense linear algebra computations are discussed and applied to matrix equations in reduced (quasi-triangular) form. Section 4 reviews parallel algorithms for reduced matrix equations based on the explicitly blocked and the recursively blocked algorithms discussed in the previous section. Condition estimation of matrix equations and related topics are discussed in Sect. 5. In Sect. 6, a brief account of some software issues for the RECSY and SCASY libraries are given. Section 7 presents some performance results with focus on the hybrid parallelization model including message passing and multi-threading. Finally, Sect. 8 is devoted to some CACSD applications, namely condition estimation of invariant subspaces of Hamiltonian matrices and periodic matrix equations.

Table 1 Considered standard and generalized matrix equations. CT and DT denote the *continuous-time* and *discrete-time* variants, respectively.

Name	Matrix equation	Acronym
Standard CT Sylvester	$AX - XB = C \in \mathbb{R}^{m\times n}$	SYCT
Standard CT Lyapunov	$AX + XA^T = C \in \mathbb{R}^{m\times m}$	LYCT
Generalized Coupled Sylvester	$(AX - YB, DX - YE) = (C, F) \in \mathbb{R}^{(m\times n)\times 2}$	GCSY
Standard DT Sylvester	$AXB - X = C \in \mathbb{R}^{m\times n}$	SYDT
Standard DT Lyapunov	$AXA^T - X = C \in \mathbb{R}^{m\times m}$	LYDT
Generalized Sylvester	$AXB^T - CXD^T = E \in \mathbb{R}^{m\times n}$	GSYL
Generalized CT Lyapunov	$AXE^T + EXA^T = C \in \mathbb{R}^{m\times m}$	GLYCT
Generalized DT Lyapunov	$AXA^T - EXE^T = C \in \mathbb{R}^{m\times m}$	GLYDT

2 Variants of Bartels–Stewart's Schur Method

Our algorithms and software rely on the standard Schur method proposed already in 1972 by Bartels and Stewart [6]. The generic standard method consists of four major steps: (1) initial transformation of the left-hand coefficient matrices to Schur form; (2) updates of the right-hand-side matrix with respect to the orthogonal transformation matrices from the first step; (3) computation of the solution of the reduced matrix equation from the first two steps in a combined forward and backward substitution process; (4) a retransformation of the solution from the third step, in terms of the orthogonal transformations from the first step, to get the solution of the original matrix equation.

Let us demonstrate the solution process by considering the SYCT equation $AX - XB = C$. Step 1 produces the Schur factorizations $T_A = Q^T AQ$ and $T_B = P^T BP$, where $Q \in \mathbb{R}^{m\times m}$ and $P \in \mathbb{R}^{n\times n}$ are orthogonal and T_A and T_B are upper quasi-triangular, i.e., having 1×1 and 2×2 diagonal blocks corresponding to real and complex conjugate pairs of eigenvalues, respectively. Reliable and efficient algorithms for the Schur reduction step can be found in LAPACK [4] and in ScaLAPACK [26, 7] for distributed memory (DM) environments.

Steps 2 and 4 are typically conducted by two consecutive matrix multiply (GEMM) operations [12, 33, 34]: $\tilde{C} = Q^T CP$ and $X = Q\tilde{X}P^T$, where $\tilde{X}$ is the solution to the triangular equation

$$T_A\tilde{X} - \tilde{X}T_B = \tilde{C},$$

resulting from steps 1 and 2, and solved in step 3. Similar Schur-based methods are formulated for the standard equations LYCT, LYDT, and SYDT.

It is also straightforward to extend the generic Bartels–Stewart method to the generalized matrix equations GCSY, GSYL, GLYCT, and GLYDT. Now, we must rely on robust and efficient algorithms and software for the generalized Schur reduction (Hessenberg-triangular reduction and the QZ algorithm) [40, 39, 10, 1, 2, 3, 32], and algorithms for solving triangular forms of generalized matrix equations.

For illustration we consider GCSY, the generalized coupled Sylvester equation $(AX - YB, DX - YE) = (C, F)$ [37, 36]. In step 1, (A, D) and (B, E) are transformed to generalized real Schur form, i.e., $(T_A, T_D) = (Q_1^T AZ_1, Q_1^T DZ_1)$, $(T_B, T_E) = (Q_2^T BZ_2, Q_2^T EZ_2)$, where T_A and T_B are upper quasi-triangular, T_D and T_E are upper triangular, and $Q_1, Z_1 \in \mathbb{R}^{m\times m}$, $Q_2, Z_2 \in \mathbb{R}^{n\times n}$ are orthogonal. Step 2 updates the right-hand-side matrices $(\tilde{C}, \tilde{F}) = (Q_1^T CZ_2, Q_1^T FZ_2)$ leading to the reduced triangular GCSY

$$(T_A\tilde{X} - \tilde{Y}T_B, T_D\tilde{X} - \tilde{Y}T_E) = (\tilde{C}, \tilde{F}),$$

which is solved in step 3. The solution (X, Y) of the original GCSY is obtained in step 4 by the orthogonal equivalence transformation $(X, Y) = (Z_1\tilde{X}Z_2^T, Q_1\tilde{Y}Q_2^T)$.

Notice that care has to be taken to preserve the symmetry of the right-hand-side C in the Lyapunov equations LYCT, LYDT, GLYCT, and GLYDT (see Table 1) during each step of the different variants of the Bartels–Stewart method. This is important for reducing the complexity in steps 2–4 (for step 3, see Sect. 3) and to ensure that the computed solution X of the corresponding Lyapunov equation is guaranteed to be symmetric on output.

In general, the computed solutions of the matrix equations in Table 1 overwrite the corresponding right-hand-side matrices of the respective equation.

3 Blocking Strategies for Reduced Matrix Equations

In this section, we review two ways of introducing blocking for solving matrix equations in reduced (quasi-triangular) form. Most common is to perform a fix block partitioning of the matrices involved and rearrange the loop nests of a single-element algorithm so that the computations are performed on submatrices (matrix blocks). This means that the operations in the inner-most loops are expressed in matrix-matrix operations that can deliver high performance via calls to optimized level 3 BLAS. Indeed, this *explicit blocking* approach is extensively used in LAPACK [4].

Another successful approach is to combine recursion and blocking, leading to an automatic variable blocking with the potential for matching the deep memory hierarchies of today's HPC systems. *Recursive blocking* means that the involved matrices are split in the middle (by columns, by rows, or both) and a number of smaller problems are generated. In turn, the subarrays involved are split again generating a number of even smaller problems. The divide phase proceeds until the size of the controlling recursion blocking parameter becomes smaller than some predefined value. This termination criteria ensures that all leaf operations in the recursion tree are substantial level 3 (matrix-matrix) computations. The conquer phase mainly consists of increasingly sized matrix multiply and add (GEMM) operations. For an overview of recursive blocked algorithms and hybrid data structures for dense matrix computations and library software, we refer to the SIAM Review paper [13].

In the next two subsections, we demonstrate how explicit and recursive blocking are applied to solving matrix equations already in reduced form. To solve each of the Sylvester equations SYCT, SYDT, GCSY, and GSYL in quasi-triangular form is an $O(m^2n+mn^2)$ operation. Likewise, solving reduced Lyapunov equations LYCT, LYDT, GLYCT, and GLYDT are all $O(m^3)$ operations. The blocked methods described below have a similar complexity, assuming that the more complicated data dependencies of the two-sided matrix equations (SYDT, LYDT, GSYL, GLYCT, GLYDT) are handled appropriately.

3.1 Explicit Blocked Methods for Reduced Matrix Equations

We apply explicit blocking to reformulate each matrix equation problem into as much level 3 BLAS operations as possible. In the following, the (i, j)th block of a partitioned matrix, say X, is denoted X_{ij}. For illustration of the basic concepts, we consider the one-sided matrix equation SYCT and the two-sided LYDT. Let m_b and n_b be the block sizes used in an explicit block partitioning of the matrices A and B, respectively. In turn, this imposes a similar block partitioning of C and X (which overwrites C). Then $D_l = \lceil m/m_b \rceil$ and $D_r = \lceil n/n_b \rceil$ are the number of diagonal blocks in A and B, respectively. Now, SYCT can be rewritten in block-partitioned form as

$$A_{ii}X_{ij} - X_{ij}B_{jj} = C_{ij} - (\sum_{k=i+1}^{D_l} A_{ik}X_{kj} - \sum_{k=1}^{j-1} X_{ik}B_{kj}), \tag{1}$$

for $i = 1,2,\ldots,D_l$ and $j = 1,2,\ldots,D_r$. This block formulation of SYCT can be implemented using a couple of nested loops that call a node (or kernel) solver for the $D_l \cdot D_r$ small matrix equations and level 3 operations in the right-hand side [22]. Similarly, we express LYDT in explicit blocked form as

$$A_{ii}X_{ij}A_{jj}^T - X_{ij} = C_{ij} - \sum_{(k,l)=(i,j)}^{(D_l,D_l)} A_{ik}X_{kl}A_{jl}^T, \; (k,l) \neq (i,j). \tag{2}$$

Notice that the blocking of LYDT decomposes the problem into several smaller SYDT ($i \neq j$) and LYDT ($i = j$) equations. Apart from solving for only the upper or lower triangular part of X in the case of LYDT, a complexity of $O(m^3)$ can be retained by using a technique that stores *intermediate sums of matrix products* to avoid computing certain matrix products in the right-hand side of (2) several times. The same technique can also be applied to all two-sided matrix equations.

Similarly, explicit blocking and the complexity reduction technique are applied to the generalized matrix equations. Here, we illustrate with the one-sided GCSY, and the explicit blocked variant takes the form

$$\begin{cases} A_{ii}X_{ij} - Y_{ij}B_{jj} = C_{ij} - (\Sigma_{k=i+1}^{D_l} A_{ik}X_{kj} - \Sigma_{k=1}^{j-1} Y_{ik}B_{kj}) \\ D_{ii}X_{ij} - Y_{ij}E_{jj} = F_{ij} - (\Sigma_{k=i+1}^{D_l} D_{ik}X_{kj} - \Sigma_{k=1}^{j-1} X_{ik}E_{kj}), \end{cases} \tag{3}$$

where $i = 1,2,\ldots,D_l$ and $j = 1,2,\ldots,D_r$. A resulting serial level 3 algorithm is implemented as a couple of nested loops over the matrix operations defined by (3).

We remark that all linear matrix equations considered can be rewritten as an equivalent large linear system of equations $Zx = y$, where Z is the Kronecker product representation of the corresponding Sylvester-type operator. This is utilized in condition estimation algorithms and kernel solvers for small-sized matrix equations (see Sects. 5 and 6.1).

3.2 Recursive Blocked Methods for Reduced Matrix Equations

To make it easy to compare the two blocking strategies, we start by illustrating recursive blocking for SYCT.

The sizes of m and n control three alternatives for doing a *recursive splitting*. In Case 1 ($1 \le n \le m/2$), A is split by rows and columns, and C by rows only. In Case 2 ($1 \le m \le n/2$), B is split by rows and columns, and C by columns only. Finally, in Case 3 ($n/2 < m < 2n$), both rows and columns of the matrices A, B, and C are split:

$$\begin{bmatrix} A_{11} & A_{12} \\ & A_{22} \end{bmatrix} \begin{bmatrix} X_{11} & X_{12} \\ X_{21} & X_{22} \end{bmatrix} - \begin{bmatrix} X_{11} & X_{12} \\ X_{21} & X_{22} \end{bmatrix} \begin{bmatrix} B_{11} & B_{12} \\ & B_{22} \end{bmatrix} = \begin{bmatrix} C_{11} & C_{12} \\ C_{21} & C_{22} \end{bmatrix}.$$

This recursive splitting results in the following four triangular SYCT equations:

$$\begin{aligned} A_{11}X_{11} - X_{11}B_{11} &= C_{11} - A_{12}X_{21}, \\ A_{11}X_{12} - X_{12}B_{22} &= C_{12} - A_{12}X_{22} + X_{11}B_{12}, \\ A_{22}X_{21} - X_{21}B_{11} &= C_{21}, \\ A_{22}X_{22} - X_{22}B_{22} &= C_{22} + X_{21}B_{12}. \end{aligned}$$

Conceptually, we start by solving for X_{21} in the third equation above. After updating C_{11} and C_{22} with respect to X_{21}, one can solve for X_{11} and X_{22}. Both updates and the triangular Sylvester solves are independent operations and can be executed concurrently. Finally, C_{12} is updated with respect to X_{11} and X_{22}, and we can solve for X_{12}. The description above defines a recursion template that is applied to the four Sylvester sub-solves leading to a recursive blocked algorithm that terminates with calls to optimized kernel solvers for the leaf computations of the recursion tree.

We also illustrate the recursive blocking and template for GSYL, the most general of the two-sided matrix equations. Also here, we demonstrate Case 3 (Cases 1 and 2 can be seen as special cases), where (A,C), (B,D), and E are split by rows and columns:

$$\begin{aligned} &\begin{bmatrix} A_{11} & A_{12} \\ & A_{22} \end{bmatrix} \begin{bmatrix} X_{11} & X_{12} \\ X_{21} & X_{22} \end{bmatrix} \begin{bmatrix} B_{11}^T & \\ B_{12}^T & B_{22}^T \end{bmatrix} \\ -&\begin{bmatrix} C_{11} & C_{12} \\ & C_{22} \end{bmatrix} \begin{bmatrix} X_{11} & X_{12} \\ X_{21} & X_{22} \end{bmatrix} \begin{bmatrix} D_{11}^T & \\ D_{12}^T & D_{22}^T \end{bmatrix} = \begin{bmatrix} E_{11} & E_{12} \\ E_{21} & E_{22} \end{bmatrix}, \end{aligned}$$

leading to the following four triangular GSYL equations:

$$\begin{aligned} A_{11}X_{11}B_{11}^T - C_{11}X_{11}D_{11}^T &= E_{11} - A_{12}X_{21}B_{11}^T - (A_{11}X_{12} + A_{12}X_{22})B_{12}^T \\ &\quad + C_{12}X_{21}D_{11}^T + (C_{11}X_{12} + C_{12}X_{22})D_{12}^T, \\ A_{11}X_{12}B_{22}^T - C_{11}X_{12}D_{22}^T &= E_{12} - A_{12}X_{22}B_{22}^T + C_{12}X_{22}D_{22}^T, \\ A_{22}X_{21}B_{11}^T - C_{22}X_{21}D_{11}^T &= E_{21} - A_{22}X_{22}B_{12}^T + C_{22}X_{22}D_{12}^T, \\ A_{22}X_{22}B_{22}^T - C_{22}X_{22}D_{22}^T &= E_{22}. \end{aligned}$$

Now, we start by solving for X_{22} in the fourth equation above. After updating E_{12} and E_{21} with respect to X_{22}, we can solve for X_{12} and X_{21}. As for SYCT, both updates and the triangular GSYL solves are independent operations and can be executed concurrently. Finally, after updating E_{11} with respect to X_{12}, X_{21} and X_{22}, we solve for X_{11}. Some of the updates of E_{11} can be combined in larger GEMM operations, for example,

$$E_{11} = E_{11} - A_{12}X_{21}B_{11}^T + C_{12}X_{21}D_{11}^T \\ - \left(\begin{bmatrix} A_{11} & A_{12} \end{bmatrix} \begin{bmatrix} X_{12} \\ X_{22} \end{bmatrix} \right) B_{12}^T + \left(\begin{bmatrix} C_{11} & C_{12} \end{bmatrix} \begin{bmatrix} X_{12} \\ X_{22} \end{bmatrix} \right) D_{12}^T.$$

For more details about the recursive blocked algorithms for the one-sided and the two-sided matrix equations in Table 1, we refer to the ACM TOMS papers [29, 30]. A summary of all explicitly blocked algorithms can be found in [21]. Formal derivations of some of these algorithms are described in [44].

4 Parallel Algorithms for Reduced Matrix Equations

We consider parallelization of algorithms for reduced matrix equations at two levels: *globally* in a distributed memory paradigm, and *locally* on shared memory or multicore nodes as part of the distributed memory environment. The motivation for this hybrid approach is twofold:

- For large-scale problems arising in real applications, it is simply not possibly to rely on single processor machines because of limitations in storage and/or computing power; using high-performance parallel computers is many times the only feasible way of conducting the computations in a limited (at least reasonable) amount of time.
- On computing clusters with SMP-like nodes, a substantial performance gain can be achieved by local parallelization of distributed memory algorithm subtasks that do not need to access remote memory modules through the network; this observation has increased in importance with the arrival of the multicore (and future manycore) processors.

Despite the simple concept of this hybrid parallelization approach, it is a true challenge to design algorithms and software that are able to exploit the performance fully without lots of less portable hand-tuning efforts. Therefore, it is important to rely on programming paradigms that are either (more or less) self-tuning or offer high portable performance by design.

The automatic variable blocking in the RECSY algorithms has the potential for self-tuning and matching of a node memory hierarchy. This will be even more important in multiple-threads programming for Intel or AMD-style multicore chips, where the cores share different levels of the cache memory hierarchy; data locality is still the issue!

Moreover, SCASY is based on the current ScaLAPACK design, which in many respects is the most successful industry standard programming model for linear algebra calculations on distributed memory computers. This standard ensures portability, good load balance, the potential for high node performance via calls to level 3

BLAS operations, and enables efficient internode communication by the BLACS-MPI API.

We remark that the basic building blocks of ScaLAPACK are under reconsideration [11]. Alternative programming models include programming paradigms with a higher level of abstraction than that of the current mainstream model, e.g., Co-arrays [46].

4.1 Distributed Wavefront Algorithms

By applying the explicit blocking concept (see Sect. 3) and two-dimensional (2D) block cyclic distribution of the matrices over a rectangular $P_r \times P_c$ mesh (see, e.g., [17]), the solution to the reduced triangular systems can be computed block (pair) by block (pair) using a *wavefront* traversal of the block diagonals (or anti-diagonals) of the right-hand-side matrix (pair). Each computed subsolution block (say X_{ij}) is used in level 3 updates of the currently unsolved part of the right-hand side.

A maximum level of parallelism is obtained for the one-sided matrix equations by observing that all subsystems associated by any block (anti-)diagonal are independent. This is also true for the two-sided matrix equations if the technique of rearranging the updates of the right-hand side using *intermediate sums of matrix products* [19, 20] is utilized. Given $k = \min(P_r, P_c)$, it is possible to compute k subsolutions on the current block (anti-)diagonal in parallel and to perform level 3 updates of the right-hand side using k^2 processors in parallel and independently, except for the need to communicate for the subsolutions (which is conducted by high-performance *broadcast* operations) and other blocks from the left-hand-side coefficient matrices (see, e.g., the equation (2)).

In Fig. 1, we illustrate the block wavefront of the two-sided GLYCT equation $AXE^T + EXA^T = C$. The subsolution blocks X_{ij} of GLYCT are computed anti-diagonal by anti-diagonal, starting at the south-east corner of C, and each computed X_{ij} overwrites the corresponding block of C. Because the pair (A, E) is in generalized Schur form with A upper quasi-triangular and E upper triangular, A^T and E^T are conceptually represented as blocked lower triangular matrices.

Under some mild simplifying assumptions, the data dependencies in the considered matrix equations imply that the theoretical limit of the *scaled speedup* S_p of the parallel wavefront algorithms is bounded as

$$S_p \leq \frac{p}{k} \quad \text{with} \quad k = 1 + \frac{1}{n_b^3} \cdot \frac{t_s}{t_a} + \frac{1}{n_b} \cdot \frac{t_w}{t_a}, \tag{4}$$

where p is the number of utilized nodes in the distributed memory environment, n_b is the data distribution block size, and t_a, t_s, and t_w denote the time for performing

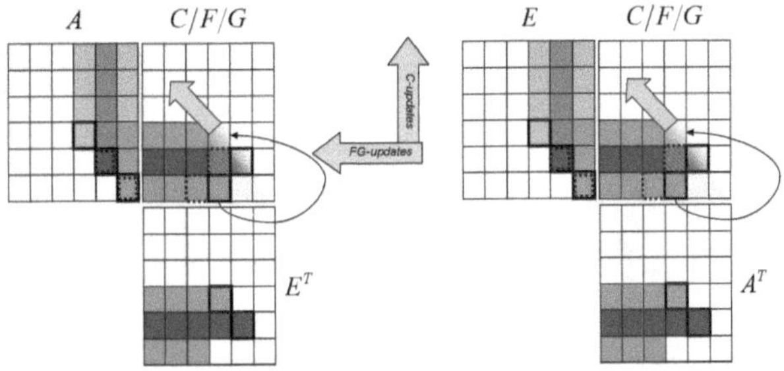

Fig. 1 The GLYCT wavefront: generalized, two-sided, symmetric. The three white blocks in the south-east corner of the right-hand-side C contain already computed subsolutions X_{ij} (the super-diagonal block obtained by symmetry). Bold bordered blocks of C mark subsolutions on the current anti-diagonal. Blocks with the same level of gray tone of A, E, and C are used together in subsystem solves, GEMM updates, or preparations for the next anti-diagonal. Dashed-bold bordered blocks are used in several operations. (F and G are buffers for intermediate results.)

an arithmetic operation, the node latency, and the inverse of the bandwidth of the interconnection network (see [20] for details).

4.2 Parallelization of Recursive Blocked Algorithms

Parallelism is invoked in two ways in the recursive blocked algorithms [29, 30, 31]. The simplest way is so called *implicit data parallelization*, which is obtained by using SMP-aware implementations of level 3 BLAS. This has especially good effects on the large and squarish GEMM updates in the conquer phase of the recursive algorithms.

Complementary to this, there is also the possibility of invoking *task parallelism*. This is done by explicit parallelization of independent tasks in a recursion tree using OpenMP [42], which typically includes calls to kernels solvers and some level 3 BLAS operations. We remark that the hybrid approach including implicit data parallelization as well as explicit task parallelization presumes that the SMP-aware BLAS and OpenMP work together. In addition, if there is more than two processors on the SMP-node, the OpenMP compiler must support nested parallelism, which most modern compilers do.

5 Condition Estimation

As briefly mentioned in Sect. 3.1, all linear matrix equations can be rewritten as an equivalent large linear system of equations $Zx = y$, where Z is the Kronecker product representation of the corresponding Sylvester-type operator, see Table 2. For

Table 2 Kronecker product matrix representations Z_{ACRO} and Z^T_{ACRO} of the Sylvester-type operators in Table 1, both conceptually used in condition estimation of the standard and generalized matrix equations.

Acronym (ACRO)	Z_{ACRO}	Z^T_{ACRO}
SYCT	$I_n \otimes A - B^T \otimes I_m$	$I_n \otimes A^T - B \otimes I_m$
LYCT	$I_m \otimes A + A \otimes I_m$	$I_m \otimes A^T - A^T \otimes I_m$
GCSY	$\begin{bmatrix} I_n \otimes A & -B^T \otimes I_m \\ I_n \otimes D & -E^T \otimes I_m \end{bmatrix}$	$\begin{bmatrix} I_n \otimes A^T & I_n \otimes D^T \\ -B \otimes I_m & -E \otimes I_m \end{bmatrix}$
SYDT	$B \otimes A - I_{m \cdot n}$	$B^T \otimes A^T - I_{m \cdot n}$
LYDT	$A \otimes A - I_{m^2}$	$A^T \otimes A^T - I_{m^2}$
GSYL	$B \otimes A - D \otimes C$	$B^T \otimes A^T - D^T \otimes C^T$
GLYDT	$A \otimes A - E \otimes E$	$A^T \otimes A^T - E^T \otimes E^T$
GLYCT	$E \otimes A + A \otimes E$	$E^T \otimes A^T + A^T \otimes E^T$

example, the size of the Z_{ACRO}-matrices for the standard and generalized Lyapunov equations are $n^2 \times n^2$. Consequently, these formulations are only efficient to use explicitly when solving small-sized problems in kernel solvers, e.g., see LAPACK's `DLASY2` and `DTGSY2` for solving SYCT and GCSY, and the kernels of the RECSY library (see Sect. 6.1 and [29, 30, 31]).

However, the linear system formulations allow us to perform condition estimation of the matrix equations by utilizing a general method [23, 27, 35] for estimating $\|A^{-1}\|_1$ of a square matrix A using reverse communication of $A^{-1}x$ and $A^{-T}x$, where $\|x\|_2 = 1$. In particular, for SYCT this approach is based on the linear system $Z_{\text{SYCT}}x = y$, where $x = \text{vec}(X)$ and $y = \text{vec}(C)$, for computing a lower bound of the inverse of the *separation between the matrices A and B* [50]:

$$\begin{aligned} \text{sep}(A,B) &= \inf_{\|X\|_F=1} \|AX - XB\|_F \\ &= \sigma_{\min}(Z_{\text{SYCT}}) = \|Z^{-1}_{\text{SYCT}}\|_2^{-1}, \end{aligned} \tag{5}$$

$$\frac{\|x\|_2}{\|y\|_2} = \frac{\|X\|_F}{\|C\|_F} \leq \|Z^{-1}_{\text{SYCT}}\|_2 = \frac{1}{\sigma_{\min}(Z_{\text{SYCT}})} = \text{sep}^{-1}(A,B). \tag{6}$$

The quantity (5) appears frequently in perturbation theory and error bounds (see, e.g., [28, 50]). Using the SVD of Z_{SYCT}, the exact value can be computed to the cost of $O(m^3n^3)$ flops. Such a direct computation of $\text{sep}(A,B)$ is only appropriate for small- to moderate-sized problems. However, its inverse can be estimated much cheaper by applying the 1-norm-based estimation technique and solving a few (normally around five) triangular SYCT equations to the cost of $O(m^2n + mn^2)$ flops [35].

This condition estimation method is applied to all matrix equations in Table 1 by considering $Z_{\text{ACRO}}x = y$, where Z_{ACRO} is the corresponding Kronecker product matrix representation of the associated Sylvester-type operator in Table 2. By choosing right-hand sides y and solving the associated reduced matrix equation for x, we obtain reliable lower bounds $\|x\|_2/\|y\|_2$ on $\|Z^{-1}_{\text{ACRO}}\|_2$.

5.1 Condition Estimation in RECSY

The condition estimation functionality in RECSY is invoked via the included LAPACK-SLICOT software wrappers (see Sect. 6.1), which means that the solution of the reduced matrix equations $Z_{\mathrm{ACRO}}x = y$ are performed using the recursive blocked algorithms. For example, the LAPACK routines `DTRSEN` and `DTGSEN` call quasi-triangular matrix equation solvers for condition estimation.

5.2 Condition Estimation in SCASY

The parallel condition estimators compute P_c different estimates independently and concurrently, one for each process column by taking advantage of the fact that the utilized ScaLAPACK estimation routine `PDLACON` requires a column vector distributed over a single process column as right-hand side, and the global maximum is formed by a scalar *all-to-all reduction* [17] in each process row (which is negligible in terms of execution time). The column vector y in each process column is constructed by performing an *all-to-all broadcast* [17] of the local pieces of the right-hand-side matrix or matrices in each process row, forming P_c different right-hand-side vectors y_i. Altogether, we compute P_c different estimates (lower bounds of the associated sep^{-1}-function) and choose the largest value $\max_i \|x_i\|_2/\|y_i\|_2$, almost to the same cost in time as computing only one estimate.

6 Library Software Highlights

In this section, we give a brief account of some software issues for the RECSY and SCASY libraries. For more details, including calling sequences and parameter settings, we refer to the library web pages [45, 47] and references therein.

6.1 The RECSY Library

RECSY includes eight native recursive blocked routines for solving the reduced (quasi-triangular) matrix equations named `REC[ACRO]`, where "`[ACRO]`" is replaced by the acronym for the matrix equation to be solved (see Table 1). In addition to the routines listed above, a similar set of routines parallelized for OpenMP platforms are provided. The signatures for these routines are `REC[ACRO]_P`. All routines in RECSY are implemented in Fortran 90 for double precision real data.

It is also possible to use RECSY via wrapper routines that overload SLICOT [48] and LAPACK [4] routines that call triangular matrix equation solvers. In Fig. 2, the routine hierarchy of RECSY is illustrated.

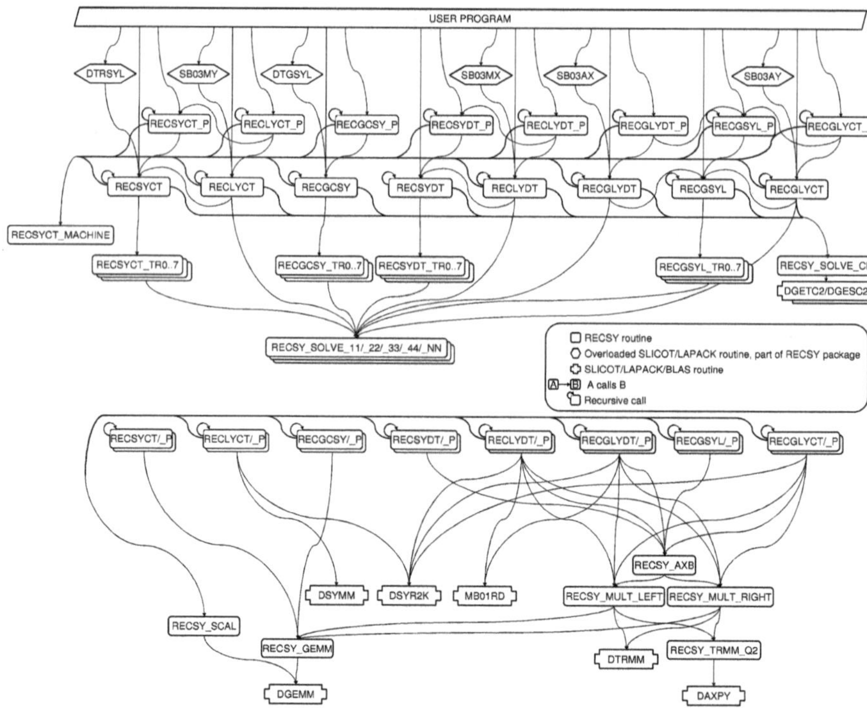

Fig. 2 Subroutine call graph of RECSY. Overloaded SLICOT and LAPACK routines are shown as hexagons. Level 3 BLAS routines and auxiliary BLAS, LAPACK, and SLICOT routines used by RECSY are displayed as crosses. RECSY routines that use recursive calls are embedded in a box with a self-referencing arrow.

6.2 The SCASY Library

SCASY includes ScaLAPACK-style general matrix equation solvers implemented as eight basic routines called `PGE[ACRO]D`, where "`P`" stands for *parallel*, "`GE`" stands for *general*, "`D`" denotes *double precision*, and "`[ACRO]`" is replaced by the acronym for the matrix equation to be solved. All parallel algorithms implemented are explicitly blocked variants of the Bartels-Stewart mehod (see [20]). These routines invoke the corresponding triangular solvers `PTR[ACRO]D`, where "`TR`" stands for *triangular*. Condition estimators `P[ACRO]CON` associated with each matrix equation are built on top of the triangular solvers accessed through the general solvers using a parameter setting that avoids the reduction part of the general algorithm.

In total, SCASY consists of 47 routines whose design depends on the functionality of a number of external libraries. The call graph in Fig. 3 shows the routine hierarchy of SCASY. The following external libraries are used in SCASY:

- ScaLAPACK [7, 49], including the PBLAS [43] and BLACS [8],
- LAPACK and BLAS [4],

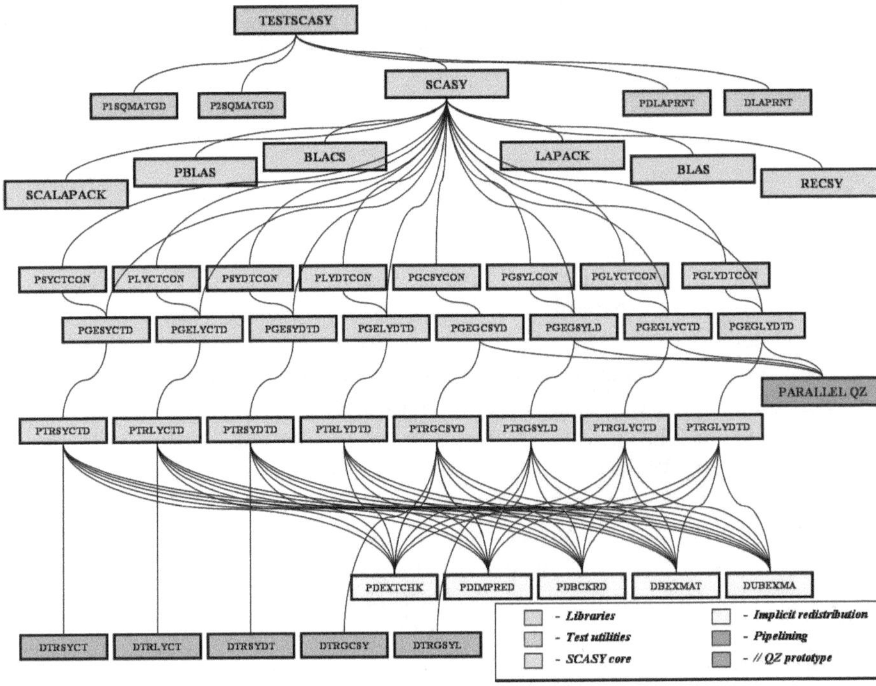

Fig. 3 Subroutine call graph of SCASY. The top three levels show testing routines and libraries called. The next three levels display the SCASY core, including routines for condition estimation, general and triangular solvers. The last two levels show routines for implicit redistribution of local data due to conjugate pairs of eigenvalues and pipelining for the one-sided routines.

- RECSY [31], which provides almost all node solvers except for one transpose case of the GCSY equation. Notice that RECSY in turn calls a small set of subroutines from SLICOT (*Software Library in Control*) [48, 14].

For example, the routines for the standard matrix equations utilize the ScaLAPACK routines PDGEHRD (performs an initial parallel Hessenberg reduction), PDLAHQR (implementation of the parallel unsymmetric QR algorithm presented in [26]), and PDGEMM (the PBLAS parallel implementation of the level-3 BLAS GEMM-operation). The triangular solvers employ the RECSY node solvers [29, 30, 31] and LAPACK's DTGSYL for solving (small) matrix equations on the nodes and the BLAS for the level-3 updates (DGEMM, DTRMM and DSYR2K operations). To perform explicit communication and coordination in the triangular solvers, we use the BLACS library.

SCASY may be compiled including node solvers from the OpenMP version of RECSY by defining the preprocessor variable OMP. By linking with a multi-threaded version of the BLAS, SCASY supports parallelization on both a global and on a node level on distributed memory platforms with SMP-aware nodes (see Sect. 4). The number of threads to use in the RECSY solvers and the threaded version of the BLAS is controlled by the user via environment variables, e.g., via `OMP_NUM_THREADS` for OpenMP and `GOTO_NUM_THREADS` for the threaded version of the GOTO-BLAS [16].

7 Experimental Results

In this section, we show some performance results of the RECSY and SCASY libraries. The results presented below illustrate the two levels of parallelization discussed in Sect. 4, i.e., the message passing model and the multithreading model for parallel computing. Two distributed memory platforms are used in the experiments, which all are conducted in double precision arithmetic ($\varepsilon_{\text{mach}} \approx 2.2 \times 10^{-16}$).

First, the Linux Cluster *seth*, which consists of 120 dual AMD Athlon MP2000+ nodes (1.667MHz, 384KB on-chip cache). Most nodes have 1GB memory, and a small number of nodes have 2GB memory. The cluster is connected with a Wolfkit3 SCI high-speed interconnect having a peak bandwidth of 667 MB/sec. The network connects the nodes in a 3-dimensional torus organized as a $6 \times 4 \times 5$ grid. We use the Portland Group's pgf90 6.0-5 compiler using the recommended compiler flags -O2 -tp athlonxp -fast and the following software libraries: ScaMPI (MPICH 1.2), LAPACK 3.0, ATLAS 3.5.9, ScaLAPACK / PBLAS 1.7.0, BLACS 1.1, RECSY 0.01alpha, and SLICOT 4.0.

Our second parallel target machine is the 64-bit Opteron Linux Cluster *sarek* with 192 dual AMD Opteron nodes (2.2GHz), 8Gb RAM per node and a Myrinet-2000 high-performance interconnect with 250 MB/sec bandwidth. We use the Portland Group's `pgf77 1.2.5 64-bit` compiler, the compiler flag `-fast` and the following software: MPICH-GM 1.5.2 [41], LAPACK 3.0 [39], GOTO-BLAS r0.94 [16], ScaLAPACK 1.7.0 and BLACS 1.1patch3 [7], and RECSY 0.01alpha [45].

The SCASY library provides parallel routines with high node performance, mainly due to RECSY, and good scaling properties. However, the inherent data dependencies of the matrix equations limit the level of concurrency and impose a lot of synchronization throughout the execution process. In Fig. 4, we display some measured parallel performance results keeping constant memory load (1.5GB) per processor. For large-scale problems, the scaled parallel speedup approaches $O(p/k)$ as projected by the theoretical analysis, where in this case $k = 3.18$ is used. Some parallel performance results of SCASY are also presented in Sect. 8, and for more extensive testings we refer to [20, 21].

In Fig. 5, we display the performance of the RECSY and SCASY solvers for the triangular SYCT using one processor (top graph), message passing with two processors on one node and one processor each on two nodes, respectively (two graphs in the middle), and multithreading using both processors on a *sarek* node (bottom graph). The results demonstrate that local node multithreading of distributed memory subtasks is often more beneficial than is parallelization using message passing.

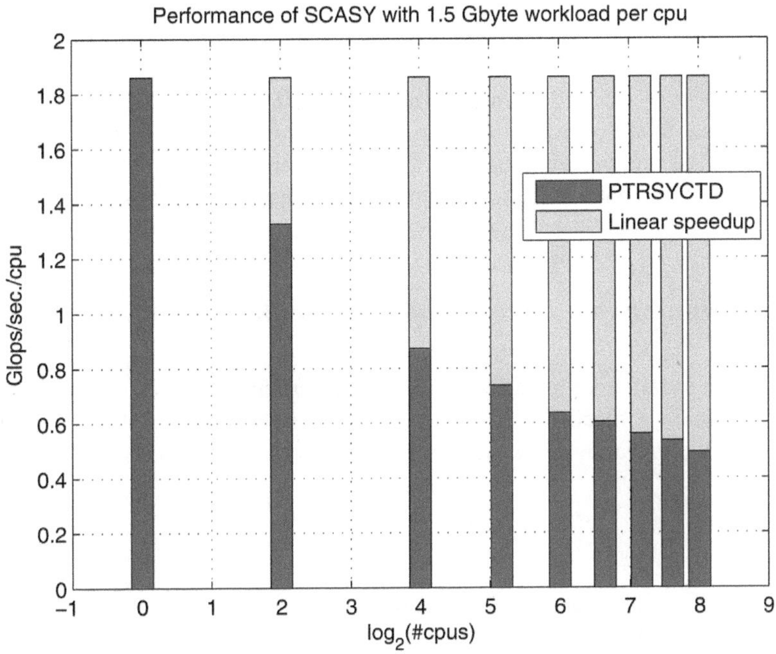

Fig. 4 Experimental performance of PTRSYCTD using $m_b = n_b = 64$, a constant memory load per processor of 1.5GB on *sarek*. The number of processors varies between 1 and 256.

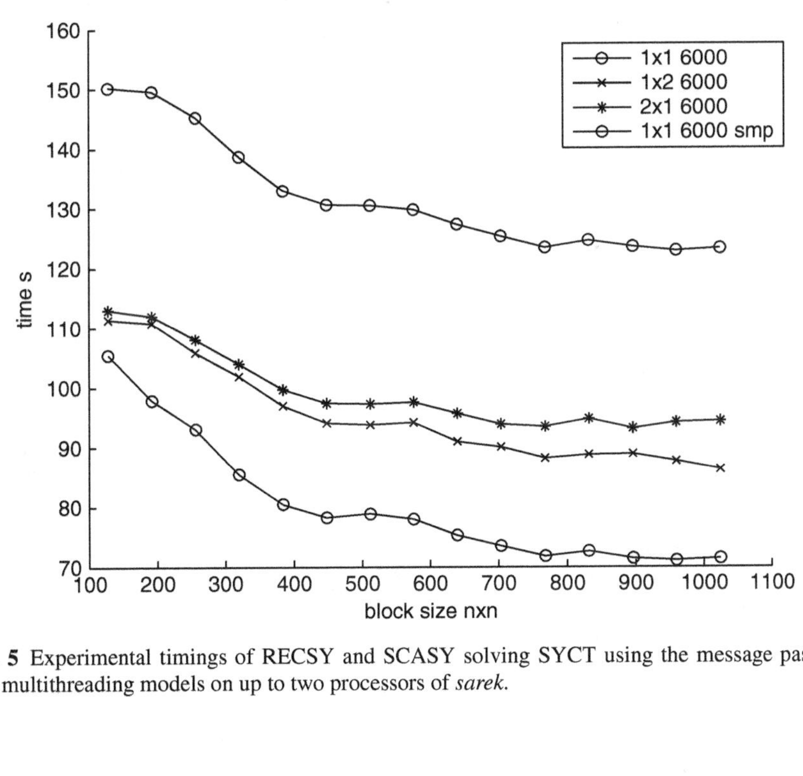

Fig. 5 Experimental timings of RECSY and SCASY solving SYCT using the message passing and multithreading models on up to two processors of *sarek*.

8 Some Control Applications and Extensions

In this section, we first demonstrate the use of RECSY and SCASY in parallel condition estimation of invariant subspaces of Hamiltonian matrices. Then, we describe how the recursive and explicit blocking strategies of RECSY and SCASY, respectively, are extended and applied to periodic matrix equations.

8.1 Condition Estimation of Subspaces with Specified Eigenvalues

We consider parallel computation of *c-stable* invariant subspaces corresponding to the eigenvalues $\{\lambda : \mathrm{Re}(\lambda) < 0\}$ and condition estimation of the selected cluster of eigenvalues and invariant subspaces of *Hamiltonian* matrices of the form

$$H = \begin{bmatrix} A & bb^T \\ cc^T & -A^T \end{bmatrix},$$

where $A \in \mathbb{R}^{\frac{n}{2} \times \frac{n}{2}}$ is a random diagonal matrix and $b, c \in \mathbb{R}^{\frac{n}{2} \times 1}$ are random vectors with real entries of uniform or normal distribution. For such matrices, $m = n/2 - k_{\mathrm{imag}}$ of the eigenvalues are c-stable, where k_{imag} is the number of eigenvalues that lie strictly on the imaginary axis. Solving such a Hamiltonian eigenvalue problem for the stable invariant subspaces can be very hard because the eigenvalues tend to cluster closely around the imaginary axis, especially when n gets large [38], leading to a very ill-conditioned *separation* problem.

A typical stable subspace computation includes the following steps:

1. Compute the real Schur form $T = Q^T H Q$ of H.
2. Reorder the m stable eigenvalues to the top left corner of T, i.e., compute an ordered Schur decomposition $\tilde{T} = \tilde{Q}^T H \tilde{Q}$, such that the m first columns of $\tilde{Q}$ span the stable invariant subspace of H; the m computed stable eigenvalues may be distributed over the diagonal of T in any order before reordering starts.
3. Compute condition estimates for (a) the selected cluster of eigenvalues and (b) the invariant subspaces.

For the last step, we utilize the condition estimation functionality and the corresponding parallel Sylvester-type matrix equation solvers provided by SCASY [20, 21, 47].

We show some experimental results in Table 3, where the following performance, output, and accuracy quantities are presented:

- The parallel runtime measures t_{3a}, t_{3b}, t_{tot} for the different condition estimation steps, the total execution time of steps 1–3 above, and the corresponding parallel speedup. We remark that the execution time is highly dominated by the current ScaLAPACK implementation of the non-symmetric QR algorithm `PDLAHQR`.
- The outputs m, s, and *sep* corresponding to the dimension of the invariant subspace and the reciprocal condition numbers of the selected cluster of eigenvalues

Table 3 Experimental parallel results from computing stable invariant subspaces of random Hamiltonian matrices on *seth*.

		Output			Timings			
n	$P_r \times P_c$	m	s	sep	t_{3a}	t_{3b}	t_{tot}	S_p
3000	1×1	1500	0.19E-03	0.14E-04	13.8	107	2497	1.00
3000	2×2	1503	0.27E-04	0.96E-06	6.79	70.2	1022	2.44
3000	4×4	1477	0.30E-04	0.40E-06	3.16	18.3	308	8.11
3000	8×8	1481	0.48E-03	0.57E-06	2.53	14.8	198	12.6
6000	2×2	2988	0.52E-03	0.32E-13	47.4	568	8449	1.00
6000	4×4	3015	0.57E-04	0.11E-17	19.7	267	2974	2.84
6000	8×8	2996	0.32E-04	0.25E-12	8.16	99.2	935	9.04

and the stable invariant subspace, respectively. The condition numbers s and sep are computed as follows:

- Solve the Sylvester equation $\tilde{T}_{11}X - X\tilde{T}_{22} = -\gamma\tilde{T}_{12}$, where

$$\tilde{T} = \begin{bmatrix} \tilde{T}_{11} & \tilde{T}_{12} \\ 0 & \tilde{T}_{22} \end{bmatrix}, \quad \text{and } \tilde{T}_{11} \in \mathbb{R}^{m\times m},$$

  compute the Frobenius norm of X in parallel, and set $s = 1/\sqrt{(1+\|X\|_F)}$.
- Compute a lower bound est of $\text{sep}^{-1}(\tilde{T}_{11}, \tilde{T}_{22})$ in parallel using the matrix norm estimation technique outlined in Sect. 5 and compute $sep = 1/est$ taking care of any risk for numerical overflow.

Both quantities $\|P\|_2$ and $\text{sep}(\tilde{T}_{11}, \tilde{T}_{22})$ appear in error bounds for a computed cluster of eigenvalues and the associated invariant subspace (see, e.g., [50]). Small values on s and sep signal ill-conditioning meaning that the computed cluster and the associated subspace can be sensitive for small perturbations in the data. If we compute a large-norm solution X, then $\|P\|_2 = \sqrt{(1+\|X\|_F)}$, i.e., the norm of the spectral projector P associated with $\tilde{T}_{11}$, becomes large and s becomes small.

8.2 Periodic Matrix Equations in CACSD

We have also extended and applied both the recursive blocking and the explicit blocking techniques to the solution of periodic matrix equations [18, 5]. Consider the periodic continuous-time Sylvester (PSYCT) equation

$$A_k X_k - X_{k+1} B_k = C_k,\ k = 0, 1, \ldots, K-1, \tag{7}$$

where $A_k \in \mathbb{R}^{m\times m}$, $B_k \in \mathbb{R}^{n\times n}$, and $C_k, X_k \in \mathbb{R}^{m\times n}$ are K-cyclic general matrices with real entries. A K-cyclic matrix is characterized by that it repeats itself in a sequence of matrices every Kth time, e.g., $A_K = A_0$, $A_{K+1} = A_1$, etc. Matrix equations of the form (7) have applications in, for example, computation and condition estimation of *periodic invariant subspaces* of square matrix products of the form

$$\mathscr{A}_{K-1}\cdots\mathscr{A}_1\mathscr{A}_0 \in \mathbb{R}^{l\times l}, \tag{8}$$

and in periodic systems design and analysis [51] related to *discrete-time periodic systems* of the form

$$\begin{aligned} x_{k+1} &= \mathscr{A}_k x_k + \mathscr{B}_k u_k \\ y_k &= \mathscr{C}_k x_k + \mathscr{D}_k u_k, \end{aligned} \tag{9}$$

with K-periodic system matrices $\mathscr{A}_k$, $\mathscr{B}_k$, $\mathscr{C}_k$, and $\mathscr{D}_k$, and the study of the *state transition matrix* of (9), defined as the square matrix $\Phi_{\mathscr{A}}(j,i) = \mathscr{A}_{j-1}\mathscr{A}_{j-2}\dots\mathscr{A}_i$, where $\Phi_{\mathscr{A}}(i,i)$ is the identity matrix. The state transition matrix over one period $\Phi_{\mathscr{A}}(j+K,j)$ is called the *monodromy matrix* of (9) at time j (its eigenvalues are called the *characteristic multipliers* at time j) and is closely related to the matrix product (8).

We use script notation for the matrices in (8) and the system matrices in (9), as the matrices that appear in the periodic matrix equations, like PSYCT, can be subarrays (blocks) of the scripted matrices in *periodic Schur form* [9, 25].

In the following, we assume that the periodic matrices A_k and B_k of PSYCT are already in periodic real Schur form (see [9, 25]). This means that $K-1$ of the matrices in each sequence are upper triangular and one matrix in each sequence, say A_r and B_s, $0 \leq r,s \leq K-1$, is quasi-triangular. The products of conforming diagonal blocks of the matrix sequences A_k and B_k contain the eigenvalues of the matrix products $A_{K-1}A_{K-2}\cdots A_0$ and $B_{K-1}B_{K-2}\cdots B_0$, respectively, where the 1×1 and 2×2 blocks on the main block diagonal of A_r and B_s correspond to real and complex conjugate pairs of eigenvalues of the corresponding matrix products. In other words, we consider a reduced (triangular) periodic matrix equation as in the non-periodic cases.

Recursive Blocking

As for the non-periodic matrix equations, we consider three ways of recursive splitting of the matrix sequences involved: A_k is split by rows and columns and C_k by rows only; B_k is split by rows and columns and C_k by columns only; all three matrix sequences are split by rows and columns. No matter which alternative is chosen, the number of flops is the same. Performance may differ greatly, though. Our algorithm picks the alternative that keeps matrices as "squarish" as possible, i.e., $1/2 < m/n < 2$, which guarantees a good ratio between the number of flops and the number of elements referenced.

Now, small instances of periodic matrix equations have to be solved at the end of the recursion tree. Each such matrix equation can be represented as a linear system $Zx = c$, where Z is a Kronecker product representation of the associated periodic Sylvester-type operator, and it belongs to the class of bordered almost block diagonal (BABD) matrices [15]. For example, the PSYCT matrix equation can be expressed as $Z_{\mathrm{PSYCT}}x = c$, where the matrix Z of size $mnK \times mnK$ is

$$Z_{\mathrm{PSYCT}} = \begin{bmatrix} B_{K-1}^T \otimes I_m & & & & I_n \otimes A_{K-1} \\ I_n \otimes A_0 & B_0^T \otimes I_m & & & \\ & \ddots & \ddots & & \\ & & & I_n \otimes A_{K-2} & B_{K-2}^T \otimes I_m \end{bmatrix},$$

and

$$x = [\mathrm{vec}(X_0), \mathrm{vec}(X_1), \cdots, \mathrm{vec}(X_{K-1})]^T, \quad c = [\mathrm{vec}(C_0), \cdots, \mathrm{vec}(C_{K-1})]^T.$$

In the algorithm, recursion proceeds down to problem sizes of 1×1 to 2×2. For these problems, a compact form of Z_{PSYCT} that preserves the sparsity structure is used, and the linear system is solved using Gaussian elimination with partial pivoting (GEPP). These solvers are based on the superscalar kernels that were developed for the RECSY library [45]. For more details, e.g., relating to ill-conditioning and the storage layout of the periodic matrix sequences, we refer to [18].

Explicit Blocking

If A_k and B_k are partitioned by square $m_b \times m_b$ and $n_b \times n_b$ blocks, respectively, we can rewrite (7) in block partitioned form

$$A_{ii}^{(k)} X_{ij}^{(k)} - X_{ij}^{(k+1)} B_{jj}^{(k)} = C_{ij}^{(k)} - \Big(\sum_{k=i+1}^{D_A} A_{ik}^{(k)} X_{kj}^{(k)} - \sum_{k=1}^{j-1} X_{ik}^{(k+1)} B_{kj}^{(k)} \Big), \tag{10}$$

where $D_A = \lceil m/m_b \rceil$. The summations in (10) can be implemented as a serial blocked algorithm using a couple of nested loops. For high performance and

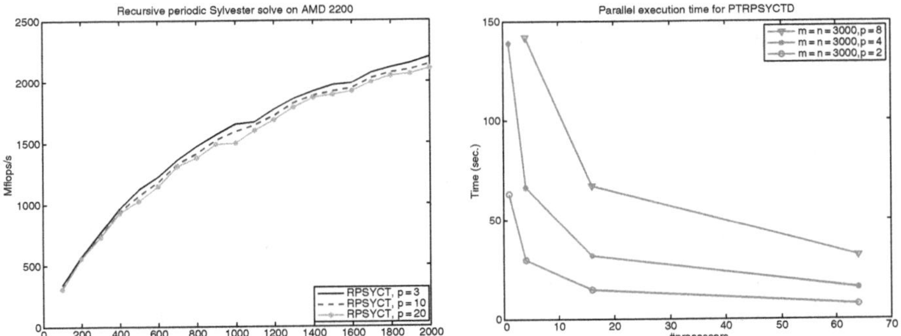

Fig. 6 *Left:* Performance results for the recursive blocked PSYCT solver with $m = n$ on AMD Opteron for varying periodicity K $(= 3, 10, 20)$. *Right:* Execution time results for the parallel DM solver `PTRPSYCTD` with $m = n = 3000$ and $K = 2, 4$, and 8 on *sarek* using the block sizes $m_b = n_b = 64$. ($K = p$ in the legends.)

portability, level 3 BLAS (mostly GEMM operations) should be utilized for the periodic right-hand-side updates.

Just as in the non-periodic case, the explicit blocking of PSYCT in (10) reveals that all subsolutions $X_{ij}^{(k)}$ located on the same block diagonal of each X_k are independent and can be computed in parallel. As for the non-periodic matrix equations, we use recursive blocked node solvers (briefly described above). In addition, all subsequent updates are mutually independent and can be performed in parallel.

Some performance results are displayed in Fig. 6. The left part displays results for the recursive blocked PSYCT algorithm. The problem size ($m = n$) ranges from 100 to 2000, and the periods $K = 3, 10$, and 20. For large enough problems, the performance approaches 70% of the DGEMM performance, which is comparable with the recursive blocked SYCT solver in RECSY [29, 45]. For an increasing period K, the performance decreases only marginally. The right part of Fig. 6 shows timing results for the explicitly blocked parallel DM solver `PTRPSYCTD` with $m = n = 3000$ and the periods $K = 2, 4$, and 8. A general observation is that an increase of the number of processors by a factor of 4 cuts down the parallel execution time by roughly a factor of 2. This is consistent with earlier observations for the non-periodic case.

Acknowledgments The research was conducted using the resources of the High Performance Computing Center North (HPC2N). Financial support was provided by the *Swedish Research Council* under grant VR 621-2001-3284 and by the *Swedish Foundation for Strategic Research* under grant SSF A3 02:128.

References

1. Adlerborn, B., Dackland, K., Kågström, B.: Parallel two-stage reduction of a regular matrix pair to Hessenberg-triangular form. In: T. Sørvik et al. (eds.) Applied Parallel Computing: New Paradigms for HPC Industry and Academia, *Lecture Notes in Computer Science*, vol. 1947, pp. 92–102. Springer (2001)
2. Adlerborn, B., Dackland, K., Kågström, B.: Parallel and blocked algorithms for reduction of a regular matrix pair to Hessenberg-triangular and generalized Schur forms. In: J. Fagerholm et al. (eds.) PARA 2002, *Lecture Notes in Computer Science*, vol. 2367, pp. 319–328. Springer-Verlag (2002)
3. Adlerborn, B., Kågström, B., Kressner, D.: Parallel variants of the multishift QZ algorithm with advanced deflation techniques. In: B. Kågström et al. (eds.) Applied Parallel Computing - State of the Art in Scientific Computing (PARA'06), *Lecture Notes in Computer Science*, vol. 4699, pp. 117–126. Springer (2007)
4. Anderson, E., Bai, Z., Bischof, C., Blackford, S., Demmel, J.W., Dongarra, J.J., Du Croz, J., Greenbaum, A., Hammarling, S., McKenney, A., Sorensen, D.C.: LAPACK Users' Guide, 3rd edn. SIAM, Philadelphia, PA (1999)
5. Andersson, P., Granat, R., Jonsson, I., Kågström, B.: Parallel algorithms for triangular periodic Sylvester-type matrix equations. In: E. Luque et al. (eds.) Euro-Par 2008 — Parallel Processing, *Lecture Notes in Computer Science*, vol. 5168, pp. 780–789. Springer (2008)
6. Bartels, R.H., Stewart, G.W.: Algorithm 432: the solution of the matrix equation $AX - BX = C$. Communications of the ACM **8**, 820–826 (1972)
7. Blackford, L.S., Choi, J., Cleary, A., D'Azevedo, E., Demmel, J.W., Dhillon, I., Dongarra, J.J., Hammarling, S., Henry, G., Petitet, A., Stanley, K., Walker, D., Whaley, R.C.: ScaLAPACK Users' Guide. SIAM, Philadelphia, PA (1997)

8. BLACS - Basic Linear Algebra Communication Subprograms. URL http://www.netlib.org/ blacs/index.html
9. Bojanczyk, A., Golub, G.H., Van Dooren, P.: The periodic Schur decomposition; algorithm and applications. In: *Proc. SPIE Conference*, vol. 1770, pp. 31–42 (1992)
10. Dackland, K., Kågström, B.: Blocked algorithms and software for reduction of a regular matrix pair to generalized Schur form. ACM Trans. Math. Software **25**(4), 425–454 (1999)
11. Demmel, J., Dongarra, J., Parlett, B., Kahan, W., Gu, M., Bindel, D., Hida, Y., Li, X., Marques, O., Riedy, J., Vömel, C., Langou, J., Luszczek, P., Kurzak, J., Buttari, A., Langou, J., Tomov, S.: Prospectus for the next LAPACK and ScaLAPACK libraries. In: B. Kågström et al. (eds.) Applied Parallel Computing - State of the Art in Scientific Computing (PARA'06), *Lecture Notes in Computer Science*, vol. 4699, pp. 11–23. Springer (2007)
12. Dongarra, J.J., Du Croz, J., Duff, I.S., Hammarling, S.: A set of level 3 basic linear algebra subprograms. ACM Trans. Math. Software **16**, 1–17 (1990)
13. Elmroth, E., Gustavson, F., Jonsson, I., Kågström, B.: Recursive blocked algorithms and hybrid data structures for dense matrix library software. SIAM Review **46**(1), 3–45 (2004)
14. Elmroth, E., Johansson, P., Kågström, B., Kressner, D.: A web computing environment for the SLICOT library. In: The Third NICONET Workshop on Numerical Control Software, pp. 53–61 (2001)
15. Fairweather, G., Gladwell, I.: Algorithms for almost block diagonal linear systems. SIAM Review **44**(1), 49–58 (2004)
16. GOTO-BLAS - High-Performance BLAS by Kazushige Goto. URL http://www.cs.utexas.edu/users/flame/goto/
17. Grama, A., Gupta, A., Karypsis, G., Kumar, V.: Introduction to Parallel Computing, 2nd edn. Addison-Wesley (2003)
18. Granat, R., Jonsson, I., Kågström, B.: Recursive blocked algorithms for solving periodic triangular Sylvester-type matrix equations. In: B. Kågström et al. (eds.) Applied Parallel Computing - State of the Art in Scientific Computing (PARA'06), *Lecture Notes in Computer Science*, vol. 4699, pp. 531–539. Springer (2007)
19. Granat, R., Kågström, B.: Parallel algorithms and condition estimators for standard and generalized triangular Sylvester-type matrix equations. In: B. Kågström et al. (eds.) Applied Parallel Computing - State of the Art in Scientific Computing (PARA'06), *Lecture Notes in Computer Science*, vol. 4699, pp. 127–136. Springer (2007)
20. Granat, R., Kågström, B.: Parallel solvers for Sylvester-type matrix equations with applications in condition estimation, Part I: theory and algorithms. Report UMINF 07.15, Dept. of Computing Science, Umeå University, Sweden. Submitted to ACM Trans. Math. Software (2007)
21. Granat, R., Kågström, B.: Parallel solvers for Sylvester-type matrix equations with applications in condition estimation, Part II: the SCASY software. Report UMINF 07.16, Dept. of Computing Science, Umeå University, Sweden. Submitted to ACM Trans. Math. Software (2007)
22. Granat, R., Kågström, B., Poromaa, P.: Parallel ScaLAPACK-style algorithms for solving continuous-time Sylvester equations. In: H. Kosch et al. (eds.) Euro-Par 2003 Parallel Processing, *Lecture Notes in Computer Science*, vol. 2790, pp. 800–809. Springer (2003)
23. Hager, W.W.: Condition estimates. SIAM J. Sci. Statist. Comput. **3**, 311–316 (1984)
24. Hammarling, S.J.: Numerical solution of the stable, non-negative definite Lyapunov equation. IMA Journal of Numerical Analysis **2**, 303–323 (1982)
25. Hench, J.J., Laub, A.J.: Numerical solution of the discrete-time periodic Riccati equation. IEEE Trans. Automat. Control **39**(6), 1197–1210 (1994)
26. Henry, G., Watkins, D.S., Dongarra, J.J.: A parallel implementation of the nonsymmetric QR algorithm for distributed memory architectures. SIAM J. Sci. Comput. **24**(1), 284–311 (2002)
27. Higham, N.J.: Fortran codes for estimating the one-norm of a real or complex matrix, with applications to condition estimation. ACM Trans. Math. Software **14**(4), 381–396 (1988)
28. Higham, N.J.: Perturbation theory and backward error for $AX - XB = C$. BIT **33**(1), 124–136 (1993)

29. Jonsson, I., Kågström, B.: Recursive blocked algorithms for solving triangular systems. I. One-sided and coupled Sylvester-type matrix equations. ACM Trans. Math. Software **28**(4), 392–415 (2002)
30. Jonsson, I., Kågström, B.: Recursive blocked algorithms for solving triangular systems. II. Two-sided and generalized Sylvester and Lyapunov matrix equations. ACM Trans. Math. Software **28**(4), 416–435 (2002)
31. Jonsson, I., Kågström, B.: RECSY - a high performance library for solving Sylvester-type matrix equations. In: H. Kosch et al. (eds.) Euro-Par 2003 Parallel Processing, *Lecture Notes in Computer Science*, vol. 2790, pp. 810–819. Springer (2003)
32. Kågström, B., Kressner, D.: Multishift variants of the QZ algorithm with aggressive early deflation. SIAM Journal on Matrix Analysis and Applications **29**(1), 199–227 (2006)
33. Kågström, B., Ling, P., Van Loan, C.: GEMM-based level 3 BLAS: high-performance model implementations and performance evaluation benchmark. ACM Trans. Math. Software **24**(3), 268–302 (1998)
34. Kågström, B., Ling, P., Van Loan, C.: Algorithm 784: GEMM-based level 3 BLAS: portability and optimization issues. ACM Trans. Math. Software **24**(3), 303–316 (1998)
35. Kågström, B., Poromaa, P.: Distributed and shared memory block algorithms for the triangular Sylvester equation with sep^{-1} estimators. SIAM J. Matrix Anal. Appl. **13**(1), 90–101 (1992)
36. Kågström, B., Poromaa, P.: LAPACK-style algorithms and software for solving the generalized Sylvester equation and estimating the separation between regular matrix pairs. ACM Trans. Math. Software **22**(1), 78–103 (1996)
37. Kågström, B., Westin, L.: Generalized Schur methods with condition estimators for solving the generalized Sylvester equation. IEEE Trans. Autom. Contr. **34**(4), 745–751 (1989)
38. Kressner, D.: Numerical methods and software for general and structured eigenvalue problems. Ph.D. thesis, TU Berlin, Institut für Mathematik, Berlin, Germany (2004)
39. LAPACK - Linear Algebra Package. URL http://www.netlib.org/lapack/
40. Moler, C.B., Stewart, G.W.: An algorithm for generalized matrix eigenvalue problems. SIAM J. Numer. Anal. **10**, 241–256 (1973)
41. MPI - Message Passing Interface. URL http://www-unix.mcs.anl.gov/mpi/
42. OpenMP - Simple, Portable, Scalable SMP Programming. URL http://www.openmp.org/
43. PBLAS - Parallel Basic Linear Algebra Subprograms. URL http://www.netlib.org/scalapack/pblas
44. Quintana-Ortí, E.S., van de Geijn, R.A.: Formal derivation of algorithms: The triangular Sylvester equation. ACM Transactions on Mathematical Software **29**(2), 218–243 (2003)
45. RECSY - High Performance library for Sylvester-type matrix equations. URL http://www.cs.umu.se/research/parallel/recsy
46. Reid, J., Numrich, R.W.: Co-arrays in the next Fortran standard. Scientific Programming **15**(1), 9–26 (2007)
47. SCASY - ScaLAPACK-style solvers for Sylvester-type matrix equations. URL http://www.cs.umu.se/research/parallel/scasy
48. SLICOT Library In The Numerics In Control Network (Niconet). URL http://www.win.tue.nl/niconet/index.html
49. ScaLAPACK Users' Guide. URL http://www.netlib.org/scalapack/slug/
50. Stewart, G.W., Sun, J.-G.: Matrix Perturbation Theory. Academic Press, New York (1990)
51. Varga, A., Van Dooren, P.: Computational methods for periodic systems - an overview. In: Proc. of IFAC Workshop on Periodic Control Systems, Como, Italy, pp. 171–176. International Federation of Automatic Control (IFAC) (2001)

Parallelization of Linear Algebra Algorithms Using ParSol Library of Mathematical Objects

Alexander Jakušev, Raimondas Čiegis, Inga Laukaitytė, and Vyacheslav Trofimov

Abstract The linear algebra problems are an important part of many algorithms, such as numerical solution of PDE systems. In fact, up to 80% or even more of computing time in this kind of algorithms is spent for linear algebra tasks. The parallelization of such solvers is the key for parallelization of many advanced algorithms. The mathematical objects library ParSol not only implements some important linear algebra objects in C++, but also allows for semiautomatic parallelization of data parallel and linear algebra algorithms, similar to High Performance Fortran (HPF). ParSol library is applied to implement the finite difference scheme used to solve numerically a system of PDEs describing a nonlinear interaction of two counter-propagating laser waves. Results of computational experiments are presented.

1 Introduction

The numerical solution of PDEs and systems of PDEs is the basis of mathematical modeling and simulation of various important industrial, technological, and engineering problems. The requirements for the size of the discrete problem and speed of its solution are growing every day, and parallelization of PDE solvers is one of the ways to meet those requirements. All the most popular PDE solvers today have parallel versions available. However, new methods and tools to solve PDEs and systems of PDEs appear constantly, and parallelization of these solvers becomes a very important task. Such a goal becomes even more urgent as multicore computers and clusters of such computers are used practically in all companies and universities.

Alexander Jakušev · Raimondas Čiegis · Inga Laukaitytė
Vilnius Gediminas Technical University, Saulėtekio al. 11, LT–10223, Vilnius, Lithuania
e-mail: {alexj · rc · Inga.Laukaityte}@fm.vgtu.lt

Vyacheslav Trofimov
M. V. Lomonosov Moscow State University, Vorob'evy gory, 119992, Russia
e-mail: vatro@cs.msu.su

Let us consider parallelization details of such well-known PDE solvers as Clawpack [13], Diffpack [12], DUNE [4], OpenFOAM [16], PETSc [1], and UG [3]. All of them share common features:

1. Parallelization is implemented on discretization and linear algebra layers, where parallel versions of linear algebra data structures and algorithms are provided. Data parallel parallelization model is used in all cases.
2. Using the PDE solvers, the user can produce parallel version of the code semi-automatically, however, it is required to specify how data is divided among processors and to define a stencil used in the discretization step.
3. The parallelization code is usually tightly connected with the rest of the solver, making it practically impossible to reuse the code for the parallelization of another PDE solver. However, widespread low-level parallelization libraries, especially MPI [17], are used for actual interprocess communications.

All these features, except for the tight connection to the rest of the code, are important and are used in our tool. However, the tight integration with the solvers prevents the reuse of parallelization code. For quick and efficient parallelization of various PDE solvers, a specific library should be used.

Many parallelization libraries and standards exist; they may be grouped by parallelization model they are using. The following models are the most popular ones:

Multithreading. This model is best suited for the Symmetric Multiprocessing (SMP) type of computers. OpenMP standard is a good representative of this model. However, this model is not suited for computer clusters and distributed parallel memory computers.

Message passing. This model is best suited for systems with distributed memory, such as computer clusters. MPI [17] and PVM [8] standards are good representatives of this model. However, this is a low-level model and it is difficult to use it efficiently.

Data parallel. This model is ideal if we want to achieve semi-automatic parallelization quickly. HPF [10] and FreePOOMA are good representatives of this model. However, in its pure form this model is pretty restricted in functionality.

Global memory. This parallel programming model may be used for parallelization of almost any kind of algorithms. PETSc [2] and Global Arrays [14] libraries are good representatives of this model. However, the tools implementing this model are difficult to optimize, which may produce inefficient code.

The conclusions above allow us to say that modern parallelization tools are still not perfect for parallelization of PDE solvers. In this chapter, the new parallelization library is described. Its main goal is parallelization of PDE solvers.

Our chapter is organized as follows. In Sect. 2, we describe main principles and details of the implementation of the library. The finite difference scheme for solution of one problem of nonlinear optics is described in Sect. 3. This algorithm is implemented by using ParSol tool. The obtained parallel algorithm is tested on the cluster of PCs in Vilnius Gediminas Technical University (VGTU). Results of computational experiments are presented. Some final conclusions are made in Sect. 4.

2 The Principles and Implementation Details of ParSol Library

The ParSol library was designed to overcome some problems of the models mentioned above. It employs the elements of both *data parallel* and *global memory* parallel programming models. From the *data parallel* model, the library takes parallelizable objects and operations with them. For linear algebra, these objects are arrays, vectors, matrices. It is important to note that BLAS level 2 compatible operations are implemented in sequential and parallel modes. The same functionality was previously implemented in HPF and PETSc tools.

In order to overcome the lack of features of the *data parallel* model, ParSol library utilizes some elements of *global memory* model. Like in *global memory* model, arrays and vectors in the library have global address space. However, there are strict limits for the interprocess communications. Any given process can communicate with only a specified set of the other processes, and to exchange with them elements defined by some a priori stencil. Such a situation is typical for many solvers of PDEs. It is well-known that for various difference schemes, only certain neighbour elements have to be accessed. The position of those neighbor elements is described by *stencil* of the grid.

Rather than trying to guess the shape of stencil from the program code, as HPF does, the library asks the programmer to specify the stencil, providing convenient means for that. While requiring some extra work from the programmer, it allows one to specify the exact necessary stencil to reduce the amount of transferred data to the minimum.

2.1 *Main Classes of ParSol*

Next we list the main classes of the ParSol library:

Arrays are the basic data structure, which may be automatically parallelized.

Vectors are arrays with additional mathematical functionality of linear algebra. Vectors are more convenient for numerical algorithms, whereas arrays are good for parallelization of algorithms where the data type of arrays does not support linear algebra operations.

Stencils are used to provide information about which neighbor elements are needed during computations.

Matrices are used in matrix vector multiplications and other operations that arise during solution of linear systems.

Array elements are internally stored in 1D array; however, the user is provided with multidimensional element indexes. The index transformations are optimized thus the given interface is very efficient. Also, arrays implement dynamically calculated boundaries, which are adjusted automatically in parallel versions.

Cyclic arrays are implemented in a specific way, using additional shadow elements where the data is copied from the opposite side of the array. This increases array footprint slightly and requires user to specify data exchange directives even in a sequential version; however, the element access does not suffer, as no additional calculations are necessary.

ParSol arrays are implemented as template classes, so it is possible to have arrays of various data types, such as integers, floats or complex numbers.

Vectors provide additional functionality on top of the arrays, such as global operations, multiplication by a constant, and calculation of various norms. All these operations make a basis of many popular numerical algorithms used to solve PDEs and systems of such equations.

The library provides the user with both dense and sparse matrices. The dense matrices are stored in 2D array, whereas sparse matrices are stored in the CSR (Compressed Sparse Row) format. We note that the same format is used in PETSc library [2, p. 57]. For sparse matrices, it is possible to estimate a priori the number of preallocated elements if the stencil of the vector is known. The matrix dimensions are defined at the creation stage, using a vector given in the constructor of the matrix (thus matrix inherits a lot of its properties that the matrix will be used in conjunction with).

The parallelization scheme is shown in Fig. 1. The arrays (and thus vectors) are divided among processors depending on the given topology of processors. Similar to the case of stencils, the library provides an interface for setting the topology through special *Topology* classes and the stencil. When data exchange starts, neighbour elements are copied to the shadow area, where they must be read-only. Several methods of data exchange are implemented in ParSol: all-at-once, pair-by-pair, and pair-by-pair-ordered.

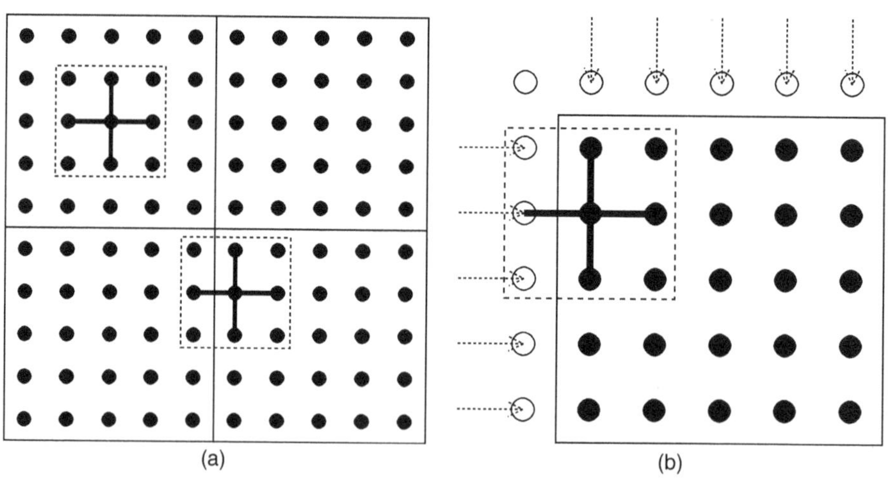

Fig. 1 Transition from sequential to parallel array: (a) the sequential array and its stencil, (b) part of the parallel array for one processor.

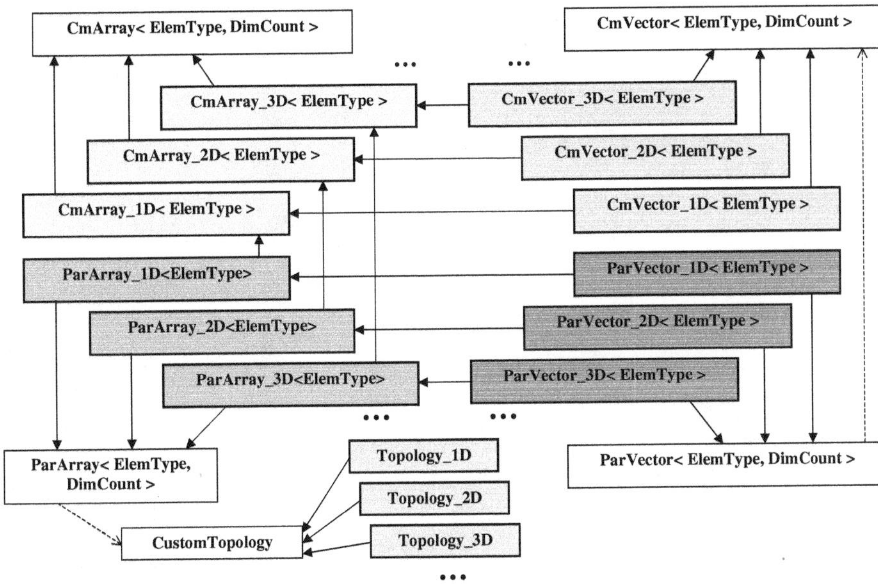

Fig. 2 Class diagram of ParSol arrays and vectors.

2.2 Implementation of ParSol

The library has been implemented in C++ [18], using such C++ features as OOP, operator overloading, and template metaprogramming [9]. This implementation is similar to such numerical libraries as Boost++ and FreePOOMA. Only standard C++ features are used, resulting in high portability of the library. As the library uses some advanced C++ features, modern C++ compilers are required.

MPI-1.1 standard is used for the parallel version of the library. The parallel versions of the classes are implemented as children of sequential analogical classes, ensuring that the code should not be changed much during parallelization. The size of the library code is ≈ 20000 lines, with ≈ 10000 more lines for regression tests. ParSol consists of more than 70 various classes, a subset of which is shown in Fig. 2. The library is available on the Internet at http://techmat.vgtu.lt/~alexj/ParSol/.

3 Parallel Algorithm for Simulation of Counter-propagating Laser Beams

In the domain (see Fig. 3)

$$D = \{0 \le z \le L_z, \quad 0 \le x_k \le L_x, \ k = 1,2\}$$

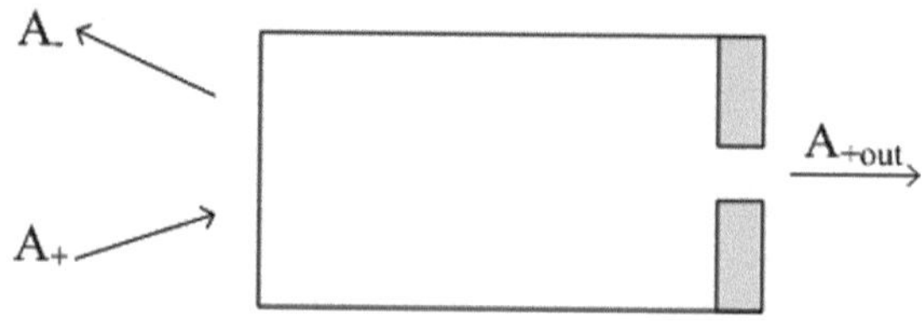

Fig. 3 Scheme of the interaction of counter-propagating laser pulses.

dimensionless equations and boundary conditions describing the interaction of two counter-propagating laser beams are given by [15]:

$$\frac{\partial A^+}{\partial t} + \frac{\partial A^+}{\partial z} + i\sum_{k=1}^{2} D_k \frac{\partial^2 A^+}{\partial x_k^2} + i\gamma(0.5|A^+|^2 + |A^-|^2)A^+ = 0,$$

$$\frac{\partial A^-}{\partial t} - \frac{\partial A^-}{\partial z} + i\sum_{k=1}^{2} D_k \frac{\partial^2 A^-}{\partial x_k^2} + i\gamma(0.5|A^-|^2 + |A^+|^2)A^- = 0,$$

$$A^+(z=0,x_1,x_2,t) = A_0(t)\exp\Big(-\sum_{k=1}^{2}\frac{(x_k - x_{ck})^{m_k}}{r_{pk}}\Big),$$

$$A^-(z=L_z,x_1,x_2,t) = A^+(z=L_z,x_1,x_2,t)R_0$$
$$\times\Big(1-\exp\Big(-\sum_{k=1}^{2}\frac{(x_k - x_{mk})^{q_k}}{R_{ak}}\Big)\Big)\exp\Big(i\sum_{k=1}^{2}\frac{(x_k - x_{mk})^2}{R_{mk}}\Big).$$

Here $A^{\pm}$ are complex amplitudes of counter-propagating pulses, γ characterizes the nonlinear interaction of laser pulses, x_{ck} are coordinates of the beam center, r_{pk} are radius of input beam on the transverse coordinates, and $A_0(t)$ is a temporal dependence of input laser pulses. In the boundary conditions, R_0 is the reflection coefficient of the mirror, R_{ak} are the radius of the hole along the transverse coordinates, x_{mk} are coordinates of the hole center, and R_{mk} characterize curvature of the mirror.

At the initial time moment, the amplitudes of laser pulses are equal to zero

$$A^{\pm}(z,x_1,x_2,0) = 0, \quad (z,x_1,x_2) \in D.$$

Boundary conditions along transverse coordinates are equal to zero.

We note that in previous papers, the mathematical model was restricted to 1D transverse coordinate case.

3.1 Invariants of the Solution

It is well-known that the solution of the given system satisfies some important invariants [19]. Here we will consider only one of such invariants. Multiplying differential equations by $(A^{\pm})^*$ and integrating over

$$Q(z,t,h) = [z-h,z] \times [0,L_x] \times [0,L_x] \times [t-h,t],$$

we prove that the full energy of each laser pulse is conserved during propagation along the characteristic directions $z \pm t$:

$$\|A^+(z,t)\| = \|A^+(z-h,t-h)\|, \quad \|A^-(z-h,t)\| = \|A^-(z,t-h)\|, \qquad (1)$$

where the L_2 norm is defined as

$$\|A(z,t)\|^2 = \int_0^{L_x} \int_0^{L_x} |A|^2 \, dx_1 dx_2.$$

This invariant describes a very important feature of the solution and therefore it is important to guarantee that the discrete analogs are satisfied for the numerical solution. In many cases, this helps to prove the existence and convergence of the discrete solution.

3.2 *Finite Difference Scheme*

In the domain $[0,T] \times D$, we introduce a uniform grid $\Omega = \Omega_t \times \Omega_z \times \Omega_x$, where

$$\Omega_t = \{t^n = nh_t, \ n = 0,\ldots,N\}, \quad \Omega_z = \{z_j = jh_z, \ j = 0,\ldots,J\},$$

$$\Omega_x = \{(x_{1l}, x_{2m}), \ x_{km} = mh_x, \ k = 1,2, \ m = 0,\ldots,M\}.$$

In order to approximate the transport part of the differential equations by using the finite differences along the characteristics $z \pm t$, we take $h_t = h_z$.

Let us denote discrete functions, defined on the grid Ω by

$$E^{\pm,n}_{j,lm} = E^{\pm}(z_j, x_{1l}, x_{2m}, t^n).$$

We also will use the following operators: $\beta(E,W) = \gamma(0.5|E|^2 + |W|^2)$,

$$\bar{E}^+ = \frac{E_j^{+,n} + E_{j-1}^{+,n-1}}{2}, \quad \bar{E}^- = \frac{E_{j-1}^{-,n} + E_j^{-,n-1}}{2},$$

$$\mathscr{D}E_{j,kl} = D_1 \frac{E_{j,l+1,m} - 2E_{j,lm} + E_{j,l-1,m}}{h_x^2} + D_2 \frac{E_{j,l,m+1} - 2E_{j,lm} + E_{j,l,m-1}}{h_x^2}.$$

Then the system of differential equations is approximated by the following finite difference scheme

$$\frac{E_j^{+,n} - E_{j-1}^{+,n-1}}{h_t} + i\mathscr{D}\bar{E}_j^+ + i\beta(\bar{E}_j^+, \bar{E}_j^-)\bar{E}_j^+ = 0, \qquad (2)$$

$$\frac{E_{j-1}^{-,n} - E_j^{+,n-1}}{h_t} + i\mathscr{D}\bar{E}_j^- + i\beta(\bar{E}_j^-, \bar{E}_j^+)\bar{E}_j^- = 0$$

with corresponding boundary and initial conditions.

We will prove that this scheme is conservative, i.e., the discrete invariants are satisfied for its solution. Let us define the scalar product and L_2 norm of the discrete functions as

$$(U,V) = \sum_{l=1}^{M-1}\sum_{m=1}^{M-1} U_{lm}V_{lm}^* h_x^2, \quad \|U\|^2 = (U,U).$$

Taking a scalar product of equations (2) by $\bar{E}^+$ and $\bar{E}^-$ respectively and considering the real parts of the equations, we get that the discrete analogs of the invariants (1) are satisfied

$$\|E_j^{+,n}\| = \|E_{j-1}^{+,n-1}\|, \quad \|E_{j-1}^{-,n}\| = \|E_j^{-,n-1}\|, \quad j = 1,\dots,J.$$

Finite difference scheme is nonlinear, thus at each time level t^n and for each $j = 1,\dots,J$ the simple linearization algorithm is applied:

$$\frac{E_j^{+,n,s+1} - E_{j-1}^{+,n-1}}{h_t} + i\mathscr{D}\bar{E}_j^{+,s+1} + i\beta(\bar{E}_j^{+,s}, \bar{E}_j^{-,s})\bar{E}_j^{+,s} = 0, \qquad (3)$$

$$\frac{E_{j-1}^{-,n,s+1} - E_j^{+,n-1}}{h_t} + i\mathscr{D}\bar{E}_j^{-,s+1} + i\beta(\bar{E}_j^{-,s}, \bar{E}_j^{+,s})\bar{E}_j^{-,s} = 0,$$

where the following notation is used

$$\bar{E}^{+,s} = \frac{E_j^{+,n,s} + E_{j-1}^{+,n-1}}{2}, \quad \bar{E}^{-,s} = \frac{E_{j-1}^{-,n,s} + E_j^{-,n-1}}{2}.$$

The iterations are continued until the convergence criterion

$$\max_{lm} |E_{lm}^{\pm,s+1} - E_{lm}^{\pm,s}| < \varepsilon_1 \max_{lm} |E_{lm}^{\pm,s}| + \varepsilon_2, \quad \varepsilon_1, \varepsilon_2 > 0$$

is valid. It is important to note that iterative scheme also satisfies the discrete invariant

$$\|E_j^{+,n,s}\| = \|E_{j-1}^{+,n-1}\|, \quad \|E_{j-1}^{-,n,s}\| = \|E_j^{-,n-1}\|, \quad s \geq 1.$$

For each iteration defined by (3) two systems of linear equations must be solved to find vectors $E_j^{+,n,s+1}$ and $E_{j-1}^{-,n,s+1}$. This is done by using the 2D FFT algorithm.

3.3 Parallel Algorithm

The finite difference scheme, which is constructed in the previous section, uses the structured grid, and the complexity of computations at each node of the grid is approximately the same (it depends on the number of iterations used to solve a nonlinear discrete problem for each z_j). The parallelization of such algorithms can be done by using domain decomposition (DD) paradigm [11], and ParSol is exactly targeted for such algorithms.

In this chapter, we apply the 1D block domain decomposition algorithm, decomposing the grid only in z direction. Such a strategy enables us to use a sequential version of the FFT algorithm for solution of the 2D linear systems with respect to (x,y) coordinates.

This parallel algorithm is generated semiautomatically by ParSol. The parallel vectors, which are used to store discrete PDE solutions, are created by specifying three main attributes:

- the dimension of the parallel vector is 3D;
- the topology of processors is 1D and only z coordinate is distributed;
- the 1D grid stencil is defined by the points (z_{j-1}, z_j, z_{j+1}).

Thus in order to implement the computational algorithm, the kth processor ($k = 0, 1, \ldots, p$) defines its subgrid as well as its ghost points $\Omega(k)$, where

$$\Omega(k) = \{(z_j, x_{1l}, x_{2m}),\ z_j \in \Omega_z(k), \quad (x_{1l}, x_{2m}) \in \Omega_x\},$$

$$\Omega_z(k) = \{z_j :\ \widetilde{j}_L(k) \le j \le \widetilde{j}_R(k)\},$$

$$\tilde{j}_L(k) = \max(j_L(k) - 1, 0),\ \tilde{j}_R = \min(j_R(k) + 1, J).$$

At each time step t^n and for each $j = 1, 2, \ldots, J$, the processors must exchange some information for ghost points values. As the computations move along the characteristics $z \pm t$, only a half of the full data on ghost points is required to be exchanged. Thus the kth processor

- sends to $(k+1)$th processor vector $E^{+,n}_{j_R,\cdot}$ and receives from it vector $E^{-,n}_{\tilde{j}_R,\cdot}$,
- sends to $(k-1)$th processor vector $E^{-,n}_{j_L,\cdot}$ and receives from it vector $E^{+,\cdot}_{\tilde{j}_L,\cdot}$.

Obviously, if $k = 0$ or $k = (p-1)$, then a part of communications is not done. In ParSol, such an optimized communication algorithm is obtained by defining temporal reduced stencils for vectors E^+ and E^-; they contain ghost points only in the required directions but not in both.

Next we present a simple scalability analysis of the developed parallel algorithm. The complexity of the serial algorithm for one time step is given by $W = KJ(M+1)^2(\gamma_1 + \gamma_2 \log M)$, where $\gamma_1 (M+1)^2$ estimates the CPU time required to compute all coefficients of the finite-difference scheme, $\gamma_2 (M+1)^2 \log M$ defines the costs of FFT algorithm, and K is the averaged number of iterations done at one time step.

Let us assume that p processors are used. Then the computational complexity of parallel algorithm depends on the size of the largest local grid part given to one processor. It is equal to

$$T_{p,\text{comp}} = \max_{0 \le k < p} K(k)\big(\lceil (J+1)/p \rceil + 1\big)(M+1)^2(\gamma_1 + \gamma_2 \log M),$$

where $\lceil x \rceil$ denotes a smallest integer number larger than or equal to x. This formula includes costs of extra computations involving ghost points.

Data communication time is given by $T_{p,\text{comm}} = 2\big(\alpha + \beta(M+1)^2\big)$; here α is the message startup time and β is the time required to send one element of data. We assume that communication between neighbour processors is done in parallel. Thus the total complexity of the parallel algorithm is given by

$$T_p = \max_{0 \le k < p} K(k)\big(\lceil (J+1)/p \rceil + 1\big)(M+1)^2(\gamma_1 + \gamma_2 \log M) + 2\big(\alpha + \beta(M+1)^2\big).$$

3.4 Results of Computational Experiments

The parallel code was tested on the cluster of PCs at Vilnius Gediminas Technical University. It consists of Pentium 4 processors (3.2GHz, level 1 cache 16kB, level 2 cache 1MB) interconnected via Gigabit Smart Switch (http://vilkas.vgtu.lt). Obtained performance results are presented in Table 1. Here for each number of processors p, the coefficients of the algorithmic speed up $S_p = T_1/T_p$ and efficiency $E_p = S_p/p$ are presented. The size of the discrete problem is $M = 31$ and $J = 125$.

More applications of the developed parallelization tool ParSol are described in [5, 6, 7].

4 Conclusions

The new tool of parallel linear algebra objects ParSol employs elements of both data parallel and global memory parallel programming models. In the current version the main objects, i.e., arrays, vectors and matrices, are targeted for structured grids. The interface for 1D, 2D, and 3D linear algebra objects is done. The algorithm

Table 1 Results of computational experiments on Vilkas cluster.

	M	J	$p=1$	$p=2$	$p=4$	$p=8$	$p=12$
S_p	31	125	1.0	1.87	3.46	5.86	7.83
E_p	31	125	1.0	0.94	0.87	0.73	0.65

implemented by ParSol objects is parallelized semi-automatically, by specifying the stencil of the grid, topology of processors, and data communication points in the algorithm.

ParSol is applied to parallelize one numerical algorithm, which is developed to simulate the nonlinear interaction of two counter-propagating nonlinear optical beams. It is shown how the specific features of the algorithm can be taken into account to minimize the amount of data communicated between processors. The results of numerical experiments show a good efficiency of the obtained parallel numerical algorithm.

Acknowledgments R. Čiegis and I. Laukaitytė were supported by the Lithuanian State Science and Studies Foundation within the project on B-03/2007 "Global optimization of complex systems using high performance computing and GRID technologies".

References

1. Akcelik, V., Biros, G., Ghattas, O., Hill, J. et al.: Frontiers of parallel computing. In: M. Heroux, P. Raghaven, H. Simon (eds.) Parallel Algorithms for PDE-Constrained Optimization. SIAM, Philadelphia (2006)
2. Balay, S., Buschelman, K., Eijkhout, V., Gropp, W.D., Kaushik, D., Knepley, M.G., Curfman McInnes, L., Smith, B.F., Zhang, H.: PETSc user manual. ANL-95/11 - Revision 2.1.5. Argonne National Laboratory (2004)
3. Bastian, P., Birken, K., Johannsen, K., Lang, S. et al.: A parallel software-platform for solving problems of partial differential equations using unstructured grids and adaptive multigrid methods. In: W. Jage, E. Krause (eds.) High Performance Computing in Science and Engineering, pp. 326–339. Springer, New York (1999)
4. Blatt, M., Bastian, P.: The iterative solver template library. In: B. Kågström, E. Elmroth, J. Dongarra, J. Wasniewski (eds.) Applied Parallel Computing: State of the Art in Scientific Computing, *Lecture Notes in Scientific Computing*, vol. 4699, pp. 666–675. Springer, Berlin Heidelberg New York (2007)
5. Čiegis, Raim, Čiegis, Rem., Jakušev, A., Šaltenienė, G.: Parallel variational iterative algorithms for solution of linear systems. Mathematical Modelling and Analysis **12**(1), 1–16 (2007)
6. Čiegis, R., Jakušev, A., Krylovas, A., Suboč, O.: Parallel algorithms for solution of nonlinear diffusion problems in image smoothing. Mathematical Modelling and Analysis **10**(2), 155–172 (2005)
7. Čiegis, R., Jakušev, A., Starikovičius, V.: Parallel tool for solution of multiphase flow problems. In: R. Wyrzykowski, J. Dongarra, N. Meyer, J. Wasniewski (eds.) Sixth International conference on Parallel Processing and Applied Mathematics. Poznan, Poland, September 10-14, 2005, *Lecture Notes in Computer Science*, vol. 3911, pp. 312–319. Springer, Berlin Heidelberg New York (2006)
8. Geist, A., Beguelin, A., Dongarra, J. et al.: PVM: Parallel Virtual Machine. A User's Guide and Tutorial for Networked Parallel Computing. MIT Press, Cambridge, MA (1994)
9. Jakušev, A.: Application of template metaprogramming technologies to improve the efficiency of parallel arrays. Mathematical Modelling and Analysis **12**(1), 71–79 (2007)
10. Koelbel, C.H., Loveman, D.B., Schreiber, R.S., Steele, G.L., Zosel, M.E.: The High Performance Fortran Handbook. The MIT Press, Cambridge, MA (1994)
11. Kumar, V., Grama, A., Gupta, A., Karypis, G.: Introduction to Parallel Computing: Design and Analysis of Algorithms. Benjamin/Cummings, Redwood City (1994)

12. Langtangen, H.P.: Computational Partial Differential Equations — Numerical Methods and Diffpack Programming, *Lecture Notes in Computational Science and Engineering*. Springer-Verlag, New York (1999)
13. Le Veque, R.: Finite Volume Methods for Hyperbolic Problems. Cambridge University Press, Cambridge, UK (2002)
14. Nieplocha, J., Palmer, B., Tipparaju, V., Krishnan, M., Trease, H., Apra, E.: Advances, applications and performance of the Global Arrays shared memory programming toolkit. International Journal of High Performance Computing Applications **20**(2), 203-231 (2006)
15. Nikitenko, K.Y., Trofimov, V.A.: Optical bistability based on nonlinear oblique reflection of light beams from a screen with an aperture on its axis. Quantum Electronics **29**(2), 147–150 (1999)
16. OpenFOAM: The Open Source CFD Toolbox. URL http://www.opencfd.co.uk/openfoam/
17. Snir, M., Oto, S., Hus-Lederman, S., Walker, D., Dongarra, J.: MPI. The Complete Reference. The MIT Press, Cambridge, Reading, MA (1995)
18. Stroustrup, B.: The C++ Programming Language. Addison-Wesley, MA (1997)
19. Tereshin, E.B., Trofimov, V.A.: Conservative finite difference scheme for the problem of propagation of a femtosecond pulse in a photonic crystal with combined nonlinearity. Comput. Math. and Mathematical Physics **46**(12), 2154–2165 (2006)

The Development of an Object-Oriented Parallel Block Preconditioning Framework

Richard L. Muddle, Jonathan W. Boyle, Milan D. Mihajlović, and Matthias Heil

Abstract The finite-element-based solution of partial differential equations often requires the solution of large systems of linear equations. Krylov subspace methods are efficient solvers when combined with effective preconditioners. We consider block preconditioners that are applicable to problems that have different types of degree of freedom (e.g., velocity and pressure in a fluid simulation). We discuss the development of an object-oriented parallel block preconditioning framework within OOMPH-LIB, the object-oriented, multi-physics, finite-element library, available as open-source software at `http://www.oomph-lib.org`. We demonstrate the performance of this framework for problems from non-linear elasticity, fluid mechanics, and fluid-structure interaction.

1 Introduction

Numerical solution techniques for partial differential equations (PDEs) typically require the solution of large sparse systems of linear equations. Krylov subspace methods are efficient solvers when combined with effective preconditioners. We present a block preconditioning framework developed in OOMPH-LIB [6], the object-oriented, multi-physics, finite-element library, available as open-source software at `http://www.oomph-lib.org`.

Richard L. Muddle · Milan D. Mihajlović
School of Computer Science, University of Manchester
Oxford Road, Manchester, M13 9PL, UK
e-mail: rlm@cs.man.ac.uk · milan@cs.man.ac.uk

Jonathan W. Boyle · Matthias Heil
School of Mathematics, University of Manchester
Oxford Road, Manchester, M13 9PL, UK
e-mail: j.boyle@manchester.ac.uk · m.heil@maths.man.ac.uk

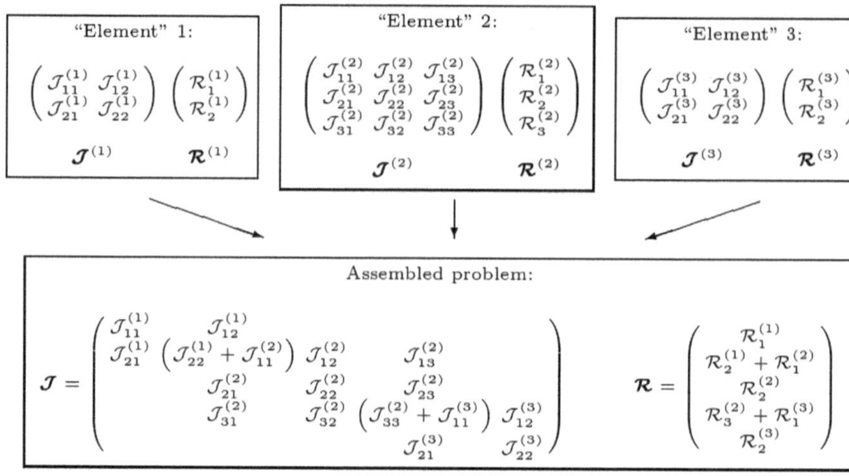

Fig. 1 Element-by-element assembly of a system of linear equations.

A key design goal of OOMPH-LIB is to provide a framework for the discretization and solution of coupled multi-physics problems, such as those from fluid-structure interation (FSI) [1]. The temporal and spatial discretization of these problems produces a system of non-linear algebraic equations, which is solved by Newton's method. This requires the repeated solution of linear systems of the form $\mathscr{J}\,\delta\mathbf{x} = -\mathbf{r}$ for the Newton correction $\delta\mathbf{x}$ where $\mathscr{J}$ is the Jacobian matrix. OOMPH-LIB employs a finite-element-like framework in which the linear systems are constructed in an element-by-element assembly procedure as illustrated in Fig. 1.

OOMPH-LIB's definition of an element is sufficiently general to allow the representation of finite elements, finite difference stencils, or algebraic constraints – the only requirement is that each element must provide its contribution to the global system of equations. OOMPH-LIB provides a large number of single-physics elements which, via templating and inheritance, can be (re-)used to construct multi-physics elements.

Within this framework, the three most computationally intensive components of the computation are

- the repeated assembly of the Jacobian matrix $\mathscr{J}$ and the residual vector $\mathbf{r}$;
- the repeated solution of system $\mathscr{J}\,\delta\mathbf{x} = -\mathbf{r}$;
- the application of the fully automated mesh adaptation procedures, which refine (or unrefine) the mesh, based on an estimated error.

Here we shall focus on the solution of the systems of linear equations. By default OOMPH-LIB employs the parallel and serial versions of the `SuperLU` direct solver

[2]; however, for large problems, direct solvers are too expensive in terms of memory and CPU time. The solution of such problems therefore requires the use of iterative methods. We consider Krylov subspace methods [9] (such as GMRES), which are only efficient when used together with an effective preconditioner. A preconditioner $\mathscr{P}$ is a linear operator that is spectrally close to $\mathscr{J}$, but computationally cheap to assemble and apply. Left preconditioning represents a transformation of the original linear system to

$$\mathscr{P}^{-1}\mathscr{J}\,\delta\mathbf{x} = -\mathscr{P}^{-1}\mathbf{r}.$$

To apply the preconditioner, we have to solve a linear system of the form $\mathscr{P}\mathbf{z} = \mathbf{y}$ for $\mathbf{z}$ at each Krylov iteration.

2 Block Preconditioning

Block preconditioners are a class of preconditioners that are applicable to problems with more than one type of degree of freedom (DOF); for example, in a fluid mechanics simulation, the fluid velocities and the pressures are different types of DOF. For problems of this type, the linear system can be reordered to group together the equations associated with each type of DOF. The system matrix can then be considered to have been decomposed into a matrix of sub-matrices, or blocks. For example, the reordered linear system for a problem with two types of DOF is

$$\begin{bmatrix} \mathscr{J}_{11} & \mathscr{J}_{12} \\ \mathscr{J}_{21} & \mathscr{J}_{22} \end{bmatrix} \begin{bmatrix} \delta\mathbf{x_1} \\ \delta\mathbf{x_2} \end{bmatrix} = -\begin{bmatrix} \mathbf{r}_1 \\ \mathbf{r}_2 \end{bmatrix}. \tag{1}$$

A block preconditioner is a linear operator assembled from the block matrices of the reordered coefficient matrix. For example, the block diagonal preconditioner associated with the linear system (1) is

$$\mathscr{P}_{diag} = \begin{bmatrix} \mathscr{J}_{11} & 0 \\ 0 & \mathscr{J}_{22} \end{bmatrix}, \tag{2}$$

the application of which requires the solution of two (smaller) subsidiary linear systems involving $\mathscr{J}_{11}$ and $\mathscr{J}_{22}$. It is often sufficient to only compute an approximate solution to these systems, for example, by applying only a few sweeps of algebraic mulitgrid (AMG) [11].

This methodology can be considered to be a two-stage preconditioning strategy comprising the approximate solution of the subsidiary problems and a global update via block matrix computations.

OOMPH-LIB's block preconditioning framework facilitates the development and application of block preconditioners. Because the order of the DOFs in $\mathscr{J}$ is arbitrary, the application of a block preconditioner generally requires the following steps:

(i) the DOFs must be classified according to their type – this is implemented at the element level;

(ii) using this classification, the relevant blocks must be identified and extracted from the system matrix $\mathscr{J}$;

(iii) for some block preconditioners (e.g., the LSC Navier–Stokes preconditioner discussed in Sect. 3), certain additional matrices may have to be generated;

(iv) the (approximate) linear solvers for the subsidiary linear systems must be defined;

(v) finally, to apply the preconditioner, the relevant sub-vectors must be extracted from the global vectors $\mathbf{y}$ and $\mathbf{z}$ and the subsidiary linear system must be solved (approximately).

This block preconditioning framework is implemented in both serial and parallel using `MPI`.

3 The Performance of the Block Preconditioning Framework

In this section, we evaluate the performance of the block preconditioning framework in a number of test problems and consider its parallel scaling performance on multiple processors. We performed the experiments on up to 8 nodes of a Beowulf cluster, in which each node has a 3.60GHz Intel Xeon processor with 2GB of memory. The compute nodes are connected via a gigabit switch with one network connection per node. In each case, we used the distributed GMRES solver implemented in the Trilinos AztecOO package [10] as the global linear solver. When required, we utilized the Hypre library [7] implementation of AMG.

3.1 Reference Problem : 2D Poisson

To put the parallel scaling results into context, we begin by considering the parallel scaling of the solution of a simple 2D Poisson problem on a unit squarewith a conjugate gradient (CG) Krylov solver, preconditioned with Jacobi smoothed AMG. Given that this solution method yields an optimal solution (i.e., the computational cost scales linearly with problem size), and the Hypre and Trilinos libraries are known to scale efficiently, this problem forms an effective benchmark for the block preconditioners investigated in this paper.

Table 1 shows performance data for this problem where n is the size of the linear system and p is the number of processors. We keep n/p approximately constant so that the work per processor remains approximately constant. We observe that the iteration count remains constant with respect to the size of the problem, and the number of processors employed. As expected, optimal scaling in setup and solution time with respect to problem size is observed: for instance, on one processor, a

Table 1 Parallel performance of the Poisson preconditioner.

n/p	∼20000				∼40000				∼80000			
p	1	2	4	8	1	2	4	8	1	2	4	8
Number of iterations	8	9	11	10	7	9	11	10	8	10	10	11
Average setup & solve time (s)	0.16	0.28	0.35	0.46	0.33	0.48	0.62	0.80	0.72	0.99	1.07	1.56
Parallel efficiency (%)	100	60.6	47.51	37.1	100	67.4	53.6	41.3	100	73.1	67.7	46.6

problem with 40000 DOFs takes approximately twice as long to solve as does one with 20000 DOFs. This is due to the almost constant iteration count and the application of AMG as the preconditioner. The parallel scaling efficiency (for a fixed n/p) is the execution time on a single processor divided by the execution time on p processors ($\times 100\%$). The parallel scaling performance is inherited from the third party libraries employed and is limited by the problem sizes considered and the speed of the cluster interconnects.

We use the parallel scaling performance observed in this problem as a benchmark against which we evaluate our block preconditioners.

3.2 Non-linear Elasticity

To evaluate the block-preconditioning framework for a non-linear elasticity problem, we consider the model problem shown in Fig. 2: an elastic cantilever beam, modeled as a 2D solid, is loaded by a uniform pressure load on its upper face and undergoes large deflections.

We discretized the governing equations (the principle of virtual displacements) with bi-quadratic (Q2) solid mechanics elements and employed a block-diagonal preconditioner [8], subdividing the degrees of freedom into horizontal and

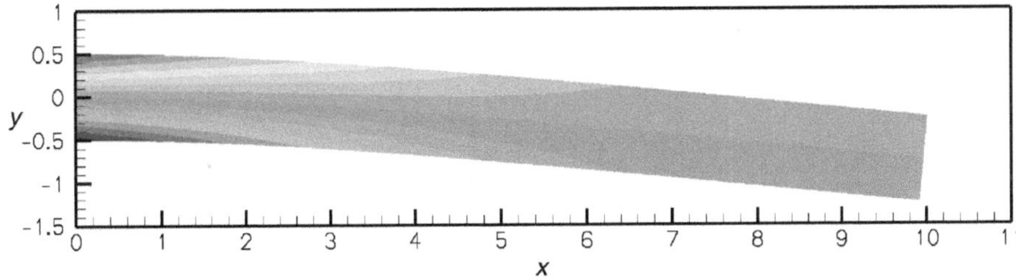

Fig. 2 Axial stress field in the cantilever beam.

Table 2 Parallel performance of the non-linear elasticity preconditioner.

n/p	~20000				~40000				~80000			
p	1	2	4	8	1	2	4	8	1	2	4	8
Average number of iterations	41.7	42.2	42.7	43.5	42.2	42.5	43.3	43.2	42.2	43	43.2	43.7
Average setup & solve time (s)	1.7	3.5	4.9	7.5	4.3	6.3	8.2	11.8	9.1	11.9	14.5	20.0
Parallel efficiency (%)	100	52.4	37.6	25.2	100	67.1	51.8	36.8	100	76.7	64.2	46.4

vertical displacement components. Each subsidiary linear system within the preconditioner was (approximately) solved using two V(2,2) cycles of undamped Jacobi smoothed AMG. In our experiments, we only considered V-cycles in AMG; other cycles (such as F(1,1)) were considered too expensive for preconditioning. In our experiments, we observed that V(2,2) cycles performed more robustly.

Table 2 shows performance data for this problem where n is the size of the linear system and p is the number of processors. The iteration count remains approximately constant with respect to the size of the problem and the number of processors. We observe a near optimal scaling in setup and solution time with respect to problem size. The scaling of the parallel efficiency is comparable with that of the Poisson reference problem.

3.3 Fluid Mechanics

Next we consider the classic driven-cavity problem in which viscous fluid is contained inside the unit square and a flow generated by a moving lid on the lower wall. The discretization of the Navier–Stokes equations with LBB-stable Taylor–Hood (Q2-Q1) elements leads to linear systems of the following block form

$$\begin{bmatrix} \mathscr{F} & \mathscr{G} \\ \mathscr{D} & 0 \end{bmatrix} \begin{bmatrix} \delta\mathbf{u} \\ \delta\mathbf{p} \end{bmatrix} = \begin{bmatrix} \mathbf{r}_u \\ \mathbf{r}_p \end{bmatrix}. \tag{3}$$

We precondition these with the least squares commutator (LSC) preconditioner [4]

$$\begin{bmatrix} \mathscr{F} & \mathscr{G} \\ 0 & -\tilde{\mathscr{M}}_s \end{bmatrix} \begin{bmatrix} \mathbf{z}_u \\ \mathbf{z}_p \end{bmatrix} = \begin{bmatrix} \mathbf{e}_u \\ \mathbf{e}_p \end{bmatrix}, \tag{4}$$

where $\tilde{\mathscr{M}}_s$ is an approximation to the pressure Schur-complement $\mathscr{M}_s = \mathscr{D}\mathscr{F}^{-1}\mathscr{G}$. The inverse of the pressure Schur-complement is approximated by

$$\tilde{\mathscr{M}}_s^{-1} = \left(\mathscr{D}\hat{\mathscr{Q}}^{-1}\mathscr{G}\right)^{-1}\left(\mathscr{D}\hat{\mathscr{Q}}^{-1}\mathscr{F}\hat{\mathscr{Q}}^{-1}\mathscr{G}\right)\left(\mathscr{D}\hat{\mathscr{Q}}^{-1}\mathscr{G}\right)^{-1},$$

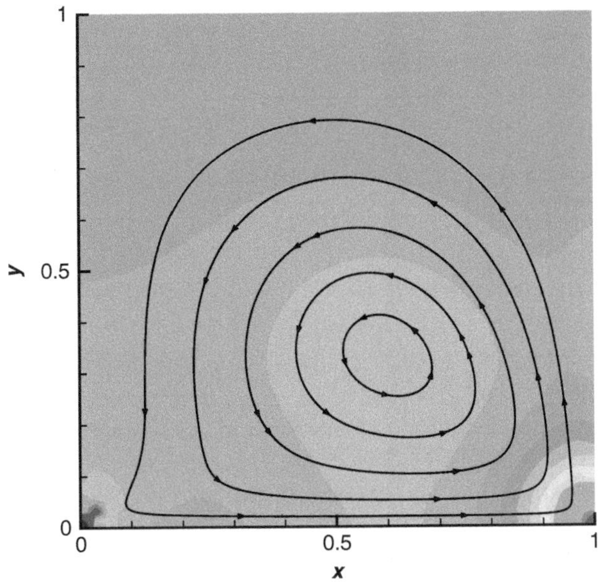

Fig. 3 Streamline and pressure contour plot of the 2D driven cavity problem.

where $\hat{\mathscr{Q}}$ is the diagonal of the velocity mass matrix. The evaluation of $\tilde{\mathscr{M}}_s^{-1}$ requires the solution of two discrete pressure Poisson operators $\mathscr{P} = \mathscr{D}\hat{\mathscr{Q}}^{-1}\mathscr{G}$.

The solution is presented in Fig. 3.

We (approximately) solve the momentum block $\mathscr{F}$ and the scaled Poisson block $\mathscr{P}$ with one V-cycle V(2,2) of AMG. Table 3 shows the performance for a Reynolds number of $Re = 200$.

We make the same observations that were made for the non-linear elasticity preconditioner; an almost constant iteration count, near optimal scaling with respect to problem size, and a parallel scaling comparable with the Poisson reference problem. The performance of our block preconditioners is comparable with similar experimental works reported in [3] and [4].

Table 3 Parallel performance of the Navier–Stokes LSC preconditioner.

n/p	∼20000				∼40000				∼80000			
p	1	2	4	8	1	2	4	8	1	2	4	8
Average number of iterations	31.3	32.8	35	39	32.6	34.8	39.4	45.8	34.8	38.2	45.2	54.8
Average setup & solve time (s)	4.8	6.8	8.3	12.3	10.1	13.5	16.4	25.3	23.1	27.5	33.9	50.8
Parallel efficiency (%)	100	70.6	58.5	39.8	100	75.9	62.9	40.8	100	84.1	68.2	45.7

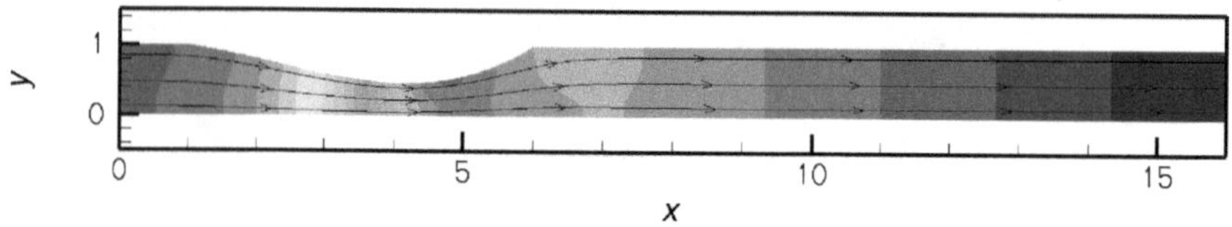

Fig. 4 Steady flow field (pressure and streamlines) of the 2D collapsible channel problem.

3.4 Fluid–Structure Interaction

Finally, we consider the classic fluid–structure interaction problem of finite-Reynolds-number flow in a 2D channel in which a finite section of the upper wall is replaced by a pre-stressed elastic membrane, modeled as a thin-walled Kirchhoff–Love beam. The fluid traction (pressure and shear stress) induces large wall deflections, which in turn affect the geometry of the fluid domain, leading to a strong two-way coupling (see Fig. 4).

We discretize the fluid and solid domains with 2D Taylor–Hood and Hermite beam elements respectively, and employ an algebraic mesh update technique [1] to update the nodal positions in the fluid mesh in response to the changes in the wall shape.

The fully coupled (monolithic) solution of the fluid and solid equations by Newton's method requires the solution of linear systems with the following block structure:

$$\left[\begin{array}{cc|c} \mathscr{F} & \mathscr{G} & \mathscr{C}_{us} \\ \mathscr{D} & 0 & \mathscr{C}_{ps} \\ \hline \mathscr{C}_{su} & \mathscr{C}_{sp} & \mathscr{S} \end{array}\right] \left[\begin{array}{c} \delta\mathbf{u} \\ \delta\mathbf{p} \\ \hline \delta\mathbf{s} \end{array}\right] = \left[\begin{array}{c} \mathbf{r}_u \\ \mathbf{r}_p \\ \hline \mathbf{r}_s \end{array}\right]. \tag{5}$$

Here $\mathscr{S}$ is the solid's tangent stiffness matrix, the matrices $\mathscr{C}_{**}$ arise from the interaction between the fluid and solid mechanics, and $\mathbf{s}$ contains the discrete wall displacements. The remaining quantities are as discussed in the Navier–Stokes problem.

Following Heil [5], we use a block-triangular approximation of the full Jacobian matrix as the preconditioner and use the LSC preconditioner discussed in the

Table 4 Parallel performance of the FSI preconditioner.

n/p	~20000				~40000				~80000			
p	1	2	4	8	1	2	4	8	1	2	4	8
Average number of iterations	16.8	14.3	14.1	12.7	14.6	14.2	12	11.6	13.1	11.3	10.9	12.6
Average setup & solve time (s)	3.2	5.2	7.3	11.5	6.8	10.1	14.1	19.7	16.1	20.9	28.0	41.5
Parallel efficiency (%)	100	66.2	48.9	31.9	100	70.9	51.8	37.6	100	79.6	60.3	41.3

previous example to approximately solve the linear systems involving the Navier–Stokes sub-block.

We again use one V-cycle V(2,2) of Jacobi smoothed AMG to (approximately) solve the Navier–Stokes blocks $\mathscr{P}$ and $\mathscr{F}$. We solve the solid block $\mathscr{S}$ exactly using serial `SuperLU`. Table 4 shows that the performance (for a Reynolds number of $Re = 50$) is comparable with the block diagonal non-linear elasticity and LSC Navier–Stokes preconditioners.

4 Conclusions

We have presented the block preconditioning framework under development in OOMPH-LIB. This facilitates the implementation of both parallel and serial block preconditioners for finite-element-like problems. A key design goal was to ensure that existing general linear algebra methods can be reused within this framework.

We have demonstrated the performance of several block preconditioners developed within the parallel framework, and we observed a near constant Krylov iteration count with respect to the problem size and the number of processors. The parallel scaling performance was comparable with the reference Poisson preconditioner.

The framework described in this paper is scheduled for inclusion in the next release of OOMPH-LIB, expected in mid-2008.

References

1. Bungartz, H.J., Schafer, M.: Fluid-Structure Interaction: Modelling, Simulation, Optimisation. Springer-Verlag, New York (2006)
2. Demmel, J.W., Eisenstat, S.C., Gilbert, J.R., Li, X.S., Liu, J.W.H.: A supernodal approach to sparse partial pivoting. SIAM J. Matrix Analysis and Applications **20**(3), 720–755 (1999)
3. Elman, H., Howle, V.E., Shadid, J., Shuttleworth, R., Tuminaro, R.: A taxonomy and comparison of parallel block multi-level preconditioners for the incompressible Navier-Stokes equations. J. Comput. Phys. **227**(3), 1790–1808 (2008)
4. Elman, H., Silvester, D., Wathen, A.: Finite Elements and Fast Iterative Solvers with Applications in Incompressible Fluid Dynamics. Oxford University Press, Oxford (2005)
5. Heil, M.: An efficient solver for the fully coupled solution of large-displacement fluid-structure interaction problems. Computer Methods in Applied Mechanics and Engineering **193**(1-2), 1–23 (2004)
6. Heil, M., Hazel, A.: OOMPH-LIB, the object-oriented multi-physics finite-element library. URL http://www.oomph-lib.org
7. HYPRE: High Performance Preconditioning Library, Center for Applied Scientific Computing at Lawrence Livermore National Laboratory. URL http://www.llnl.gov/CASC/hypre/
8. Mijalković, S.Z., Mihajlović, M.D.: Component-wise algebraic multigrid preconditioning for the iterative solution of stress analysis problems from microfabrication technology. Commun. Numer. Meth. Eng. **17**(10), 737–747 (2001)

9. Saad, Y.: Iterative Methods for Sparse Linear Systems. PWS, Boston (1996)
10. TRILINOS, Sandia National Laboratories. URL http://trilinos.sandia.gov
11. Wesseling, P.: Introduction to Multigrid Methods. Institute for Computer Applications in Science and Engineering (ICASE) R.T. Edwards, Philadelphia, PA (1995)

A Sparse Linear System Solver Used in a Distributed and Heterogenous Grid Computing Environment

Christophe Denis, Raphael Couturier, and Fabienne Jézéquel

Abstract Many scientific applications need to solve very large sparse linear systems in order to simulate phenomena close to the reality. Grid computing is an answer to the growing demand of computational power. In a grid computing environment, communication times are significant and the bandwidth is variable, therefore frequent synchronizations slow down performances. Thus it is desirable to reduce the number of synchronizations in a parallel direct algorithm. Inspired from multisplitting techniques, the GREMLINS (GRid Efficient Methods for LINear Systems) solver we developed consists of solving several linear problems obtained by splitting. The principle of the balancing algorithm is presented, and experimental results are given.

1 Introduction

Many scientific and industrial applications need to solve very large sparse linear systems in order to simulate phenomena close to the reality. Grid computing is an answer to the growing demand of computational power. In a grid computing environment, communication times are significant and the bandwidth is variable, therefore frequent synchronizations slow down performances. Thus it is desirable

Christophe Denis
School of Electronics, Electrical Engineering & Computer Science
The Queen's University of Belfast, Belfast BT7 1NN, UK
e-mail: C.Denis@qub.ac.uk

Raphael Couturier
Laboratoire d Informatique del Université de Franche-Comté
BP 527, 90016 Belfort Cedex, France
e-mail: Raphael.Couturier@iut-bm.univ-fcomte.fr

Christophe Denis · Fabienne Jézéquel
UPMC Univ Paris 06, Laboratoire d'Informatique LIP6
4 place Jussieu, 75252 Paris Cedex 05, France
e-mail: Christophe.Denis@lip6.fr · Fabienne.Jezequel@lip6.fr

to reduce the number of synchronizations in a parallel direct algorithm. Inspired by multisplitting techniques, the GREMLINS (GRid Efficient Methods for LINear Systems) solver we developed consists of solving several linear problems obtained by splitting the original one [13]. The design of the GREMLINS solver has been supported by the French National Research Agency (ANR). Each linear system is solved independently on a cluster by using a direct or iterative method. The GREMLINS solver can be not well balanced in terms of computing time as sparse matrices are concerned. Consequently, the parallel global computing time can be decreased by load balancing the GREMLINS solver.

The chapter is organized as follows. Sect. 1 presents the parallel multisplitting method used in the GREMLINS library. The principle of the balancing algorithm is discussed in Sect. 2, and experimental results are analyzed in Sect. 3. Finally, we present our concluding remarks and our future work.

2 The Parallel Linear Multisplitting Method Used in the GREMLINS Solver

We first define the linear multisplitting method. The parallelization of this method in the GREMLINS solver is based on a domain decomposition. Consider a sparse linear system $Ax = b$ to be solved with the GREMLINS solver. The decomposition of the matrix A is illustrated in Fig. 1.

The submatrix (denoted by A_{sub}) is the square matrix that a processor is in charge of. The part of the rectangle matrix before the submatrix represents the left dependencies, called $DepLeft$, and the part after the submatrix represents the right dependencies, called $DepRight$. Similarly, X_{sub} represents the unknown part to solve and B_{sub} the right-hand side involved in the computation. At each step, a processor computes X_{sub} by solving the following subsystem:

$$A_{\text{sub}}X_{\text{sub}} = B_{\text{sub}} - DepLeft\,X_{\text{left}} - DepRight\,X_{\text{right}}. \tag{1}$$

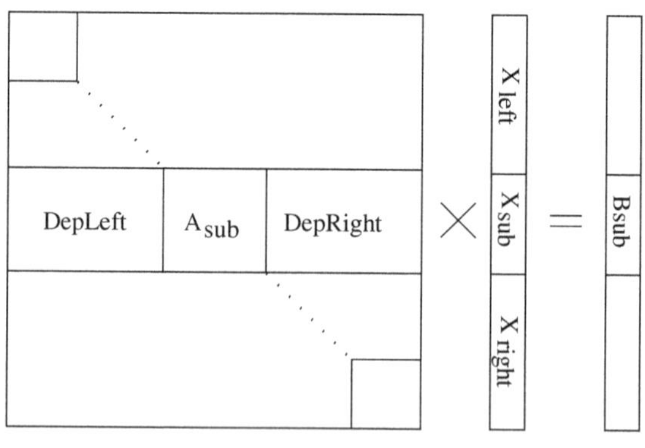

Fig. 1 Decomposition of the matrix.

Algorithm 1 The four main steps of the linear multisplitting.

1: **Initialization**: The way the matrix is loaded or generated is free. Each processor manages the load of the rectangle matrix $DepLeft + A_{\text{sub}} + DepRight$. Then until convergence, each processor iterates on:

2: **Computation**: At each iteration, each processor computes

$$B_{\text{loc}} = B_{\text{sub}} - DepLeft\, X_{\text{left}} - DepRight\, X_{\text{right}}.$$

Then, it solves X_{sub} using the *Solve (ASub,BLoc)* function with a sequential solver. The sequential solver comes from popular libraries: MUMPS[1] and SuperLU [12] (direct solvers) and Sparselib [10] (iterative solver).

3: **Data exchange**: Each processor sends its dependencies to its neighbors. When a processor receives a part of the solution vector (denoted by X_{sub}) of one of its neighbors, it should update its part of X_{left} or X_{right} vector according to the rank of the sending processor.

4: **Convergence detection**: Two methods are possible to detect the convergence. We can either use a centralized algorithm described in [3] or a decentralized version, that is, a more general version, as described in [4].

As soon as it has computed the solution of the subsystem, the processor sends this solution to all processors depending on it. The four main steps of the linear multisplitting method are described in Algorithm 1.

Algorithm 1 is similar to the block Jacobi algorithm. It actually generalizes the block Jacobi method, which is only a particular case of the multisplitting method. The two main differences of the multisplitting method are

- The multisplitting method may use asynchronous iterations. In this case, the execution time may be reduced. For a complete description of asynchronous algorithms, interested readers may consult [5].
- Some components may be overlapped and computed by more than one processor. The overlapped multisplitting methods have some similarities with overlapped block domain decomposition ones. But unlike domain decomposition methods, multisplitting methods are generated by the combination of several decompositions of matrix A (see [7]).

From a practical point of view, the use of asynchronous iterations consists in using non-blocking receptions, dissociating computations from communications using threads, and using an appropriate convergence detection algorithm. To solve a subsystem, the multisplitting algorithm uses a sequential solver. It can be a direct one or an iterative one. For the former class (direct), the most time consuming part is the factorization part, which is only achieved out at the first iteration. With large matrices, even after the decomposition process, the size of a submatrix that a processor is responsible to solve in sequence may be quite large. So, the time required to factorize a submatrix may be long. As opposed to a long factorization time, the use of a sequential direct solver allows one to compute other iterations very quickly because only the right-hand side changes at each iteration. For the latter class (iterative), all iterations require approximately the same time to be computed. The pattern of a submatrix acts on the time required to factorize it at the first iteration. Some previous works, carried out about the load balancing of the multiple front method [9], have successfully been adapted to the linear multisplitting solver. The condition for the asynchronous version to converge is slightly different and more restrictive than that of the synchronous one, i.e., in some rare practical cases, the synchronous version would converge whereas the asynchronous one would not. As the explanation for this condition is quite complex, because it relies on several mathematical tools, we invite interested readers to consult [2].

3 Load Balancing of the Direct Multisplitting Method

The performance of the direct linear multisplitting method is strongly influenced by the matrix decomposition. In particular, if the submatrix factorization computing times are not balanced, some processors will be idle.

The parallel factorization of each diagonal block A_{sub} is performed during the first iteration of the method. This iteration is the most time consuming. We use the sequential direct solver coming from the MUMPS software [1] for the load balancing algorithm. The submatrices issued from the matrix decomposition are load balanced in terms of number of rows but unfortunately not in terms of computational volume. Indeed, as sparse factorization techniques are used, the computational volume depends on the submatrix pattern and not on its size.

The aim of the load balancing method we have designed is to better distribute the factorization computational volume over diagonal blocks by predicting their computing times. A model returning an estimated computing time for a given amount of operations has been established. The input parameter of the load balancing algorithm is a matrix decomposition $\mathscr{D}_i$ issued from an external program. It returns a matrix decomposition $\mathscr{D}_c$ better load balanced in terms of computational volume. Algorithm 2 presents the outline of the load balancing algorithm.

Algorithm 2 Outline of the load balancing algorithm.

1: $NbIt \leftarrow 0$
2: Compute the estimated partial factorization computing times for each A_{sub}
3: **repeat**
4: $NbIt \leftarrow NbIt + 1$
5: Select the block A_{sub} having the maximum estimated computing time
6: Transfer locally n_t rows from A_{sub} to its nearest neighbor diagonal block and compute a new matrix decomposition $\mathscr{D}$
7: Reorder in parallel each A_{sub} with the reordering method used by MUMPS
8: Compute the estimated partial factorization computing times for each A_{sub}
9: Save the partition $\mathscr{D} \rightarrow \mathscr{D}_c$ if $\mathscr{D}$ is better load balanced than $\mathscr{D}_c$
10: **until** $NbIt < NbItMax$
11: **Return** $\mathscr{D}_c$

4 Experimental Results

We present firstly some results about the solving with GREMLINS of a sparse linear system issued from a system of 3D advection-diffusion equation discretized with a finite difference scheme [16]. Secondly, the load balancing benefit of the GREMLINS solver is highlighted. Experimentation has been conducted on the GRID'5000 architecture, a nationwide experimental grid [6]. Currently, the GRID'5000 platform is composed of an average of 1300 bi-processors that are located in 9 sites in France: Bordeaux, Grenoble, Lille, Lyon, Nancy, Orsay, Rennes, Sophia-Antipolis, Toulouse. Most of those sites have a Gigabit Ethernet Network for local machines. Links between the different sites ranges from 2.5 Gbps up to 10 Gbps. Most processors are AMD Opteron. For more details on the GRID'5000 architecture, interested readers are invited to visit the website: www.grid5000.fr. For all the experiments, we have chosen a 10^{-8} precision using the infinity norm.

4.1 Experiments with a Matrix Issued from an Advection-Diffusion Model

Transport of pollutants combined with their bio-chemical interactions can be mathematically formulated using a system of 3D advection-diffusion reaction equations having the following form:

$$\frac{\partial c}{\partial d} + A(c,a) = D(c,d) + R(c,t), \tag{2}$$

where c denotes the vector of unknown species concentrations, a contains the local fluid velocities, d is the diffusion coefficients matrix, and R includes interspecies chemical reactions and emissions or absorption from sources. The sparse linear system is issued from the discretization of the considered system with a 3D grid composed of $150 \times 150 \times 150$ discretized points, and the model contains two components per discretized point. The size of the obtained sparse matrix is 6,750,000 and the number of non-zero elements is 53,730,000. The interested reader is invited to consult [16] to find more details on this mathematical model.

The parallel solving is carried out with the GREMLINS solver with three sequential solvers in order to compare them: MUMPS and SuperLU (direct solvers) and Sparselib (iterative solver). Results are synthesized in Table 1.

We can observe that the ratio between the synchronous and the asynchronous version is of the same order as that of the generated matrices. Although SuperLU is slower than the other solvers in the synchronous case, it is equivalent to SparseLib in the asynchronous mode. This can be explained by the fact that the number of iterations is more important than in previous examples, so after the factorization step, a direct solver has less work to do at each iteration than does an iterative one. The direct sequential solver MUMPS is usually more efficient than is SuperLU. Consequently, we only use the MUMPS sequential solver on the next experiments.

Table 1 Execution times of our solver with the advec-diffu matrix with 90 machines located in 3 sites (30 in Rennes, 30 in Sophia, and 30 in Nancy).

	Synchronous		Asynchronous	
Solver	Exec. time (s)	No. iter.	Exec. time (s)	No. iter.
MUMPS	54.03	146	39.89	[293-354]
SuperLU	92.01	146	58.95	[259-312]
SparseLib	76.11	146	58.09	[250-291]

4.2 Results of the Load Balancing

The objective of the experiments is to show the benefit of the load balancing method. Results are carried out with the GREMLINS solver on synchronous mode with only the MUMPS sequential solver on the GRID'5000 architecture. The decomposition required by the multisplitting method strongly influences the efficiency of the direct linear multisplitting method. We have chosen these three following 1D block decomposition methods to underline this fact:

NAT: The initial decomposition set is built by dividing equitably the rows of the matrix on each processors.

MET: The matrix A is presented as a graph, and the multilevel graph partitioning tool METIS[11] is applied on it to obtain the initial decomposition set.

RCM: The matrix is re-ordered with the Reverse Cuthill McKee method [8] then the initial decomposition set is built by dividing equitably the rows of the re-ordered matrix on each processor.

Table 2 presents the matrices taken from from the University of Florida Sparse Matrix Collection [1]

The direct linear multisplitting method is run with and without the LB algorithm. The load balancing criterion $\Delta(\mathscr{D})$ is defined as follows: let the set $\{q_0, ..., q_i, ..., q_{L-1}\}$ be the measured factorization computing time associated with $\mathscr{D}$. The load balancing criterion $\Delta(\mathscr{D})$ is

$$\Delta(\mathscr{D}) = \frac{100}{L} \frac{\sum_{i=0}^{L-1} q_i}{\max_{0 \le i \le L-1} q_i}. \tag{3}$$

Table 2 The three matrices used in the experiments.

Matrix name	Number of rows	Number of nonzero entries
cage13	445,315	7,479,343
cage14	1,505,785	27,130,349
cage15	5,154,859	99,199,551

[1] http://www.cise.ufl.edu/research/sparse/matrices.

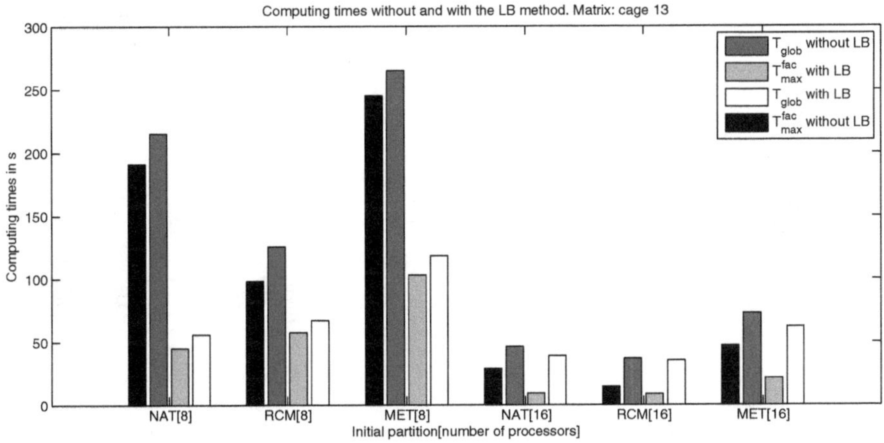

Fig. 2 Computing times with and without the use of the LB algorithm for the matrix *cage13* on 8 and 16 processors on a local cluster.

We note that $\Delta(\mathscr{D})$ of a decomposition set with ideal load balancing qualities q_i is equal to 100%.

The number of iterations *NbIt* of the load balancing (LB) algorithm is set to 5. Figures 2 and 3 show the maximum factorization computing time T_{max}^{fac} among processors and the global computing time T_{glob} with and without the use of the LB algorithm for the matrix *cage13* and *cage14*.

The relative gain obtained on the global computing time by using the LB algorithm is presented in Figs. 4 and 5. Without using the load balancing method, the amount of memory required on a processor is too high to solve the *cage14* matrix

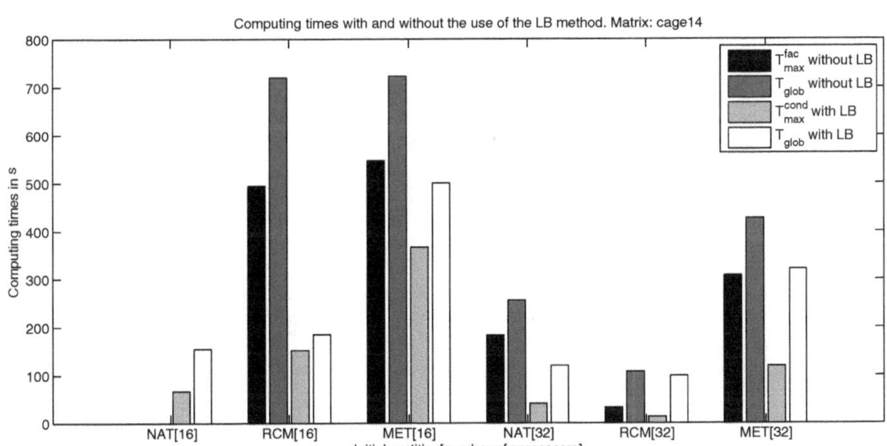

Fig. 3 Computing times with and without the use of the LB algorithm for the matrix *cage14* on 16 and 32 processors on a local cluster.

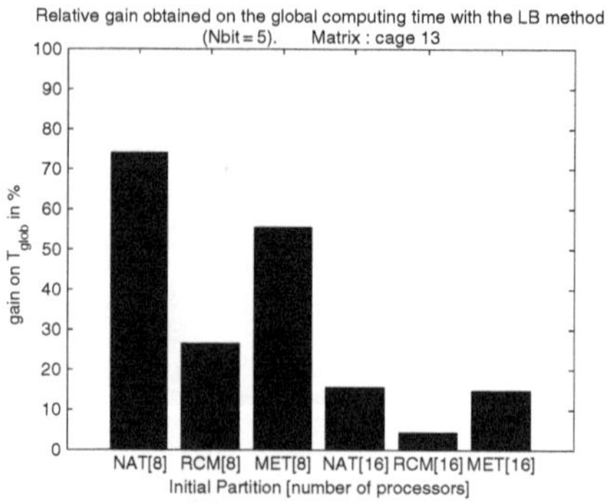

Fig. 4 The relative gain obtained on the global computing time by using the LB algorithm for the matrix *cage13* on 8 and 16 processors on a local cluster.

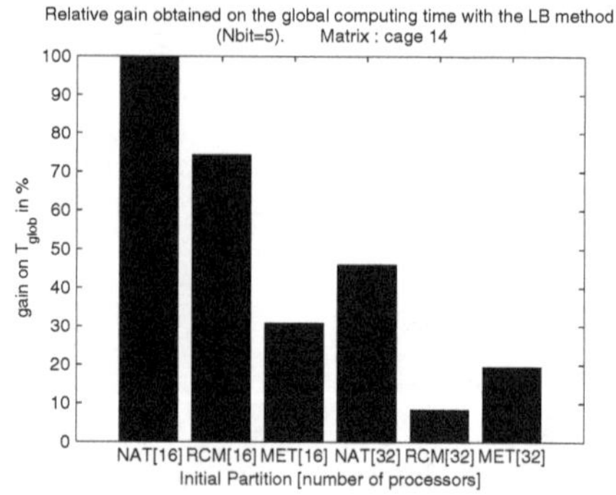

Fig. 5 The relative gain obtained on the global computing time by using the LB algorithm for the matrix *cage14* on 16 and 32 processors on a local cluster.

with the NAT partition on 16 processors. Consider now the case when the *cage14* matrix is split on 32 processors with MET as initial decomposition. The use of the load balancing method with $NbIt = 5$ permits an increase $\Delta(\mathscr{D})$ from 5.8% to 22%. So, we obtained thanks to the LB algorithm a gain on the global computing time of $g_{T_{glob}} = 19.4\%$. We observe the same phenomena over experiments.

We compare now the behavior of the load balancing algorithm on a local cluster or on two distant clusters. We also increase the size of the matrix by taking the *cage15* matrix and 128 processors. The METIS and RCM decomposition are not used on *cage15* as they require a sequential reordering on the whole matrix producing a bottleneck for larger matrices such as *cage15*. We could not use the parallel

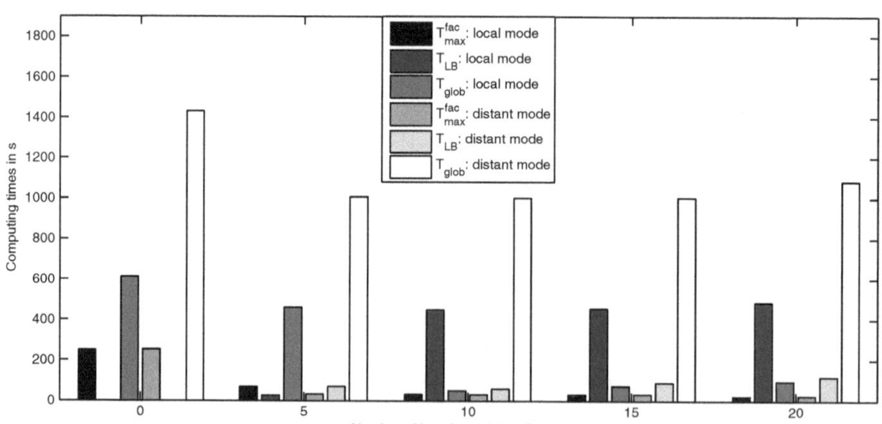

Fig. 6 The evolution of the maximum factorization, the load balancing, and the global computing times on a local cluster and on two distant clusters.

version of METIS in distributed grid computing as it requires a large number of communications and synchronizations. We used in these experiments the NAT domain decomposition, and the number of iterations of the LB algorithm varies from 0 (without the LB algorithm) to 20. Figure 6 presents the evolution of T_{max}^{fac}, the load balancing computing time T_{LB} and T_{glob} on a local cluster (local mode) and on two distant clusters (distant mode). Its computing time does not increase much when it runs on two distant clusters. The global computing time are higher in distant mode than in a local one. This is not surprising as the matrix deployment and distant communications take more time than do local ones. Consequently, the comparison between a local running and a distant one may seem irrelevant. In this case, using a distant cluster is only limited to solve large matrices that cannot be solved using a local cluster.

5 Conclusions and Future Work

We have presented some features of the GREMLINS solver. The MUMPS sequential direct solver seems to be the most powerful. However, a load balancing algorithm has been designed to decrease the global computing time of the GREMLINS solver using direct sequential solvers. An optimal number of iterations of the LB method exists, but it is difficult to predict as the decrease of the maximal factorization time is irregular due to the reordering process. The load balancing algorithm could be used on a local cluster and on distributed clusters as the overhead of computing time due to communication is not high.

Scientific computation has unavoidable approximations built into its very fabric. One important source of error that is difficult to detect and control is round-off error propagation. It is more important on a heterogeneous grid, which could have different arithmetic. Using ideas proposed in [15], we want to perform regular numerical 'health checks' on the GREMLINS solver in order to detect the cancerous effect of round-off error propagation.

The future work is to adapt our load balancing algorithm on the GREMLINS solver using sequential iterative methods. We also will study the scalability of the load balancing algorithm and, more important, think about the deployment of the matrix, which is very time consuming on distributed clusters. On the contrary, it seems that some discretization methods like the finite element method could be used with the direct multisplitting method as done with the parallel multiple front method in [14]. Comparing these two approaches in the GRID'5000 distributed grid would be an interesting perspective.

Acknowledgments This work was supported by the French Research Agency by grant ANR-JC05-41999 (Gremlins project). Experiments presented in this chapter were carried out using the Grid'5000 experimental testbed, an initiative from the French Ministry of Research through the ACI GRID incentive action, INRIA, CNRS, and RENATER and other contributing partners.

References

1. Amestoy, P.R., Duff, I.S., Koster, J., L'Excellent, J.Y.: A fully asynchronous multifrontal solver using distributed dynamic scheduling. SIAM Journal of Matrix Analysis and Applications **23**(1), 15–41 (2001)
2. Bahi, J., Couturier, R.: Parallelization of direct algorithms using multisplitting methods in grid environm ents. In: IPDPS 2005, pp. 254b, 8 pages. IEEE Computer Society Press (2005)
3. Bahi, J.M., Contassot-Vivier, S., Couturier, R.: Evaluation of the asynchronous iterative algorithms in the context of distant heterogeneous clusters. Parallel Computing **31**(5), 439–461 (2005)
4. Bahi, J.M., Contassot-Vivier, S., Couturier, R., Vernier, F.: A decentralized convergence detection algorithm for asynchronous parallel iterative algorithms. IEEE Transactions on Parallel and Distributed Systems **1**, 4–13 (2005)
5. Bertsekas, D.P., Tsitsiklis, J.N.: Parallel and Distributed Computation: Numerical Methods. Prentice Hall, Englewood Cliffs, NJ (1989)
6. Bolze, R., Cappello, F., Caron, E., Dayd, M., Desprez, F., Jeannot, E., Jgou, Y., Lanteri, S., Leduc, J., Melab, N., Mornet, G., Namyst, R., Primet, P., Quetier, B., Richard, O., Talbi, E.G., Touche, I.: Grid'5000: A large scale and highly reconfigurable experimental grid testbed. International Journal of High Performance Computing Applications **20**(4), 481–494 (2006)
7. Bruch, J.C.J.: Multisplitting and domain decomposition. In: L.C. Wrobel, C.A. Brebia (eds.) Computational Mechanics International Series On Computational Engineering. Computational methods for free and moving boundary problems in heat and fluid flow, pp. 17–36, Computational Mechanics, Inc. (1993)
8. Cuthill, E., McKee, J.: Reducing the bandwidth of sparse symmetric matrices. In: Proceedings of the 1969 24th National Conference, pp. 157–172, ACM Press, New York, NY, USA (1969)
9. Denis, C., Boufflet, J.-P., Breitkopf, P.: A load balancing method for a parallel application based on a domain decomposition. In: International Parallel and Distributed Processing Symposium, 2005. Proceedings 19th IEEE international. pp. 17a – 17a, ISBN: 0-7695-2312-9 (2005). DOI 10.1109/IPDPS.2005.36
10. Dongarra, J., Lumsdaine, A., Pozo, R., Remington, K.: A sparse matrix library in C++ for high performance architectures. In: Proceedings of the 2nd Annual Object - Oriented Numerics Conference (00N-SKI '94, Sun River, OR, Apr.), pp. 214–218 (1994)
11. Karypis, G., Kumar, V.: Multilevel k-way partitioning scheme for irregular graphs. Journal of Parallel and Distributed Computing **48**, 96–129 (1988)
12. Li, X.S., Demmel, J.W.: SuperLU_DIST: A scalable distributed-memory sparse direct solver for unsymmetric linear systems. ACM Transactions on Mathematical Software **29**(2), 110–140 (2003)
13. O'Leary, D.P., White, R.E.: Multi-splittings of matrices and parallel solution of linear systems. Journal on Algebraic and Discrete Mathematic **6**, 630–640 (1985)
14. Scott, J.A.: Parallel frontal solvers for large sparse linear systems. ACM Trans. Math. Softw. **29**(4), 395–417 (2003)
15. Scott, N.S., Jezequel, F., Denis, C., Chesneaux, J.M.: Numerical 'health check' for scientific codes: the cadna approach. Computer Physics Communications **176**, 507–521 (2007)
16. Verwer, J.G., Blom, J.G., Hundsdorfer, W.: An implicit-explicit approach for atmospheric transport-chemistry problems. Appl. Numer. Math. **20**, 191–209 (1996)

Parallel Diagonalization Performance on High-Performance Computers

Andrew G. Sunderland

Abstract Eigenvalue and eigenvector computations arise in a wide range of scientific and engineering applications. For example, in quantum chemistry and atomic physics, the computation of eigenvalues is often required to obtain electronic energy states. For large-scale complex systems in such areas, the eigensolver calculation usually represents a huge computational challenge. It is therefore imperative that suitable, highly efficient eigensolver methods are used in order to facilitate the solution of the most demanding scientific problems. This presentation will analyze the performance of parallel eigensolvers from numerical libraries such as ScaLAPACK on the latest parallel architectures using data sets derived from large-scale scientific applications.

1 Introduction

Efficient parallel diagonalization performance is essential for many scientific and engineering application codes. For example, in quantum chemistry and quantum physics, the computation of eigenvalues may be required in order to calculate electronic energy states. Computations often involve matrices of dimension of tens or even hundreds of thousands that need to be solved quickly with manageable memory requirements on the latest large-scale high-performance computing platforms.

This paper analyzes the performance of parallel eigensolver library routines across a range of applications, problem sizes, and architectures. New developments of particular note include a pre-release ScaLAPACK implementation of the Multiple Relatively Robust Representations (MRRR) algorithm and the next generation series of high-end parallel computers such as the Cray XT series and IBM's BlueGene. The results presented are based upon Hamiltonian matrices generated during

Andrew G. Sunderland
STFC Daresbury Laboratory, Warrington, UK
e-mail: a.g.sunderland@dl.ac.uk

electron-atom scattering calculations using the PRMAT code [16] and matrices from the CRYSTAL [3] package generated during the computation of electronic structure using Hartree-Fock theory.

2 Parallel Diagonalization Methods

The standard eigenvalue problem is described as

$$Ax = \lambda x, \tag{1}$$

where A is a matrix and λ is the eigenvalue corresponding with eigenvector x. For symmetric matrices, this equation can be rearranged to give the equation describing the diagonalization of matrix A:

$$A = Q\Lambda Q^T, \tag{2}$$

where the columns of the matrix Q are represented by the orthogonal eigenvectors of A, and the diagonal matrix Λ represents the associated eigenvalues.

2.1 Equations for Matrix Diagonalizations in PRMAT

The PRMAT code is based on the Baluja–Burke–Morgan [2] approach for solving the non-relativistic Schrödinger equation describing the scattering of an electron by an N-electron atom or ion:

$$H_{N+1}\Psi = E\Psi, \tag{3}$$

where E is the total energy in atomic units, and H_{N+1} is the $(N+1)$-electron Hamiltonian matrix.

In this approach, the representative of the Green function $(H+L-EI)^{-1}$ are diagonalized within a basis. The symmetric matrix $(H+L-EI)$ is reduced to diagonal form by the orthogonal transformation:

$$X^T(H+L-E)X = (E_k - E), \tag{4}$$

where the columns of the orthogonal matrix X^T represent the eigenvectors and E_k the eigenvalues of $(H+L)$.

2.2 Equations for Matrix Diagonalizations in CRYSTAL

The CRYSTAL package [3] performs *ab initio* calculations of the ground state energy, electronic wave function, and properties of periodic systems. Development of the software has taken place jointly by the Theoretical Chemistry Group at the

University of Torino and the Computational Materials Science Group at STFC Daresbury Laboratory (UK). The computation of the electronic structure is performed using either Hartree–Fock or Density Functional theory. In each case, the fundamental approximation made is the expansion of the single particle wave functions as a linear combination of atom-centered atomic orbitals (LCAO) based on Gaussian functions.

The computational core of the basic Hartree–Fock algorithm reduces to an iterative loop:

$$
\begin{array}{lrcll}
 & i & = & 0; & \\
\textbf{repeat} & & & & \\
 & \mathbf{H}_r^i & = & \mathbf{P}_r^i.\mathbf{I}_r^i, & \text{where } \mathbf{I}_r^i; \text{ is the sum of independent integrals} \\
 & \mathbf{H}_k^i & \Longleftarrow & \mathbf{Q}_k^{i\,T}\mathbf{H}_r^i\mathbf{Q}_k^i; & \text{... Fourier transform and matrix multipy} \\
 & \mathbf{H}_k^i\psi_k^i & = & \varepsilon_k^i\psi_k^i; & \text{... Parallel diagonalization} \\
 & \mathbf{P}_r^{i+1} & \Longleftarrow & \mid \psi_k^i \mid^2; & \\
 & \multicolumn{4}{l}{\textbf{if } \mathbf{P}_r^{i+1} - \mathbf{P}_r^i \text{ is sufficiently small then } \textbf{exit;}} \\
 & \textbf{else } i & = & i+1; & \\
\textbf{end repeat} & & & &
\end{array}
\qquad (5)
$$

where the suffixes r and k represent real-space and k-space, respectively.

2.3 Symmetric Eigensolver Methods

The solution to the real or Hermitian dense symmetric eigensolver problem usually takes place via three main steps:

1. Reduction of the matrix to tri-diagonal form, typically using the Householder Reduction,
2. Solution of the real symmetric tri-diagonal eigenproblem via one of the following methods:
   - Bisection for the eigenvalues and inverse iteration for the eigenvectors [6, 19],
   - QR algorithm [5],
   - Divide & Conquer method (D&C) [17],
   - Multiple Relatively Robust Representations (MRRR algorithm) [4],
3. Back transformation to find the eigenvectors for the full problem from the eigenvectors of the tridiagonal problem.

For an $n \times n$ matrix, the reduction and back transformation phases each require $O(n^3)$ arithmetic operations. Until recently, all algorithms for the symmetric tridiagonal eigenproblem also required $O(n^3)$ operations in the worst case and associated memory overheads of $O(n^2)$. However, for matrices with clustered eigenvalues, the Divide and Conquer method takes advantage of a process known as *deflation* [17], which often results in an reduced operation count. The potential advantages of the MRRR algorithm are both that theoretically only $O(kn)$ operations are required, where k is the number of desired eigenpairs, and the additional memory requirements are only $O(n)$.

2.4 Eigensolver Parallel Library Routines

Several eigensolver routines for solving standard and generalized dense symmetric or dense Hermitian problems are available in the current release of ScaLAPACK [12]. These include:

- *PDSYEV* based on the QR Method,
- *PDSYEVX* based on Bisection and Inverse Iteration,
- *PDSYEVD* based on the Divide and Conquer method,
- Also tested here is a new routine *PDSYEVR* [1] based on the MRRR algorithm. At the time of this analysis, this routine is a pre-release version and is still undergoing testing and development by the ScaLAPACK developers.

PDSYEV and *PDSYEVD* only calculate *all* the eigenpairs of a matrix. However, *PDSYEVX* and the new *PDSYEVR* have the functionality to calculate subsets of eigenpairs specified by the user. For reasons of conciseness, the performance results reported in this chapter will focus on the latest parallel solvers *PDSYEVD* and *PDSYEVR*. For a comparison of a fuller range of eigensolvers, readers are recommended to consult the HPCx Technical Report HPCxTR0608 [15].

3 Testing Environment

The matrices analyzed here are derived from external sector Hamiltonian Ni3+ and Fe+ scattering calculations using the PRMAT code. They are all real symmetric matrices with dimensions ranging from 5280 to 20064. The eigenvalue distribution is fairly well-spaced with comparatively few degeneracies, though some clustering does exist. For the main cross-platform comparisons, diagonalizations using matrices obtained from the CRYSTAL package have been measured. The eigenvalue distribution of these real symmetric matrices is typically much more clustered than are those obtained from the PRMAT code.

The majority of the parallel timings presented are from runs undertaken on the current National Supercomputing facility HPCx [9] at STFC [14] comprising 160 IBM p5-575 nodes, totalling 2536 processors. Timings are also shown that map the evolution of the HPCx system over the past five years. The original Phase 1 configuration consisted of p690 processors with the colony (SP) switch. Figures also show timing comparisons of runs taken on HPCx with runs undertaken on other contemporary HPC platforms: an IBM Blue Gene/L and Blue Gene/P machine [10], also sited at STFC; a Cray XT3 machine sited at the Swiss Supercomputing Centre CSCS [13] with AMD 2.6GHz Opteron processors; and the new 'HECToR' Cray XT4 machine with 11,328 AMD 2.8GHz Opteron cores sited at the University of Edinburgh [8].

The results include comparisons for dual-core processors (Cray XT4, BG/L), quad-core processors (BG/P), and 16-way and 32-way shared-memory processors (SMPs) (IBM p5-575 and IBM p690). For reasons of consistency, throughout the performance analysis charts, "Number of Processors" is taken to be equivalent to "Number of Cores".

4 Results

Figures 1 and 2 show the scaling of performance for the diagonalization routines PDSYEVR and PDSYEVD for a range of problem sizes on the current configuration of HPCx (IBM p5-575) and HECToR (Cray XT4). The relative performance reported is the time taken to solution for the algorithm and HPC platform with respect to the CPU time of PDSYEVD on 16 processors of HPCx. The performance for two different versions of PDSYEVR are shown: an older version from 2007 and a very recent version from early 2008. It is shown that the parallel performance of PDSYEVR (2008) is very close to that of PDSYEVD for the two problem sizes, though the performance on the highest processor count can degrade, possibly due to slightly uneven distributions of the eigenvalue representation tree among processors [1]. Figure 2 shows how performance increases up to a maximum on

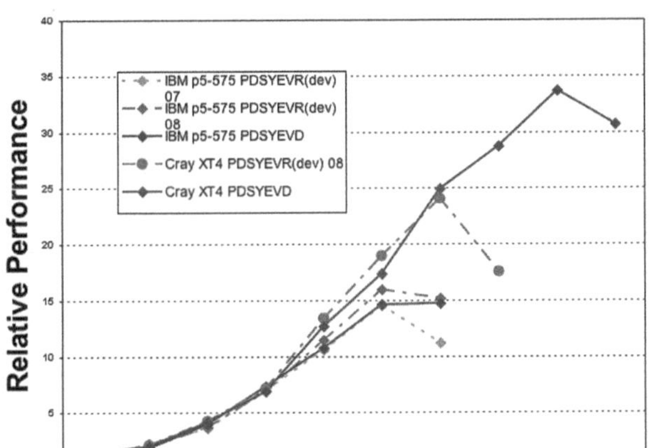

Fig. 1 Parallel performance of PDSYEVD and PDSYEVR for PRMAT sector Hamiltonian matrix, $n = 20064$.

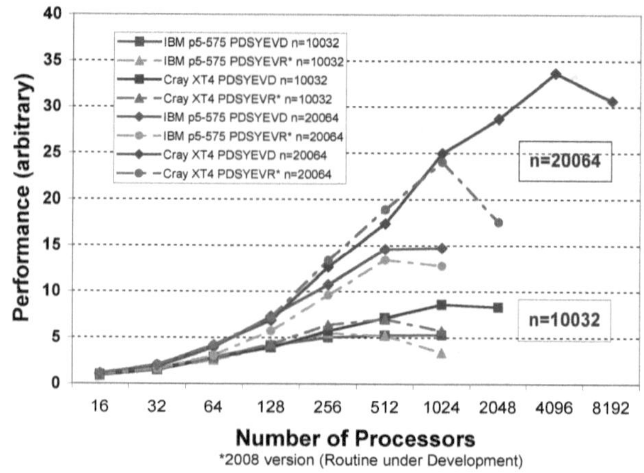

Fig. 2 Relative scaling of PDSYEVD and PDSYEVR for different Hamiltonian matrix sizes.

4096 processors for the larger problem size. However, performance is only around 34 times faster on 4096 processors than on 16 processors. This is much lower that the theroretical perfect parallel scaling, which would result in a performance improvement factor of 256 (4096/16).

Figures 3 and 4 show how the PDSYEVD routine scales with processor count on the high-end computing platforms detailed in Sect. 3. Parallel performance is best on

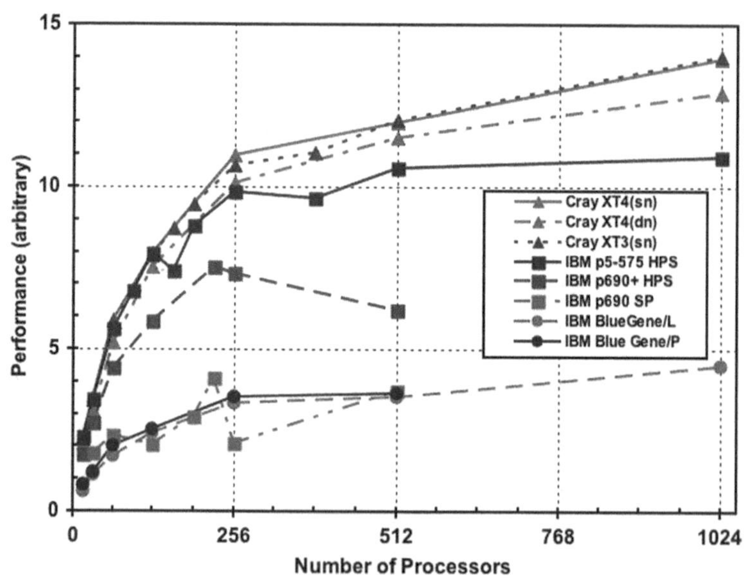

Fig. 3 Performance of PDSYEVD on the latest HPC architectures (CRYSTAL matrix, $n = 7194$).

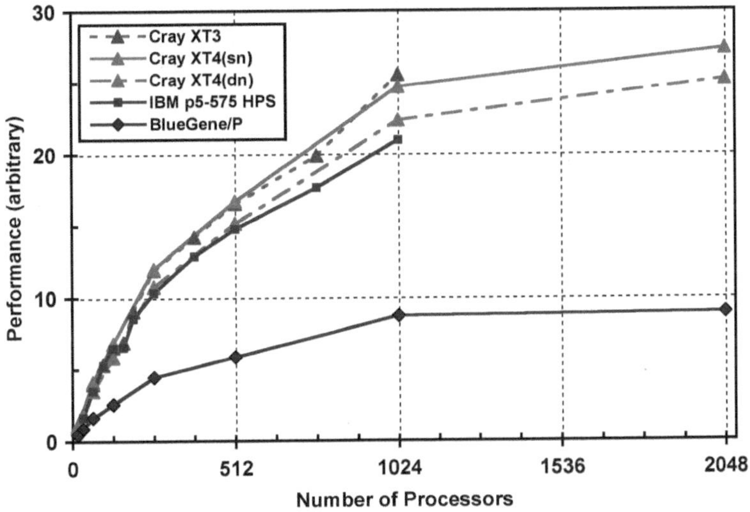

Fig. 4 Performance of PDSYEVD on the latest HPC architectures (CRYSTAL matrix, $n = 20480$).

the Cray XT machines for both matrices tested here, relatively closely matched by the current configuration of HPCx (IBM p5-575 with the High Performance Switch). Different sets of results are provided for single core usage (sn) and dual core usage (dn) of the dual core nodes of the Cray XT4. At the time these tests were undertaken, the Cray XT3 testing platform consisted of single core processors only. As is often the case, the higher clock speed of the XT4 processors relative to the XT3 results in a negligible improvement to parallel performance. The performance of the IBM BlueGene/L and BlueGene/P is around three times slower than that of the Cray XTs, roughly matching the performance of the orginal HPCx system (p690 SP).

The advantage of the BlueGene/P machine is demonstrated most clearly in Fig. 5, where the power consumption for parallel matrix eigensolves relative to that undertaken on 16 processors of BlueGene/P is shown. At present, BlueGene/P at STFC Daresbury Laboratory in the United Kingdom is the most efficent supercomputer (with respect to flops per Watt) in the world [7].

One characteristic of new parallel eigensolvers that has become evident during the course of the tests is that the tridiagonal eigensolver may no longer be the primary computational bottleneck in the full symmetric eigensolve. Figure 6 shows how the balance between reduction, tri-diagonal eigensolve, and back transformation has changed significantly with different eigensolver methods. It can be seen that the time taken in the tri-diagonal eigensolve using Divide-and-Conquer is relatively small compared with the time taken in reducing the full matrix to tri-diagonal form. Although the back-transformation calculation scales very well to large numbers of processors, the relative computational costs of the reduction phase remain high. This contrasts markedly with the traditional QR-based approach where the tri-diagonal eigensolve dominates the overall time taken to solution.

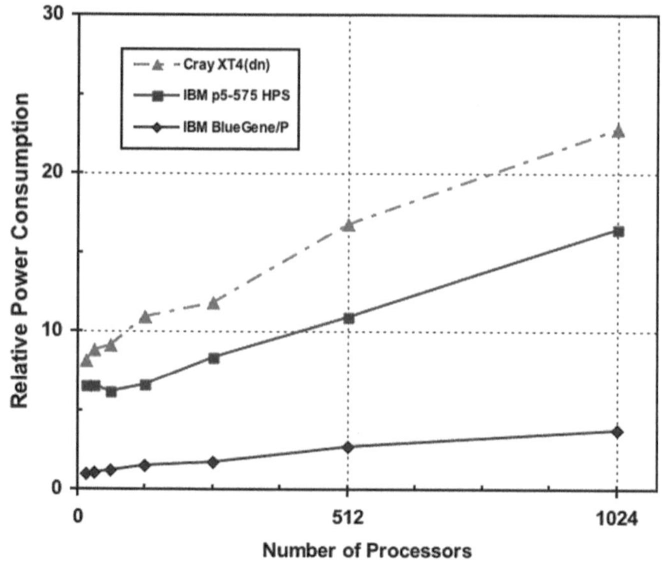

Fig. 5 Relative power consumption for eigensolve on different platforms (CRYSTAL matrix, $n = 20480$).

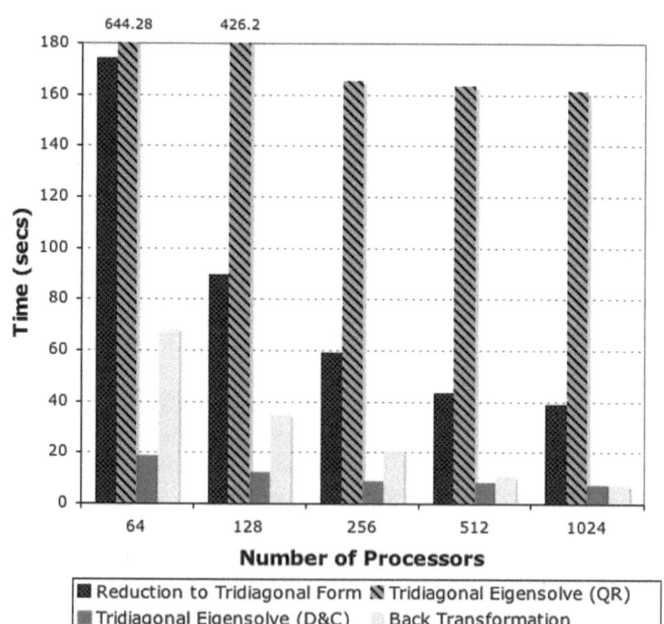

Fig. 6 Timing breakdown of the three phases of the eigensolve (CRYSTAL matrix, $n = 20480$).

5 Conclusions

The latest ScaLAPACK eigensolvers are generally reliable and perform well for the applications tested in this paper. Typically, the parallel scaling improves for the larger problem sizes on all the platforms, as the computation to communication ratio increases. In other reports, it has been established that both solvers generally perform preferably to the original ScaLAPACK solvers *PDSYEV* and *PDSYEVX* for the matrices under test here (see [15]). The parallel performance of the pre-release version of the MRRR-based solver *PDSYEVR* obtained from the developers for testing performs comparably with the Divide-and Conquer based *PDSYEVD* over a range of problem sizes. On large processor counts where the division of the problem is relatively thin, the performance of PDSYEVR appears to degrade somewhat. This problem is addressed in [18]. It remains to be seen if the "holy grail" properties of $O(kn)$ operations and memory overheads of $O(n)$ will be achieved for the final release of *PDSYEVR* in a future release of ScaLAPACK.

Timings from the IBM p5-575 and the new Cray XT series machines show that good parallel scaling can be achieved for larger matrices up to a few thousands of processors. The results from the new BlueGene architectures show that it is generally two to three times slower than are equivalent parallel runs on the the Cray XT4 for large-scale parallel diagonalizations. This ratio pretty much matches the respective processor clock speed (2.8GHz vs 850MHz). on the two machines. However, it is now of increasing importance that parallel architectures are power efficient (flops/Watt) in addition to being performance efficient (flops/sec). Figure 6 shows that the power consumption of the BlueGene/P is around six times lower than that of the Cray XT4 for a corresponding matrix diagonalization.

When timings for the full symmetric eigenproblem are broken down into the three constituent phases – reduction, tri-diagonal eigensolve and back transformation (Fig. 6) – it is shown that the tri-diagonal eigensolve may no longer dominate timings. Moreover, the Householder reduction is both relatively slow and scales poorly on large processor counts. This has been recognized by parallel numerical routine developers, and new methods are now under investigation to improve the parallel performance of this phase of the calculation [11].

To meet the challenges of petascale architectures, where runs may involve tens of thousands of processing cores, it is evident that new parallelization strategies will be required. For example, the PRMAT Hamiltonian matrices represent the wavefunction for a sector of external configuration space defined when calculating the electron-atom scattering problem. A typical problem contains multiple sectors, and with a little reorganization, the sector Hamiltonian matrix diagonalizations can be calculated concurrently by sub-groups of processors divided up from the global processor population. Thus good parallel scaling of the overall scientific problem could be achieved on many of thousands of processors by utilizing parallel diagonalization methods that perform efficiently on processor counts of mere hundreds.

References

1. Antonelli, D., Vömel, C.: PDSYEVR: ScaLAPACKs parallel MRRR algorithm for the symmetric eigenvalue problem. Tech. Rep. UCB/CSD-05-1399, Lapack working note (2005). URL http://www.netlib.org/lapack/lawnspdf/lawn168.pdf
2. Baluja, K.L., Burke, P.G., Morgan, L.A.: R-matrix propagation program for solving coupled second-order differential equations. Computer Physics Communications **27**(3), 299–307 (1982)
3. CRYSTAL: A computational tool for solid state chemistry and physics. URL http://www.crystal.unito.it
4. Dhillon, I.S.: Algorithm for the symmetric tridiagonal eigenvalue/eigenvector problem. Ph.D. thesis, University of California, Berkeley, California (1997)
5. Francis, J.G.F.: The QR transformation, parts I & II. The Computer J. **4**, 265–271 and 332–345 (1961–1962)
6. Givens, W.J.: The numerical computation of the characteristic values of a real symmetric matrix. Tech. Rep. ORNL-1574, Oak Ridge National Laboratory (1954)
7. The Green500 list. URL http://www.green500.org
8. HECTOR - the UK supercomputing service. URL http://www.hector.ac.uk
9. The HPCx Supercomputing Facility. URL http://www.hpcx.ac.uk
10. The IBM BlueGene. URL http://www.research.ibm.com/bluegene/
11. Kaya, D.: Parallel algorithms for reduction of a symmetric matrix to tridiagonal form on a shared memory multiprocessor. Applied Mathematics and Computation **169**(2), 1045–1062 (2005)
12. The ScaLAPACK Project. URL www.netlib.org/scalapack/
13. The Swiss National Supercomputing Centre. URL http://www-users.cscs.ch/xt4/
14. The Science and Technology Facilities Council. URL http://www.stfc.ac.uk
15. Sunderland, A.G.: Performance of a new parallel eigensolver PDSYEVR on HPCx. HPCx Technical Report (2006). URL http://www.netlib.org/scalapack/scalapack_home.html
16. Sunderland, A.G., Noble, C.J., Burke, V.M., Burke, P.G.: A parallel R-matrix program PRMAT for electron-atom and electron-ion scattering calculations. Computer Physics Communications **145**, 311–340 (2002)
17. Tisseur, F., Dongarra, J.: The QR transformation, parts I & II. SIAM J. Comput. **20**(6), 2223–2236 (1999)
18. Vömel, C.: A refined representation tree for MRRR. Tech. Rep. 194, Lapack Working Note (2007). URL http://www.netlib.org/lapack/lawnspdf/lawn194.pdf
19. Wilkinson, J., Reinsch, C.: Contribution II/18: The calculation of specified eigenvectors by inverse iteration, vol. II - Linear Algebra, 2nd edn. Springer-Verlag, Berlin (1971)

Part II
Parallel Optimization

Parallel Global Optimization in Multidimensional Scaling

Julius Žilinskas

Abstract Multidimensional scaling is a technique for exploratory analysis of multidimensional data, whose essential part is optimization of a function possessing many adverse properties including multidimensionality, multimodality, and non-differentiability. In this chapter, global optimization algorithms for multidimensional scaling are reviewed with particular emphasis on parallel computing.

1 Introduction

Many problems in engineering, physics, economics, and other fields are reduced to global minimization with many local minimizers. Mathematically the problem is formulated as

$$f^* = \min_{\mathbf{x} \in \mathbf{D}} f(\mathbf{x}),$$

where $f(\mathbf{x})$ is a nonlinear function of continuous variables $f : \mathbb{R}^N \to \mathbb{R}$, $\mathbf{D} \subseteq \mathbb{R}^N$ is a feasible region, and N is the number of variables. Besides the global minimum f^*, one or all global minimizers $\mathbf{x}^* : f(\mathbf{x}^*) = f^*$ should be found. No assumptions on unimodality are included into formulation of the problem [10, 20, 28].

Global optimization problems are classified difficult in the sense of the algorithmic complexity theory. Therefore, global optimization algorithms are computationally intensive, and solution time crucially depends on the dimensionality of a problem. Large practical problems, unsolvable with available computers, always exist. When computing power of usual computers is not sufficient to solve a practical problem, the high-performance parallel computers may be helpful. An algorithm is more applicable in case its parallel implementation is available, because larger practical problems may be solved by means of parallel computing. Therefore,

Julius Žilinskas
Institute of Mathematics and Informatics, Akademijos 4, LT-08663 Vilnius, Lithuania
e-mail: julius.zilinskas@ktl.mii.lt

implementation and investigation of parallel versions of algorithms for global optimization is important.

Multidimensional scaling (MDS) is a technique for exploratory analysis of multidimensional data widely usable in different applications [5, 9]. An essential part of the technique is optimization of a function possessing many optimization adverse properties: it is high dimensional, normally it has many local minima, it is invariant with respect to translation and mirroring, and it can be non-differentiable.

2 Global Optimization

A point $\mathbf{x}^*$ is a local minimum point of the function f if $f(\mathbf{x}^*) \leq f(\mathbf{x})$ for $\mathbf{x} \in \mathbf{N}$, where $\mathbf{N}$ is a neighborhood of $\mathbf{x}^*$. A local minimum point can be found using local optimization, e.g., by stepping in the direction of steepest descent of the objective function. Without additional information, one cannot say if the local minimum is global. Global optimization methods not only try to find a good function value fast, but also try to explore the whole feasible region by evaluating function values at sampling points or investigating sub-regions of the feasible region.

Classification of global optimization methods is given in [28]:

- Methods with guaranteed accuracy:
  - Covering methods;
- Direct methods:
  - Random search methods,
  - Clustering methods,
  - Generalized descent methods;
- Indirect methods:
  - Methods approximating level sets,
  - Methods approximating objective function.

Theoretically, covering methods can solve global optimization problems of some classes with guaranteed accuracy. Covering methods detect the sub-regions not containing the global minimum and discard them from further search. The partitioning of the sub-regions stops when the global minimizers are bracketed in small sub-regions guaranteeing the prescribed accuracy. A lower bound of the objective function over the sub-region is used to indicate the sub-regions that can be discarded. Some methods are based on a lower bound constructed as convex envelope of an objective function [10]. Lipschitz optimization is based on assumption that the slope of an objective function is bounded [20]. Interval methods estimate the range of an objective function over a multidimensional interval using interval arithmetic [19]. A branch and bound technique can be used for managing the list of sub-regions and the process of discarding and partitioning. Although covering, selection, branching and bounding rules differ in different branch and bound algorithms, the structure of the algorithm remains the same. This allows implementation of generalized branch and bound templates [2, 3]. Standard parts of branch and bound algorithms are implemented in the template, only specific rules should be implemented by the user. Templates ease implementation of branch and bound algorithms for combinatorial optimization and for covering methods of continuous global optimization [4]. Parallel versions can be obtained automatically using sequential program implemented using the template.

Random search methods may be adaptive and non-adaptive. Non-adaptive methods generate random trial points with predefined distribution. The trial points can be used as starting points of local searches. For example, pure random search does not contain local searches at all. Single start performs a single local search starting from the best trial point (with the smallest value of the objective function). Multi-start performs local searches starting from all trial points, and the global minimum is the smallest minimum found. These methods are very simple but inefficient. Normally the probability of finding the global minimum approaches one when the number of observations of the objective function approaches infinity. Usually non-experts use these methods to solve practical problems because the methods are simple and easy to implement. Sometimes these methods are used by the researchers to extract the characteristic of a problem: the global minimum, the number of global and local minimizers, the probability that a local search started from a random point would reach the global minimum. The parallelization of non-adaptive methods is obvious. Each process independently runs the same algorithm on equal sub-regions of or on the overall feasible region. There is no need to communicate between processes. The speedup is equal to the number of processes, and the efficiency of parallelization is equal to one.

The main idea of adaptive random search is to distribute the trial points non-uniformly in the feasible region with greater density in the most promising sub-regions. The best function values found indicate the promising sub-regions. Genetic algorithms [13, 25, 27] simulate evolution (selection, mutation, crossover) in which a population of solutions evolves improving function values. Genetic algorithms are suitable to parallelization [7], for example implementing multiple populations. Simulated annealing [8, 21] replaces the current solution by a random solution with a probability that depends on the difference of the function values and a temperature parameter. In the beginning, the temperature parameter is large allowing non-improving changes. Gradually temperature is decreased and the search becomes descent.

In clustering methods, the trial points are grouped into clusters identifying the neighborhoods of the local minimizers, and just one local search is started from every cluster. The repeated descent to a local minimizer is prevented. The trial points may be sampled using a grid or randomly.

Generalized descent methods are the generalization of the local search methods to global optimization. In the trajectory methods, the differential equation describing the local descent is modified. In the penalty methods, the local search algorithm is repeatedly applied to a modified objective function preventing the descent to known local minima. Tabu search [11, 12] modifies neighborhood definition to avoid repeated descent to the known local minimizers.

In methods approximating the objective function, statistical models of the objective function are used. The unknown values of the objective function are modeled using random variables. The auxiliary computations to determine the next trial point are expensive, therefore these methods are reasonable for expensive objective functions. The optimization technique based on a stochastic function model and minimization of the expected deviation of the estimate from the real global optimum is called Bayesian [26].

Criteria of performance of global optimization algorithms are speed, best function value found, and reliability. The speed is measured using time of optimization or the number of objective function (and sometimes gradient, bounding, and other functions) evaluations. Both criteria are equivalent when objective function is expensive—its evaluation takes much more time than do auxiliary computations of an algorithm. When an algorithm does not guarantee the global solution, best function value found and reliability, showing how often problems are solved with prescribed accuracy, are used to compare the performance of algorithms.

In general, parallelization of algorithms for global optimization is not straightforward. For example, independent adaptive search cannot be performed efficiently in different parts of the feasible region. However, some subclasses of global optimization algorithms (e.g., random search, evolutionary strategies) are favorable to parallelization.

Efficiency of the parallelization can be evaluated using standard criteria taking into account the optimization time and the number of processes. A commonly used criterion of parallel algorithms is the speedup: $s_{\text{size}} = t_1/t_{\text{size}}$, where t_{size} is the time used by the algorithm implemented on *size* processes. The speedup divided by the number of processes is called the efficiency: $e_{\text{size}} = s_{\text{size}}/\text{size}$.

3 Multidimensional Scaling

Multidimensional scaling (MDS) is a technique for exploratory analysis of multidimensional data. Let us give a short formulation of the problem. Pairwise dissimilarities among n objects are given by the matrix (δ_{ij}), $i, j = 1, \dots, n$. A set of points in an embedding metric space is considered as an image of the set of objects. Normally, an m-dimensional vector space is used, and $\mathbf{x}_i \in \mathbb{R}^m$, $i = 1, \dots, n$, should be found whose inter-point distances fit the given dissimilarities. Images of the considered objects can be found minimizing a fit criterion, e.g., the most frequently used least squares *Stress* function:

$$S(\mathbf{x}) = \sum_{i=1}^{n} \sum_{j=1}^{n} w_{ij} \left(d\left(\mathbf{x}_i, \mathbf{x}_j\right) - \delta_{ij} \right)^2, \qquad (1)$$

where $\mathbf{x} = (\mathbf{x}_1, \dots, \mathbf{x}_n)$, $\mathbf{x}_i = (x_{i1}, x_{i2}, \dots, x_{im})$. It is supposed that the weights are positive: $w_{ij} > 0, i, j = 1, \dots, n$; $d(\mathbf{x}_i, \mathbf{x}_j)$ denotes the distance between the points $\mathbf{x}_i$ and $\mathbf{x}_j$. Usually Minkowski distances are used:

$$d_r(\mathbf{x}_i, \mathbf{x}_j) = \left(\sum_{k=1}^{m} \left| x_{ik} - x_{jk} \right|^r \right)^{1/r}. \quad (2)$$

Equation (2) defines Euclidean distances when $r = 2$, and city-block distances when $r = 1$. The most frequently used distances are Euclidean. However, MDS with other Minkowski distances in the embedding space can be even more informative than MDS with Euclidean distances [1].

Although *Stress* function is defined by an analytical formula, which seems rather simple, its minimization is a difficult global optimization problem [16]; its dimensionality is $N = n \times m$. Global optimization of *Stress* function is difficult, therefore frequently only a local minimum is sought. Although improved local search procedures are used for some applications of multidimensional scaling, certain applications can be solved only with global optimization. Two examples of such applications are described in [23]. One of the applications is the estimation of the position of GSM mobile phone using the measured powers of the 6 signals received from surrounding base stations. Another application is interpretation of the results on experimental testing of soft drinks [14]. It is shown in [23] that there are many local minima for these problems, and interpreting the data on the basis of the achieved configuration from local minima leads to different results. So it is necessary to find the global minimum and the corresponding configuration that explains the data best.

A tunneling method for global minimization was introduced and adjusted for MDS with general Minkowski distances in [17]. The tunneling method alternates a local search step, in which a local minimum is sought, with a tunneling step in which a different configuration is sought with the same value of Stress as the previous local minimum. In this manner, successively better local minima are obtained, and the last one is often the global minimum.

A method for MDS based on combining a local search algorithm with an evolutionary strategy of generating new initial points was proposed in [24]. Its efficiency is investigated by numerical experiments. The testing results in [15, 23] proved that the hybrid algorithm combining an evolutionary global search with an efficient local descent is the most reliable though the most time-consuming method for MDS. To speed up computation, parallel version of the algorithm has been implemented in [31].

A heuristic algorithm based on simulated annealing for two-dimensional city-block scaling was proposed in [6]. The heuristic starts with the partition of each coordinate axis into equally spaced discrete points. A simulated annealing algorithm is used to search the lattice defined by these points with the objective of minimizing least-squares or least absolute deviation loss function. The object permutations for each dimension of the solution obtained by the simulated annealing algorithm are used to find a locally optimal set of coordinates by quadratic programming. A multivariate randomly alternating simulated annealing procedure with permutation and translation phases has been applied to develop an algorithm for multidimensional scaling in any Minkowski metric in [30].

A bi-level method for city-block MDS was proposed in [33]. The method employs a piecewise quadratic structure of *Stress* with city-block distances reformulating the global optimization problem as a two-level optimization problem, where the upper level combinatorial problem is defined over the set of all possible permutations of $1,\ldots,n$ for each coordinate of the embedding space, and the lower level problem is a quadratic programming problem with a positively defined quadratic objective function and linear constraints setting the sequences of values of coordinates defined by m permutations. The lower level problems are solved using a quadratic programming algorithm. The upper level combinatorial problem can be solved by guaranteed methods for small n and using evolutionary search for larger problems. Branch and bound algorithm for the upper level combinatorial problem has been proposed in [34]. Parallel genetic algorithm for city-block MDS has been developed and investigated [29, 31, 32]. Better function values have been found [31] for Morse code confusion problem than had been published previously [6].

Let us consider visualization of 4-dimensional hyper-cube in two-dimensional space as an example of multidimensional scaling. A 4-dimensional hyper-cube may be used in network topology of 16 computing elements having four possible interconnections. Vertices of a 4-dimensional unit hyper-cube are considered as objects for the problem. The number of vertices is $n = 2^4 = 16$, and the dimensionality of the corresponding global minimization problem is $N = 2 \times 16 = 32$. The coordinates of ith vertex of a multidimensional hyper-cube are equal either to 0 or to 1, and they are defined by a binary code of $i = 0,\ldots,n-1$. Dissimilarities between vertices are measured by Euclidean or city-block distances in original multidimensional space. Influence of the type of distances in original and embedding spaces to the result of MDS has been investigated in [33], and it was shown that the results are more influenced by the distances in the embedding space. Images of 4-dimensional hyper-cube visualized using MDS with Euclidean and city-block distances are shown in Fig. 1.

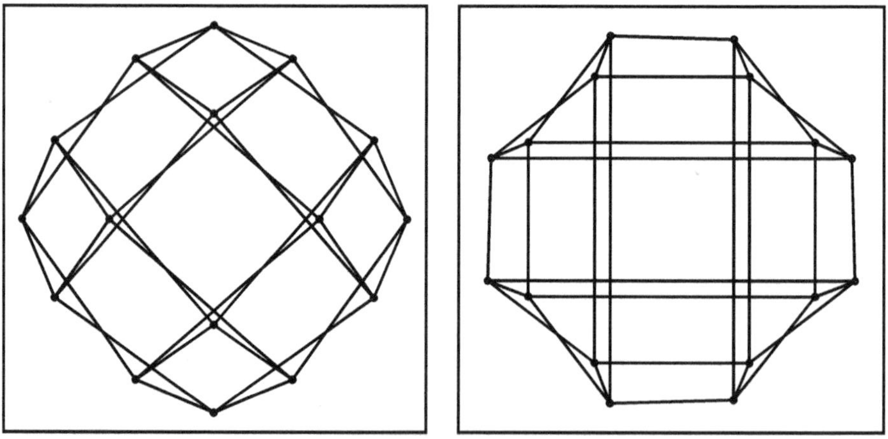

Fig. 1 Images of 4-dimensional hyper-cube visualized using MDS with Euclidean (left) and city-block (right) distances.

The vertices are shown as circles, and adjacent vertices are joined by lines to make representations more visual. Although it is difficult to imagine 4-dimensional hyper-cube, it is known that vertices of a hyper-cube are equally far from the center and compose clusters containing 2^d vertices corresponding to edges, faces, etc. When city-block distances are used, the images of vertices tend to form a rotated square, which is equivalent to a circle in Euclidean metric—all points on its edges are of the same distance to the center. The images of the hyper-cube visualized using MDS with city-block distances well visualize equal location of all vertices of the hyper-cube with respect to the center. This property is not visible in the images corresponding to Euclidean distances. In both images, it is possible to identify clusters of vertices corresponding to edges and faces.

4 Multidimensional Scaling with City-Block Distances

The case of city-block distances in MDS is different from the other cases of Minkowski metric where positiveness of distances imply differentiability of *Stress* at a local minimum point [22, 18]. In case of city-block distances, *Stress* can be non-differentiable even at a minimum point [33]. Therefore MDS with city-block distances is an especially difficult optimization problem. However, *Stress* with city-block distances is piecewise quadratic, and such a structure can be exploited in the following way. *Stress* (1) with city-block distances $d_1(\mathbf{x}_i, \mathbf{x}_j)$ can be redefined as

$$S(\mathbf{x}) = \sum_{i=1}^{n} \sum_{j=1}^{n} w_{ij} \left(\sum_{k=1}^{m} |x_{ik} - x_{jk}| - \delta_{ij} \right)^2 . \tag{3}$$

Let $A(\mathbf{P})$ be a set of $\mathbb{R}^{n \cdot m}$ such that

$$A(\mathbf{P}) = \{\mathbf{x} | x_{ik} \le x_{jk} \text{ for } p_{ki} < p_{kj},\ i, j = 1, \ldots, n,\ k = 1, \ldots, m\},$$

where $\mathbf{P} = (\mathbf{p}_1, \ldots, \mathbf{p}_m)$, $\mathbf{p}_k = (p_{k1}, p_{k2}, \ldots, p_{kn})$ is a permutation of $1, \ldots, n$; $k = 1, \ldots, m$.

For $\mathbf{x} \in A(\mathbf{P})$, (3) can be rewritten in the following form

$$S(\mathbf{x}) = \sum_{i=1}^{n} \sum_{j=1}^{n} w_{ij} \left(\sum_{k=1}^{m} (x_{ik} - x_{jk})\, z_{kij} - \delta_{ij} \right)^2 ,$$

where

$$z_{kij} = \begin{cases} 1, & p_{ki} > p_{kj}, \\ -1, & p_{ki} < p_{kj}. \end{cases}$$

Because the function $S(\mathbf{x})$ is quadratic over polyhedron $A(\mathbf{P})$, the minimization problem

$$\min_{\mathbf{x} \in A(\mathbf{P})} S(\mathbf{x}) \tag{4}$$

can be reduced to the quadratic programming problem

$$\min\Bigg(-\sum_{k=1}^{m}\sum_{i=1}^{n}x_{ik}\sum_{j=1}^{n}w_{ij}\delta_{ij}z_{kij}\ +$$

$$\frac{1}{2}\Bigg(\sum_{k=1}^{m}\sum_{l=1}^{m}\sum_{i=1}^{n}x_{ik}x_{il}\sum_{t=1,t\neq i}^{n}w_{it}z_{kit}z_{lit}-\sum_{k=1}^{m}\sum_{l=1}^{m}\sum_{i=1}^{n}\sum_{j=1,j\neq i}^{n}x_{ik}x_{jl}w_{ij}z_{kij}z_{lij}\Bigg)\Bigg),$$

$$\text{s.t. } \sum_{i=1}^{n}x_{ik}=0,\ k=1,\ldots,m, \tag{5}$$

$$x_{\{j|p_{kj}=i+1\},k}-x_{\{j|p_{kj}=i\},k}\geq 0,\ k=1,\ldots,m,\ i=1,\ldots,n-1. \tag{6}$$

Polyhedron $A(\mathbf{P})$ is defined by the linear inequality constraints (6), and the equality constraints (5) ensure centering to avoid translated solutions. A standard quadratic programming method can be applied for this problem. However, a solution of a quadratic programming problem is not necessarily a local minimizer of the initial problem of minimization of *Stress*. If a solution of a quadratic programming problem is on the border of polyhedron $A(\mathbf{P})$, a local minimizer possibly is located in a neighboring polyhedron. Therefore, a local search can be continued by solving a quadratic programming problem over the polyhedron on the opposite side of the active inequality constraints and solution of quadratic programming problems is repeated while better values are found, and some inequality constraints are active.

The structure of the minimization problem (4) is favorable to apply a two-level minimization:

$$\min_{\mathbf{P}} S(\mathbf{P}), \tag{7}$$

$$\text{s.t. } S(\mathbf{P})=\min_{\mathbf{x}\in A(\mathbf{P})} S(\mathbf{x}), \tag{8}$$

where (7) is a problem of combinatorial optimization, and (8) is a problem of quadratic programming with a positively defined quadratic objective function and linear constraints. The problem at the lower level is solved using a standard quadratic programming algorithm. Globality of search is ensured by the upper level algorithms.

The upper level (7) objective function is defined over the set of m-tuple of permutations of $1,\ldots,n$ representing sequences of coordinate values of image points. The number of feasible solutions of the upper level combinatorial problem is $(n!)^m$. A solution of MDS with city-block distances is invariant with respect to mirroring when changing direction of coordinate axes or exchanging of coordinates. The feasible set can be reduced taking into account such symmetries. The number of feasible solutions can be reduced to $(n!/2)^m$ refusing mirrored solutions changing direction of each coordinate axis. It can be further reduced to approximately $(n!/2)^m/m!$ refusing mirrored solutions with exchanged coordinates. Denoting $u=n!/2$, the number of feasible solutions is u in the case $m=1$, $(u^2+u)/2$ in the case $m=2$, and $(u^3+3u^2+2u)/6$ in the case $m=3$. The upper level combinatorial problem can be solved using different algorithms, e.g., small problems can be solved by explicit enumeration. Such a bi-level method is covering method with guaranteed accuracy. In this case, sub-regions are polyhedrons $A(\mathbf{P})$ where exact minimum can be found using convex quadratic programming. For larger dimensionalities, genetic algorithms seem prospective. In this case, the guarantee to find the exact solution is lost, but good solutions may be found in acceptable time.

5 Parallel Algorithms for Multidimensional Scaling

Although the dimensionality of MDS problems solvable by means of enumeration cannot be large because of exponentially growing number of potential solutions, it is important to implement and apply such an algorithm for the problems of highest possible dimensionality. Parallel computation enables solution of larger problems by explicit enumeration. It can be assumed that generation of the solutions to be explicitly enumerated requires much less computational time than does the explicit enumeration, which requires solution of the lower level quadratic programming problem. Therefore, it is possible to implement parallel version of explicit enumeration where each process runs the same algorithm generating feasible solutions that should be enumerated explicitly, but only each (*size*)th is explicitly enumerated on each process. The first process (which *rank* is 0) explicitly enumerates the first, $(size+1)$th and so on, generated solutions. The second process (which *rank* is 1) explicitly enumerates the second, $(size+2)$th, and so on, generated solutions. The (*size*)th (which *rank* is $size-1$) process explicitly enumerates the (*size*)th, (2*size*)th, and so on, generated solutions. The results of different processes are collected when the generation of solutions and explicit enumeration are finished. The standardized message-passing communication protocol MPI can be used for communication between parallel processes. The detailed algorithm is given in Algorithm 1. To refuse mirrored solutions with changed direction of coordinate axis, the main cycle continues while $j > 2$, which means that the coordinate values of the first object will never be smaller than corresponding coordinate values of the second object. To refuse mirrored solutions with exchanged coordinates, some restrictions on permutations are set. Let us define the order of permutations: for permutations of $1,\dots,3$ it is "123" $\prec$ "132" $\prec$ "231" and for permutations of $1,\dots,4$ it is "1234" $\prec$ "1243" $\prec$ "1342" $\prec$ "2341" $\prec$ "1324" $\prec$ "1423" $\prec$ "1432" $\prec$ "2431" $\prec$ "2314" $\prec$ "2413" $\prec$ "3412" $\prec$ "3421". A permutation $\mathbf{p}_k$ cannot precede $\mathbf{p}_l$ for $k > l$ ($l < k \Rightarrow \mathbf{p}_l \preceq \mathbf{p}_k$).

Performance of the parallel algorithm composed of explicit enumeration of combinatorial problem and quadratic programming on SUN Fire E15k high-performance computer for some test problems is shown in Fig. 2. Each line corresponds to different global optimization problems, and dimensionality of problems is $N = 14$. On a single process, optimization takes up to 20 minutes. Different numbers of processes from 1 to 24 have been used. On 24 processes, optimization takes less than one minute. The speedup is almost linear and equal to the number of processes; the efficiency of parallel algorithm is close to one. This is because decomposition

Algorithm 1 Parallel explicit enumeration algorithm for multidimensional scaling.

Input: n; m; δ_{ij}, $w_{i,j}$, $i,j = 1,\ldots,n$; $rank$; $size$

Output: S^*, $\mathbf{x}^*$

```
p_ki ← i, i = 1,...,n, k = 1,...,m; j ← n+1; k ← m+1; S* ← ∞; nqp ← 0
while j > 2 do
    if j > n then
        if nqp%size = rank then
            if min_{x∈A(P)} S(x) < S* then // Evaluate solution
                x* ← x; update S*
            end if
        end if
        j ← n; k ← m
    end if
    if j > 2 then // Form next tuple of permutations
        if p_kj = 0 then
            p_kj ← j
            if k > 1 and p_k ≺ p_{k−1} then // Detect refusable symmetries
                p_ki ← p_{k−1 i}, i = 1,...,j
            end if
            k ← k+1
        else
            p_kj ← p_kj − 1
            if p_kj = 0 then
                p_ki ← p_ki − 1, i = 1,...,j−1; k ← k−1
                if k < 1 then
                    j ← j−1; k ← m
                end if
            else
                find i: p_ki = p_kj, i = 1,...,j−1; p_ki ← p_ki + 1; k ← k+1
            end if
        end if
    end if
    nqp ← nqp+1
end while
Collect S*, x* from the different processes, keep the best.
```

in explicit enumeration leads to predictable number of independent sub-problems. Therefore the algorithm scales well.

Performance of the same parallel algorithm on a cluster of personal computers is shown in Fig. 3. Dimensionality of the global optimization problems is $N = 14$ and $N = 16$. The cluster is composed of 3 personal computers with 3GHz Pentium 4 processors and hyper-threading technology allowing simultaneous multithreading. When the number of processes is up to 3, the speedup is almost linear and equal to the number of processes, the efficiency is close to one. In this case, one process per each personal computer is used. If the number of processes is larger than 3, the efficiency of the parallel algorithm is around 0.6 as at least two processes run on one personal computer using multithreading. Because of static distribution of workload, the efficiency is determined by the slowest element of the system. Therefore, the efficiency is similar for the number of processes 4–6. The speedup of approximately

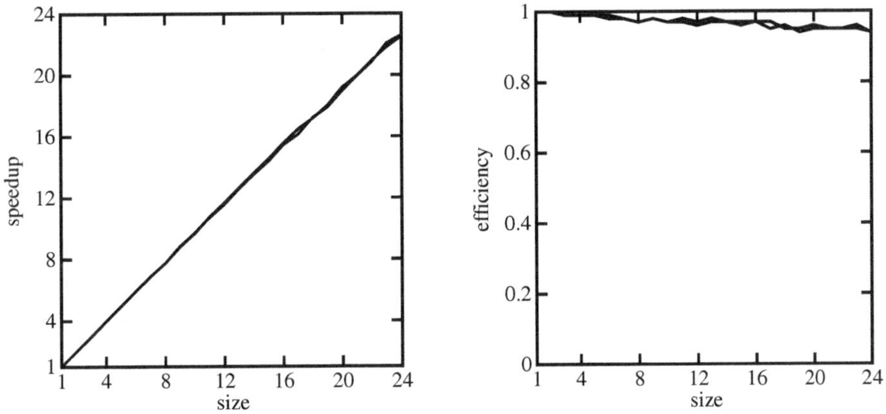

Fig. 2 Performance of parallel explicit enumeration on SUN Fire E15k parallel computer.

3.6 has been reached. Parallel computation yielded the speedup of approximately 3; hyper-threading yielded approximately 20% improvement. With the help of parallel computation, problems with $N = 18$ have been solved.

The general idea of genetic algorithm is to maintain a population of best (with respect to *Stress* value) solutions whose crossover can generate better solutions. The permutations in $\mathbf{P}$ are considered as a chromosome representing an individual. The initial population of individuals is generated randomly and improved by local search. The population evolves generating offspring from two randomly chosen individuals of the current population with the chromosomes $\mathbf{Q}$ and $\mathbf{U}$, where the first corresponds to the better fitted parent. The chromosome of the offspring is defined by 2-point crossover: $\mathbf{p}_k = (q_{k1}, \ldots, q_{k\xi_1}, v_{k1}, \ldots, v_{k(\xi_2-\xi_1)}, q_{k\xi_2}, \ldots, q_{kn})$, where $k = 1, \ldots, m$; ξ_1, ξ_2 are two integer random numbers with uniform

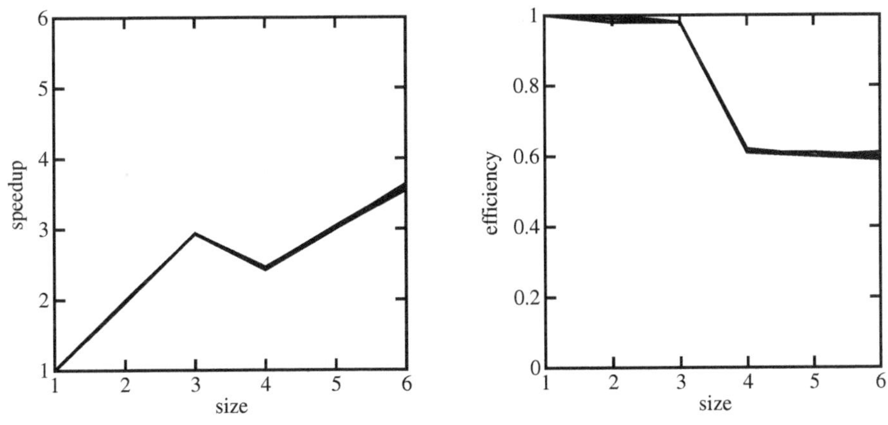

Fig. 3 Performance of explicit enumeration on a cluster of 3 personal computers.

distribution over $1,\ldots,n$; and v_{ki} constitute the subset of $1,\ldots,n$ complementary to $q_{k1},\ldots,q_{k\xi_1},q_{k\xi_2},\ldots,q_{kn}$; the numbers v_{ki} are ordered in the same way as they are ordered in $\mathbf{u}_k$. The offspring is improved by local search, and its fitness is defined by the optimal value of the corresponding lower level problem. An elitist selection is applied: if the offspring is better fitted than the worst individual of the current population, then the offspring replaces the latter. Minimization continues generating new offspring and terminates after the predetermined computing time t_c.

A parallel version of the genetic algorithm with multiple populations can be developed as shown in Algorithm 2. Each process runs the same genetic algorithm with different sequences of random numbers. This is ensured by initializing different seeds for random number generators in each process. The results of different processes are collected when search is finished after the predefined time t_c. To make parallel implementation as portable as possible, the general message-passing paradigm of parallel programming has been chosen. The standardized message-passing communication protocol MPI can be used for communication between parallel processes.

The parallel genetic algorithm has been used to solve problems with different multidimensional data. Improvement of the reliability is significant, especially while comparing the results of single processor with results of the maximal number of processors. However, it is difficult to judge about the efficiency of parallelization. In all cases, genetic algorithm finds the same global minima as found by explicit enumeration. These minima are found with 100% reliability (100 runs out of 100) in 10 seconds, although genetic algorithm does not guarantee that the global minima are found. Let us note that genetic algorithm solves these problems in seconds, whereas the algorithm of explicit enumeration requires an hour on a cluster of 3 personal computers to solve problems with $N = 16$ and a day on a cluster of 10 personal computers to solve problems with $N = 18$. Larger problems cannot be solved in acceptable time by the algorithm with explicit enumeration, but the genetic algorithm still produces good solutions. The parallel genetic algorithm finds minima of artificial geometric test problems of up to $N = 32$ variables with 100% reliability (100

Algorithm 2 Parallel genetic algorithm for multidimensional scaling.

Input: N_{init}; pp; t_c; n; m; δ_{ij}, $w_{i,j}$, $i,j = 1,\ldots,n$; *rank*
Output: S^*, $\mathbf{x}^*$

```
1: Initialize seed for random number generator based on the number of process (rank).
2: Generate N_init uniformly distributed random vectors x of dimension n · m.
3: Perform search for local minima starting from the best pp generated vectors.
4: Form the initial population from the found local minimizers.
5: while t_c time has not passed do
6:     Randomly with uniform distribution select two parents from a current population.
7:     Produce an offspring by means of crossover and local minimization.
8:     if the offspring is more fitted than the worst individual of the current population then
9:         the offspring replaces the latter.
10:    end if
11: end while
12: Collect S*, x* from the different processes, keep the best.
```

runs out of 100 find the same minimum) in 10 seconds on a cluster of 3 personal computers. Probably a more valuable result is that better function values have been found [31] for Morse code confusion problem with $N = 72$ variables than has been published previously [6]. However in this case, the algorithm has run for 2 hours on 8 processes of SUN Fire E15k.

6 Conclusions

Parallel two-level global optimization algorithm for multidimensional scaling with city-block distances based on explicit enumeration and quadratic programming scales well. The speedup is almost equal to the number of processes, and the efficiency of parallel algorithm is close to one. Global optimization algorithm for multidimensional scaling with city-block distances based on genetic algorithm and quadratic programming finds the same global minima as found by explicit enumeration faster, although it does not guarantee the global solution. Parallel computing enables solution of larger problems.

Acknowledgments The research is partially supported by Lithuanian State Science and Studies Foundation within the project B-03/2007 “Global optimization of complex systems using high performance computing and GRID technologies”.

References

1. Arabie, P.: Was Euclid an unnecessarily sophisticated psychologist? Psychometrika **56**(4), 567–587 (1991). DOI 10.1007/BF02294491
2. Baravykaitė, M., Čiegis, R.: An implementation of a parallel generalized branch and bound template. Mathematical Modelling and Analysis **12**(3), 277–289 (2007) DOI 10.3846/1392-6292.2007.12.277-289
3. Baravykaitė, M., Čiegis, R., Žilinskas, J.: Template realization of generalized branch and bound algorithm. Mathematical Modelling and Analysis **10**(3), 217–236 (2005)
4. Baravykaitė, M., Žilinskas, J.: Implementation of parallel optimization algorithms using generalized branch and bound template. In: I.D.L. Bogle, J. Žilinskas (eds.) Computer Aided Methods in Optimal Design and Operations, *Series on Computers and Operations Research*, vol. 7, pp. 21–28. World Scientific, Singapore (2006)
5. Borg, I., Groenen, P.J.F.: Modern Multidimensional Scaling: Theory and Applications, 2nd edn. Springer, New York (2005)
6. Brusco, M.J.: A simulated annealing heuristic for unidimensional and multidimensional (city-block) scaling of symmetric proximity matrices. Journal of Classification **18**(1), 3–33 (2001)
7. Cantú-Paz, E.: Efficient and Accurate Parallel Genetic Algorithms. Kluwer Academic Publishers, New York (2000)
8. Černý, V.: Thermodynamical approach to the traveling salesman problem: An efficient simulation algorithm. Journal of Optimization Theory and Applications **45**(1), 41–51 (1985). DOI 10.1007/BF00940812
9. Cox, T.F., Cox, M.A.A.: Multidimensional Scaling, 2nd edn. Chapman & Hall/CRC, Boca Raton (2001)
10. Floudas, C.A.: Deterministic Global Optimization: Theory, Methods and Applications, *Nonconvex Optimization and its Applications*, vol. 37. Kluwer Academic Publishers, New York (2000)

11. Glover, F.: Tabu search – Part I. ORSA Journal on Computing **1**(3), 190–206 (1989)
12. Glover, F.: Tabu search – Part II. ORSA Journal on Computing **2**(1), 4–32 (1990)
13. Goldberg, D.E.: Genetic Algorithms in Search, Optimization, and Machine Learning. Adison-Wesley, Reading, MA (1989)
14. Green, P., Carmone, F., Smith, S.: Multidimensional Scaling: Concepts and Applications. Allyn and Bacon, Boston (1989)
15. Groenen, P., Mathar, R., Trejos, J.: Global optimization methods for multidimensional scaling applied to mobile communication. In: W. Gaul, O. Opitz, M. Schander (eds.) Data Analysis: Scientific Modeling and Practical Applications, pp. 459–475. Springer, New York (2000)
16. Groenen, P.J.F.: The Majorization Approach to Multidimentional Scaling: Some Problems and Extensions. DSWO Press, Leiden (1993)
17. Groenen, P.J.F., Heiser, W.J.: The tunneling method for global optimization in multidimensional scaling. Psychometrika **61**(3), 529–550 (1996). DOI 10.1007/BF02294553
18. Groenen, P.J.F., Mathar, R., Heiser, W.J.: The majorization approach to multidimensional scaling for Minkowski distances. Journal of Classification **12**(1), 3–19 (1995). DOI 10.1007/BF01202265
19. Hansen, E., Walster, G.W.: Global Optimization Using Interval Analysis, 2nd edn. Marcel Dekker, New York (2003)
20. Horst, R., Pardalos, P.M., Thoai, N.V.: Introduction to Global Optimization, *Nonconvex Optimization and its Applications*, vol. 48, 2nd edn. Kluwer Academic Publishers, New York (2001)
21. Kirkpatrick, S., Gelatt, C.D.J., Vecchi, M.P.: Optimization by simulated annealing. Science **220**(4598), 671–680 (1983). DOI 10.1126/science.220.4598.671
22. Leeuw, J.D.: Differentiability of Kruskal's stress at a local minimum. Psychometrika **49**(1), 111–113 (1984). DOI 10.1007/BF02294209
23. Mathar, R.: A hybrid global optimization algorithm for multidimensional scaling. In: R. Klar, O. Opitz (eds.) Classification and Knowledge Organization, pp. 63–71. Springer, New York (1997)
24. Mathar, R., Žilinskas, A.: On global optimization in two-dimensional scaling. Acta Applicandae Mathematicae **33**(1), 109–118 (1993). DOI 10.1007/BF00995497
25. Michalewich, Z.: Genetic Algorithms + Data Structures = Evolution Programs. Springer, Berlin (1996)
26. Mockus, J.: Bayesian Approach to Global Optimization. Kluwer Academic Publishers, Boston (1989)
27. Schwefel, H.P.: Evolution and Optimum Seeking. John Wiley & Sons, New York (1995)
28. Törn, A., Žilinskas, A.: Global optimization. Lecture Notes in Computer Science **350**, 1–252 Springer-Verlag, Berlin (1989). DOI 10.1007/3-540-50871-6
29. Varoneckas, A., Žilinskas, A., Žilinskas, J.: Multidimensional scaling using parallel genetic algorithm. In: I.D.L. Bogle, J. Žilinskas (eds.) Computer Aided Methods in Optimal Design and Operations, *Series on Computers and Operations Research*, vol. 7, pp. 129–138. World Scientific, Singapore (2006)
30. Vera, J.F., Heiser, W.J., Murillo, A.: Global optimization in any Minkowski metric: a permutation-translation simulated annealing algorithm for multidimensional scaling. Journal of Classification **24**(2), 277–301 (2007). DOI 10.1007/s00357-007-0020-1
31. Žilinskas, A., Žilinskas, J.: Parallel hybrid algorithm for global optimization of problems occurring in MDS-based visualization. Computers & Mathematics with Applications **52**(1-2), 211–224 (2006). DOI 10.1016/j.camwa.2006.08.016
32. Žilinskas, A., Žilinskas, J.: Parallel genetic algorithm: assessment of performance in multidimensional scaling. In: GECCO '07: Proceedings of the 9th annual conference on Genetic and evolutionary computation, pp. 1492–1501. ACM, New York (2007). DOI 10.1145/1276958.1277229
33. Žilinskas, A., Žilinskas, J.: Two level minimization in multidimensional scaling. Journal of Global Optimization **38**(4), 581–596 (2007). DOI 10.1007/s10898-006-9097-x
34. Žilinskas, A., Žilinskas, J.: Branch and bound algorithm for multidimensional scaling with city-block metric. Journal of Global Optimization **in press** (2008). DOI 10.1007/s10898-008-9306-x

High-Performance Parallel Support Vector Machine Training

Kristian Woodsend and Jacek Gondzio

Abstract Support vector machines are a powerful machine learning technology, but the training process involves a dense quadratic optimization problem and is computationally expensive. We show how the problem can be reformulated to become suitable for high-performance parallel computing. In our algorithm, data is preprocessed in parallel to generate an approximate low-rank Cholesky decomposition. Our optimization solver then exploits the problem's structure to perform many linear algebra operations in parallel, with relatively low data transfer between processors, resulting in excellent parallel efficiency for very large-scale problems.

1 Introduction

Support vector machines (SVMs) are powerful machine learning techniques for classification and regression. They were developed by Vapnik [11] and are based on statistical learning theory. They have been applied to a wide range of applications, with excellent results, and so they have received significant interest.

Like many machine learning techniques, SVMs involve a training stage, where the machine learns a pattern in the data from a *training data set*, and a separate test or validation stage where the ability of the machine to correctly predict labels (or values in the case of regression) is evaluated using a previously unseen *test data set*. This process allows parameters to be adjusted toward optimal values, while guarding against overfitting.

The training stage for support vector machines involves at its core a dense convex quadratic optimization problem (QP). Solving this optimization problem is computationally expensive, primarily due to the dense Hessian matrix. Solving the QP with a general-purpose QP solver would result in the time taken scaling cubically with

Kristian Woodsend · Jacek Gondzio
School of Mathematics, University of Edinburgh, The King's Buildings, Edinburgh, EH9 3JZ, UK
e-mail: k.j.woodsend@sms.ed.ac.uk · j.gondzio@ed.ac.uk

the number of data points ($\mathcal{O}(n^3)$). Such a complexity result means that, in practice, the SVM training problem cannot be solved by general purpose optimization solvers.

Several schemes have been developed where a solution is built by solving a sequence of small-scale problems, where only a few data points (an *active set*) are considered at a time. Examples include decomposition [9] and sequential minimal optimization [10], and state-of-the-art software use these techniques. Active-set techniques work well when the data is clearly separable by a hyperplane, so that the separation into active and non-active variables is clear. With noisy data, however, finding a good separating hyperplane between the two classes is not so clear, and the performance of these algorithms deteriorates.

In addition, the active set techniques used by standard software are essentially sequential — they choose a small subset of variables to form the active set at each iteration, and this selection is based upon the results of the previous iteration. It is not clear how to efficiently implement such an algorithm in parallel, due to the dependencies between each iteration and the next. The parallelization schemes proposed so far typically involve splitting the training data into smaller sub-problems that are considered separately, and which can be distributed among the processors. The results are then combined in some way to give a single output [3, 5, 8].

There have been only a few parallel methods in the literature that train a standard SVM on the whole of the data set. Zanghirati and Zanni [14] decompose the QP into a sequence of smaller, though still dense, QP sub-problems and develop a parallel solver based on the variable projection method. Chang et al. [2] use interior point method (IPM) technology for the optimizer. To avoid the problem of inverting the dense Hessian matrix, they generate a low-rank approximation of the kernel matrix using partial Cholesky decomposition with pivoting. The dense Hessian matrix can then be efficiently inverted implicitly using the low-rank approximation and the Sherman–Morrison–Woodbury (SMW) formula. The SMW formula has been widely used in interior point methods; however, sometimes it runs into numerical difficulties.

This paper summarizes a parallel algorithm for large-scale SVM training using interior point methods. Unlike some previous approaches, the full data set is seen by the algorithm. Data is evenly distributed among the processors, and this allows the potential processing of huge data sets. The formulation exactly solves the linear SVM problem, using the feature matrix directly. For non-linear SVMs, the kernel matrix has to be approximated using partial Cholesky decomposition with pivoting. Unlike previous approaches that have used IPM to solve the QP, we use Cholesky decomposition rather than the SMW formula. This gives better numerical stability. In addition, the decomposition is applied to all features at once, and this allows the memory cache of the processors to be used more efficiently. By exploiting the structure of the QP optimization problem, the training itself can be achieved with near-linear parallel efficiency. The resulting implementation is therefore a highly efficient SVM training algorithm, which is scalable to large-scale problems.

This paper is structured as follows: Sect. 2 outlines the interior point method for optimizing QPs. Section 3 describes the binary classification SVM problem and how the problem can be reformulated to be more efficient for an IPM-based approach. Sections 4 and 5 describe how the Cholesky decomposition and QP optimization can be implemented efficiently in a parallel processing environment.

We now briefly describe the notation used in this paper. $\mathbf{x}_i$ is the attribute vector for the ith data point, and it consists of the observation values directly. There are n observations in the training set, and k attributes in each vector $\mathbf{x}_i$. We assume throughout this paper that $n \gg k$. X is the $n \times k$ matrix whose rows are the attribute row vectors $\mathbf{x}_i^T$ associated with each point. The classification label for each data point is denoted by $y_i \in \{-1, 1\}$. The variables $\mathbf{w} \in \mathbb{R}^k$ and $\mathbf{z} \in \mathbb{R}^n$ are used for the primal variables ("weights") and dual variables (α in SVM literature) respectively, and $w_0 \in \mathbb{R}$ for the bias of the hyperplane. Scalars are denoted using lowercase letters, column vectors in boldface lowercase, and uppercase letters denote matrices. D, S, U, V, Y, and Z are the diagonal matrices of the corresponding lowercase vectors.

2 Interior Point Methods

Interior point methods represent state-of-the-art techniques for solving linear, quadratic, and non-linear optimization programs. In this section, the key issues of implementation for QPs are discussed very briefly; for more details, see [13].

For the purposes of this chapter, we need a method to solve the box and equality-constrained convex quadratic problem

$$\begin{aligned} \min_{\mathbf{z}} \quad & \frac{1}{2}\mathbf{z}^T Q\mathbf{z} + \mathbf{c}^T\mathbf{z} \\ \text{s.t.} \quad & A\mathbf{z} = \mathbf{b}\,, \ \ \mathbf{0} \le \mathbf{z} \le \mathbf{u}\,, \end{aligned} \tag{1}$$

where $\mathbf{u}$ is a vector of upper bounds, and the constraint matrix A is assumed to have full row rank. Dual feasibility requires that $A^T\lambda + \mathbf{s} - \mathbf{v} - Q\mathbf{z} = \mathbf{c}$, where λ is the Lagrange multiplier associated with the linear constraint $A\mathbf{z} = \mathbf{b}$ and $\mathbf{s}, \mathbf{v} \ge \mathbf{0}$ are the Lagrange multipliers associated with the lower and upper bounds of $\mathbf{z}$, respectively.

At each iteration, an interior point method makes a damped Newton step toward satisfying the primal feasibility, dual feasibility, and complementarity product conditions,

$$ZS\mathbf{e} = \mu\mathbf{e}\,, \ \ (U - Z)V\mathbf{e} = \mu\mathbf{e}$$

for a given $\mu > 0$. The algorithm decreases μ before making another iteration and continues until both infeasibilities and the duality gap (which is proportional to μ) fall below required tolerances.

The Newton system to be solved at each iteration can be transformed into the *augmented system equations*:

$$\begin{bmatrix} -(Q+\Theta^{-1}) & A^T \\ A & 0 \end{bmatrix} \begin{bmatrix} \Delta\mathbf{z} \\ \Delta\lambda \end{bmatrix} = \begin{bmatrix} \mathbf{r}_c \\ \mathbf{r}_b \end{bmatrix}, \tag{2}$$

where $\Delta\mathbf{z}, \Delta\lambda$ are components of the Newton direction in the primal and dual spaces, respectively, $\Theta^{-1} \equiv Z^{-1}S + (U-Z)^{-1}V$, and $\mathbf{r}_c$ and $\mathbf{r}_b$ are appropriately defined residuals. If the block $(Q+\Theta^{-1})$ is diagonal, an efficient method to solve such a system is to form the Schur complement $C = A\,(Q+\Theta^{-1})^{-1}A^T$, solve the smaller system $C\Delta\lambda = \hat{\mathbf{r}}_b$ for $\Delta\lambda$, and back-substitute into (2) to calculate $\Delta\mathbf{z}$. Unfortunately, as we shall see in the next section, for the case of SVM training, the Hessian matrix Q is a completely dense matrix.

3 Support Vector Machines

In this section we briefly outline the standard SVM binary classification primal and dual formulations, and summarise how they can be reformulated as a separable QP; for more details, see [12].

3.1 *Binary Classification*

A support vector machine (SVM) is a classification learning machine that learns a mapping between the features and the target label of a set of data points known as the *training set*, and then uses a hyperplane $\mathbf{w}^T\mathbf{x} + w_0 = 0$ to separate the data set and predict the class of further data points. The labels are the binary values "yes" or "no", which we represent using the values $+1$ and -1. The objective is based on the structural risk minimization principle, which aims to minimize the risk functional with respect to both the empirical risk (the quality of the approximation to the given data, by minimizing the misclassification error) and maximize the confidence interval (the complexity of the approximating function, by maximizing the separation margin) [11]. A fuller description is also given in [4].

For a *linear kernel*, the attributes in the vector $\mathbf{x}_i$ for the ith data point are the observation values directly, whereas for a *non-linear kernel* the observation values are transformed by means of a (possibly infinite dimensional) non-linear mapping Φ.

3.2 *Linear SVM*

For a linear SVM classifier using a 2-norm for the hyperplane weights $\mathbf{w}$ and a 1-norm for the misclassification errors $\xi \in \mathbb{R}^n$, the QP that forms the core of training the SVM takes the form:

$$\begin{aligned} \min_{\mathbf{w},w_0,\xi} \quad & \frac{1}{2}\mathbf{w}^T\mathbf{w} + \tau\mathbf{e}^T\xi \\ \text{s.t.} \quad & Y(X\mathbf{w} + w_0\mathbf{e}) \geq \mathbf{e} - \xi\,, \quad \xi \geq \mathbf{0}\,, \end{aligned} \tag{3}$$

where $\mathbf{e}$ is the vector of all ones, and τ is a positive constant that parameterizes the problem.

Due to the convex nature of the problem, a Lagrangian function associated with (3) can be formulated, and the solution will be at the saddle point. Partially differentiating the Lagrangian function gives relationships between the primal variables ($\mathbf{w}$, w_0 and ξ) and the dual variables ($\mathbf{z} \in \mathbb{R}^n$) at optimality, and substituting these relationships back into the Lagrangian function gives the standard dual problem formulation

$$\begin{aligned} \min_{\mathbf{z}} \quad & \frac{1}{2}\mathbf{z}^T YXX^T Y\mathbf{z} - \mathbf{e}^T\mathbf{z} \\ \text{s.t.} \quad & \mathbf{y}^T\mathbf{z} = 0\,, \quad \mathbf{0} \le \mathbf{z} \le \tau\mathbf{e}\,. \end{aligned} \tag{4}$$

However, using one of the optimality relationships, $\mathbf{w} = (YX)^T\mathbf{z}$, we can rewrite the quadratic objective in terms of $\mathbf{w}$. Consequently, we can state the classification problem (4) as a separable QP:

$$\begin{aligned} \min_{\mathbf{w},\mathbf{z}} \quad & \frac{1}{2}\mathbf{w}^T\mathbf{w} - \mathbf{e}^T\mathbf{z} \\ \text{s.t.} \quad & \mathbf{w} - (YX)^T\mathbf{z} = 0\,, \; \mathbf{y}^T\mathbf{z} = 0\,, \; \mathbf{0} \le \mathbf{z} \le \tau\mathbf{e}\,. \end{aligned} \tag{5}$$

The Hessian is simplified to the diagonal matrix $Q = \operatorname{diag}\left(\begin{bmatrix}\mathbf{e}_k \\ \mathbf{0}_n\end{bmatrix}\right)$ while the constraint matrix becomes:

$$A = \begin{bmatrix} I_k & -(YX)^T \\ 0 & y^T \end{bmatrix} \in \mathbb{R}^{(k+1)\times(k+n)}\,.$$

As described in Sect. 2, the Schur complement C can be formed efficiently from such matrices and used to determine the Newton step. Building the matrix C is the most expensive operation, of order $\mathcal{O}(n(k+1)^2)$, and inverting the resulting matrix is of order $\mathcal{O}((k+1)^3)$.

3.3 Non-linear SVM

Non-linear kernels are a powerful extension to the support vector machine technique, allowing them to handle data sets that are not linearly separable. Through transforming the attribute vectors $\mathbf{x}$ into some feature space, through a non-linear mapping $\mathbf{x} \rightarrow \Phi(\mathbf{x})$, the data points can be separated by a polynomial curve or by clustering. One of the main advantages of the dual formulation is that the mapping can be represented by a kernel matrix K, where each element is given by $K_{ij} = \Phi(\mathbf{x}_i)^T\Phi(\mathbf{x}_j)$, resulting in the QP

$$\begin{aligned}\min_{\mathbf{z}} \quad & \frac{1}{2}\mathbf{z}^T YKY\mathbf{z} - \mathbf{e}^T\mathbf{z} \\ \text{s.t.} \quad & \mathbf{y}^T\mathbf{z} = 0\,,\ \ \mathbf{0} \le \mathbf{z} \le \tau\mathbf{e}\,.\end{aligned} \tag{6}$$

As the original attribute vectors appear only in terms of inner products, kernel functions allow the matrix K to be calculated without knowing $\Phi(\mathbf{x})$ explicitly.

The matrix resulting from a non-linear kernel is normally dense, but several researchers have noted (see [6]) that it is possible to make a good low-rank approximation of the kernel matrix. An efficient approach is to use partial Cholesky decomposition with pivoting to compute the approximation $K \approx LL^T$ [6]. In [12], we showed that the approximation $K \approx LL^T + \text{diag}(\mathbf{d})$ can be determined at no extra computational expense. Using a similar methodology as applied to the linear kernel above, the QP can be reformulated to give a diagonal Hessian:

$$\begin{aligned}\min_{\mathbf{w},\mathbf{z}} \quad & \frac{1}{2}(\mathbf{w}^T\mathbf{w} + \mathbf{z}^T D\mathbf{z}) - \mathbf{e}^T\mathbf{z} \\ \text{s.t.} \quad & \mathbf{w} - (YL)^T\mathbf{z} = \mathbf{0}\,,\ \ \mathbf{y}^T\mathbf{z} = 0\,,\ \ \mathbf{0} \le \mathbf{z} \le \tau\mathbf{e}\,.\end{aligned} \tag{7}$$

The computational complexity is $\mathcal{O}(n(r+1)^2 + nkr + (r+1)^3)$, where r is the number of columns in L.

4 Parallel Partial Cholesky Decomposition

By construction, a kernel matrix K will be positive semidefinite. Full Cholesky decomposition will compute L, where $LL^T := K$. Partial Cholesky decomposition produces the first r columns of the matrix L and leaves the other columns as zero, which gives a low-rank approximation of the matrix K. The advantage of this technique, compared with eigenvalue decomposition, is that its complexity is linear with the number of data points. In addition, it exploits the symmetry of K.

Algorithm 1 describes how to perform partial Cholesky decomposition in a parallel environment. As data sets may be large, the data is segmented between the processors. To determine pivot candidates, all diagonal elements are calculated (steps 1–4). Then, for each of the r columns, the largest diagonal element is located (steps 7–8). The processor p^* that owns the pivot row j^* calculates the corresponding row of L, and this row forms part of the "basis" B (steps 10–13). The basis and the original features $\mathbf{x}_{j^*}$ need to be known by all processors, so this information is broadcast (step 14). With this information, all processors can update the section of column i of L for which they are responsible and also update corresponding diagonal elements (steps 16–19).

Although the algorithm requires the processors to be synchronized at each iteration, little of the data needs to be shared among the processors: the bulk of the communication between processors (step 14) is limited to a vector of length k and a vector of at most length r. Note that matrix K is known only implicitly, through

Algorithm 1 Parallel Cholesky decomposition with partial pivoting: $LL^T + \text{diag}(\mathbf{d}) \approx K$.

Input:

n_p Number of samples on each processor p
r Required rank of approximation matrix L
X_p Processor-local features dataset

Output:

B Global basis matrix
L Partial Cholesky decomposition of processor-local data
$\mathbf{d}$ Diagonal of residual matrix $K - LL^T$

$\mathscr{J} := \{1 \ldots n_p\}$
for $j \in \mathscr{J}$ **do**
    $d_j := K_{jj}$ // Initialize the diagonal
end for
// Calculate a maximum of r columns
for $i = 1 : r$ **do**
    if $\sum_p \sum_{j=i}^{n_p} d_j > \varepsilon_{tol}$ **then**
        On each machine, find local $j_p^* : d_{j_p^*} = \max_{j \in \mathscr{J}} d_j$
        Locate global $(j^*, p^*) : j^* = \max_p d_{j_p^*}$ // e.g. using MPI_MAXLOC
        if machine is p^* **then**
            $B_{i,:} := L_{j^*}$ // Move row j^* to basis
            $\mathscr{J} := \mathscr{J} \backslash j^*$
            $B_{ii} := \sqrt{d_{j^*}}$
            $d_{j^*} := 0$
            Broadcast features $\mathbf{x}_{j^*}$, and basis row $B_{i,:}$.
        end if
        // Calculate column i on all processors. $\mathscr{J}$ is all rows not moved to basis
        $L_{\mathscr{J},i} := (K_{\mathscr{J},i} - L_{\mathscr{J},1:i}(L_{i,1:i})^T)/B_{i,i}$
        // Update the diagonal
        for $j \in \mathscr{J}$ **do**
            $d_j := d_j - (L_{j,i})^2$
        end for
    else
        $r := i - 1$
        return
    end if
end for

the kernel function, and calculating its values is an expensive process. The algorithm therefore calculates each kernel element required to form L only once, giving a complexity of $\mathscr{O}(nr^2 + nkr)$.

5 Implementing the QP for Parallel Computation

To apply formulations (5) and (7) to truly large-scale data sets, it is necessary to employ linear algebra operations that exploit the block structure of the formulations [7].

5.1 Linear Algebra Operations

We use the augmented system matrix $H = \begin{bmatrix} -Q-\Theta^{-1} & A^T \\ A & 0 \end{bmatrix}$ from (2), where Q, Θ and A were described in Sect. 2. For the formulations (5) and (7), this results in H having a symmetric bordered block diagonal structure. We can break H into blocks:

$$H = \begin{bmatrix} H_1 & & & & A_1^T \\ & H_2 & & & A_2^T \\ & & \ddots & & \vdots \\ & & & H_p & A_p^T \\ A_1 & A_2 & \dots & A_p & 0 \end{bmatrix},$$

where H_i and A_i result from partitioning the data set evenly across the p processors. Due to the "arrow-head" structure of H, a block-based Cholesky decomposition of the matrix $H = LDL^T$ will be guaranteed to have the structure:

$$H = \begin{bmatrix} L_1 & & & \\ & \ddots & & \\ & & L_p & \\ L_{A1} & \dots & L_{Ap} & L_C \end{bmatrix} \begin{bmatrix} D_1 & & & \\ & \ddots & & \\ & & D_p & \\ & & & D_C \end{bmatrix} \begin{bmatrix} L_1^T & & & L_{A1}^T \\ & \ddots & & \vdots \\ & & L_p^T & L_{Ap}^T \\ & & & L_C^T \end{bmatrix}.$$

Exploiting this structure allows us to compute the blocks L_i, D_i, and L_{Ai} in parallel. Terms that form the Schur complement can be calculated in parallel but must then be gathered and the corresponding blocks L_C and D_C computed serially. This requires the exchange of matrices of size $(r+1) \times (r+1)$ between processors.

5.2 Performance

We used the SensIT data set[1], collected on types of moving vehicles by using a wireless distributed sensor networks. It consists of 100 dense attributes, combining acoustic and seismic data. There were 78,823 samples in the training set and 19,705 in the test set. The classification task set was to discriminate class 3 from the other two. This is a relatively noisy data set — benchmark accuracy is around 85%. We used partial Cholesky decomposition with 300 columns to approximate the kernel matrix, as described in Sect. 4, which gave a test set accuracy of 86.9%.

By partitioning the data evenly across the processors, and exploiting the structure as outlined above, we get very good parallel efficiency results compared with

[1] http://www.csie.ntu.edu.tw/~cjlin/libsvmtools/datasets/.

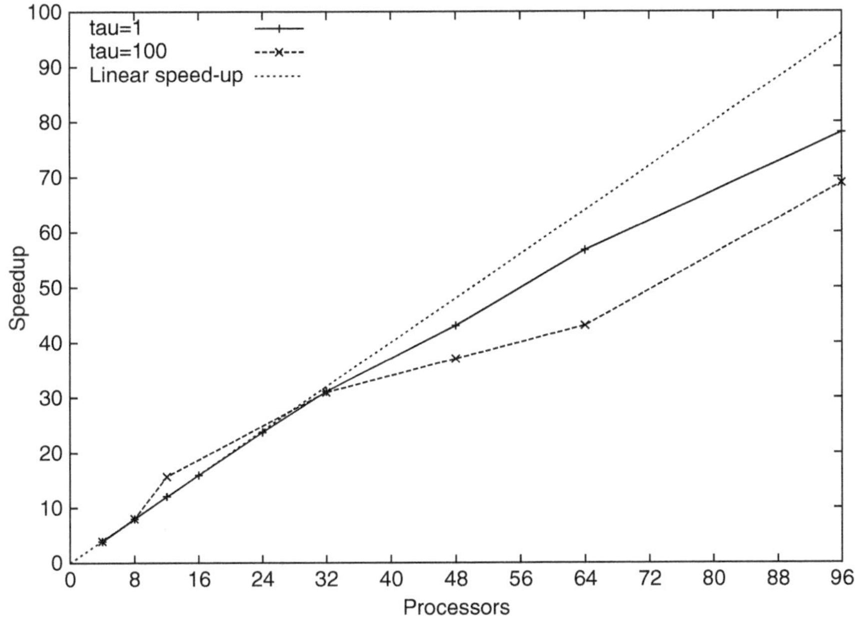

Fig. 1 Parallel efficiency of the training algorithm, using the SensIT data set. Speedup measurements are based on "wall-clock" (elapsed) time, with the performance of 8 processors taken as the baseline.

a baseline of 8 processors, as shown in Fig. 1. Training times are competitive (see Table 1): our implementation on 8 processors was 7.5 times faster than LIBSVM [1] running serially ($\tau = 100$).

Note that, despite the high parallel efficiency for 4 or more processors, the serial version of the algorithm is still significantly faster in terms of CPU time. We assume that this is due to cache inefficiencies in our implementation on multi-core processors. It is possible that a multi-mode implementation would combine the cache efficiency of the serial implementation with the scalability of the parallel implementation; this will be the subject of our future research.

Table 1 Comparison of training times of the serial and parallel implementations, and with the reference software LIBSVM ($\tau = 100$).

Name	Elapsed time (s)
Parallel implementation using 8 processors	932
Serial implementation	1982
Serial LIBSVM	6976

6 Conclusions

We have shown how to develop a high-performance parallel implementation of support vector machine training. The approach supports both linear and non-linear kernels and includes the entire data set in the problem. Non-linear kernel matrices need to be approximated, and we have described a parallel algorithm to compute the approximation. Through reformulating the optimization problem to give an implicit inverse of the kernel matrix, we are able to use interior point methods to solve it efficiently. The block structure of the augmented matrix can be exploited, so that the vast majority of linear algebra computations are distributed among parallel processors. Our implementation was able to solve problems involving large data sets, and excellent parallel efficiency was observed.

References

1. Chang, C.C., Lin, C.J.: LIBSVM – A library for support vector machines (2001). Software available at http://www.csie.ntu.edu.tw/~cjlin/libsvm
2. Chang, E., Zhu, K., Wang, H., Bai, J., Li, J., Qiu, Z., Cui, H.: Parallelizing Support Vector Machines on distributed computers. In: NIPS (2007)
3. Collobert, R., Bengio, S., Bengio, Y.: A parallel mixture of SVMs for very large scale problems. Neural Computation **14**(5), 1105–1114 (2002)
4. Cristianini, N., Shawe-Taylor, J.: An Introduction to Support Vector Machines. Cambridge University Press (2000)
5. Dong, J., Krzyzak, A., Suen, C.: A fast parallel optimization for training support vector machine. In: P. Perner, A. Rosenfeld (eds.) Proceedings of 3rd International Conference on Machine Learning and Data Mining, *Lecture Notes in Computer Science*, vol. 2734, pp. 96–105. Springer (2003)
6. Fine, S., Scheinberg, K.: Efficient SVM training using low-rank kernel representations. Journal of Machine Learning Research **2**(2), 243–264 (2002)
7. Gondzio, J., Grothey, A.: Parallel interior-point solver for structured quadratic programs: Application to financial planning problems. Annals of Operations Research **152**(1), 319–339 (2007)
8. Graf, H.P., Cosatto, E., Bottou, L., Dourdanovic, I., Vapnik, V.: Parallel support vector machines: the Cascade SVM. In: L.K. Saul, Y. Weiss, L. Bottou (eds.) Advances in Neural Information Processing Systems 17, pp. 521–528. MIT Press (2005)
9. Osuna, E., Freund, R., Girosi, F.: An improved training algorithm for support vector machines. In: J. Principe, L. Gile, N. Morgan, E. Wilson (eds.) Neural Networks for Signal Processing VII — Proceedings of the 1997 IEEE Workshop, pp. 276–285. IEEE (1997)
10. Platt, J.C.: Fast training of support vector machines using sequential minimal optimization. In: B. Schölkopf, C.J.C. Burges, A.J. Smola (eds.) Advances in Kernel Methods — Support Vector Learning, pp. 185–208. MIT Press (1999)
11. Vapnik, V.N.: Statistical Learning Theory. Wiley (1998)
12. Woodsend, K., Gondzio, J.: Exploiting separability in large-scale Support Vector Machine training. Technical Report MS-07-002, School of Mathematics, University of Edinburgh (2007). Submitted for publication. Available at http://www.maths.ed.ac.uk/~gondzio/reports/wgSVM.html
13. Wright, S.J.: Primal-Dual Interior-Point Methods. SIAM (1997)
14. Zanghirati, G., Zanni, L.: A parallel solver for large quadratic programs in training support vector machines. Parallel Computing **29**(4), 535–551 (2003)

Parallel Branch and Bound Algorithm with Combination of Lipschitz Bounds over Multidimensional Simplices for Multicore Computers

Remigijus Paulavičius and Julius Žilinskas

Abstract Parallel branch and bound for global Lipschitz minimization is considered. A combination of extreme (infinite and first) and Euclidean norms over a multidimensional simplex is used to evaluate the lower bound. OpenMP has been used to implement the parallel version of the algorithm for multicore computers. The efficiency of the developed parallel algorithm is investigated solving multidimensional test functions for global optimization.

1 Introduction

Many problems in engineering, physics, economics, and other fields may be formulated as optimization problems, where the minimum value of an objective function should be found. We aim to find at least one globally optimal solution to the problem

$$f^* = \min_{x \in D} f(x), \tag{1}$$

where an objective function $f(x)$, $f : \mathbb{R}^n \to \mathbb{R}$ is a real-valued function, $D \subset \mathbb{R}^n$ is a feasible region, and n is the number of variables.

Lipschitz optimization is one of the most deeply investigated subjects of global optimization. It is based on the assumption that the slope of an objective function

Remigijus Paulavičius
Vilnius Pedagogical University, Studentu 39, LT-08106 Vilnius, Lithuania
Institute of Mathematics and Informatics, Akademijos 4, LT-08663 Vilnius, Lithuania
e-mail: r.paulavicius@vpu.lt

Julius Žilinskas
Vilnius Gediminas Technical University, Saulėtekio 11, LT-10223 Vilnius, Lithuania
Institute of Mathematics and Informatics, Akademijos 4, LT-08663 Vilnius, Lithuania
e-mail: julius.zilinskas@ktl.mii.lt

is bounded [2]. The advantages and disadvantages of Lipschitz global optimization methods are discussed in [1, 2].

A function $f : D \to \mathbb{R}, D \subseteq \mathbb{R}^n$, is said to be Lipschitz if it satisfies the condition

$$|f(x) - f(y)| \leq L \|x - y\|, \quad \forall x, y \in D, \tag{2}$$

where $L > 0$ is a constant called Lipschitz constant, D is compact, and $\|\cdot\|$ denotes the norm. No assumptions on unimodality are included into formulation of the problem—many local minima may exist.

In Lipschitz optimization, the lower bound for the minimum is evaluated exploiting Lipschitz condition (2):

$$f(x) \geq f(y) - L\|x - y\|.$$

In [7], it has been suggested to estimate the bounds for the optimum over the simplex using function values at one or more vertices. In this chapter, the values of the function at all vertices of the simplex are used. The upper bound for the minimum is the smallest value of the function at the vertex:

$$UB(I) = \min_{x_v \in I} f(x_v),$$

where x_v is a vertex of the simplex I. A combination of Lipschitz bounds is used as lower bound for the minimum. It is the maximum of bounds found using extreme (infinite and first) and Euclidean norms over a multidimensional simplex I [5, 6]:

$$LB(I) = \max_{x_v \in I}(f(x_v) - \min\{L_1 \max_{x \in I} \|x - x_v\|_\infty, L_2 \max_{x \in I} \|x - x_v\|_2, L_\infty \max_{x \in I} \|x - x_v\|_1\}).$$

A global optimization algorithm based on branch and bound technique has been developed. OpenMP C++ version 2.0 was used to implement the parallel version of the algorithm. The efficiency of the parallelization was measured using speedup and efficiency criteria. The speedup is

$$s_p = \frac{t_1}{t_p}, \tag{3}$$

where t_p is time used by the algorithm implemented on p processes. The speedup divided by the number of processes is called the efficiency:

$$e_p = \frac{s_p}{p}. \tag{4}$$

2 Parallel Branch and Bound with Simplicial Partitions

The idea of branch and bound algorithm is to detect the subspaces (simplices) not containing the global minimum by evaluating bounds for the minimum over considered sub-regions and discard them from further search. Optimization stops when global optimizers are bracketed in small sub-regions guaranteeing the required accuracy. A general n-dimensional sequential simplex-based branch and bound algorithm for Lipschitz optimization has been proposed in [7]. The rules of selection, covering, branching and bounding have been justified by results of experimental investigations.

An n-dimensional simplex is the convex hull of a set of $n+1$ affinely independent points in the n-dimensional space. In one-dimensional space, a simplex is a segment of line, in two-dimensional space it is a triangle, in three-dimensional space it is a tetrahedron. A simplex is a polyhedron in n-dimensional space, which has the minimal number of vertices $(n+1)$. Therefore, if bounds for the minimum over a sub-region defined by polyhedron are estimated using function values at all vertices of the polyhedron, a simplex sub-region requires the smallest number of function evaluations to estimate bounds.

Usually, a feasible region in Lipschitz optimization is defined by a hyper-rectangle—intervals of variables. To use simplicial partitions, the feasible region should be covered by simplices. Experiments in [7] have shown that the most preferable initial covering is face to face vertex triangulation—partitioning of the feasible region into finitely many n-dimensional simplices, whose vertices are also the vertices of the feasible region. There are several ways to divide the simplex into sub-simplices. Experiments in [7] have shown that the most preferable partitioning is subdivision of simplex into two by a hyper-plane passing through the middle point of the longest edge and the vertices whose do not belong to the longest edge.

Sequential branch and bound algorithm with simplicial partition and combination of Lipschitz bounds was proposed in [5]. In this work, the parallel version of the algorithm was created. The rules of covering, branching, bounding and selection by parallel algorithm are the same as by the sequential algorithm:

- A feasible region is covered by simplices using face to face vertex triangulation [8]. The examples of such partition are shown in Fig. 1.

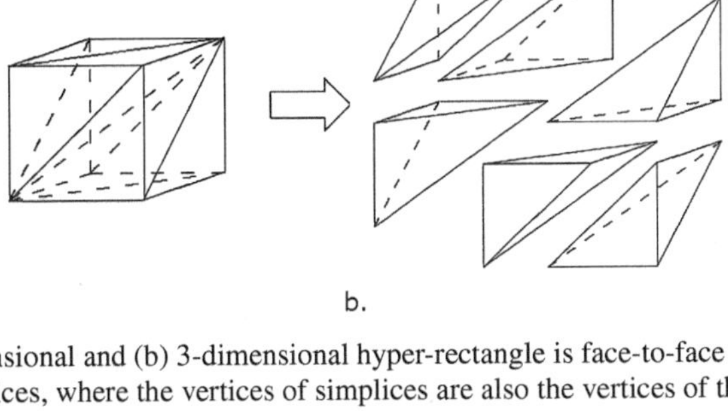

Fig. 1 (a) 2-dimensional and (b) 3-dimensional hyper-rectangle is face-to-face vertex triangulated into 2 and 6 simplices, where the vertices of simplices are also the vertices of the hyper-rectangle.

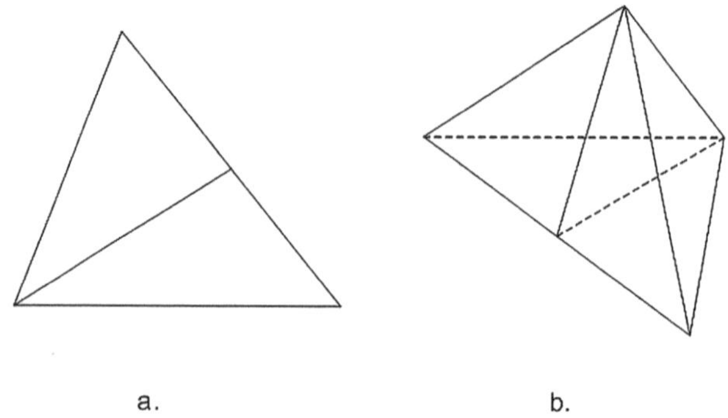

Fig. 2 (a) 2-dimensional and (b) 3-dimensional examples how simplices are divided by a hyper-plane passing through the middle point of the longest edge and the vertices not belonging to the longest edge. This ensures that the longest edge of sub-simplices is not more than two times longer than other edges.

- The simplices are branched by a hyper-plane passing through the midpoint of the longest edge and the vertices that do not belong to the longest edge. The examples of such division are shown in Fig. 2.
- The lower and upper bounds for the minimum of the function over the simplex are estimated using function values at vertices.
- The breadth first selection strategy is used.

Data parallelism is used in the developed parallel branch and bound algorithm. Feasible region D is subsequently divided into set of simplices $I = I_k$. The number of threads that will be used in the next parallel region is set using "omp_set_num_threads(p)". Directive "for" specifies that the iterations of the loop immediately following it must be executed in parallel by different threads. "schedule(dynamic, 1)" describes that iterations of the loop are dynamically scheduled among the threads and when a thread finishes one iteration another is dynamically assigned. The directive "critical" specifies a region of code that must be executed by only one thread at a time.

Each simplex is evaluated trying to find if it can contain optimal solution. For this purpose, a lower bound $LB(I_j)$ for objective function f is evaluated over each simplex and compared with upper bound $UB(D)$ for the minimum value over feasible region. If $LB(I_j) \geq UB(D) + \varepsilon$, then the simplex I_j cannot contain the function value better than the found one by more than the given precision ε, and therefore it is rejected. Otherwise it is inserted into the set of unexplored simplices I. The algorithm finalizes when the small simplex is found, which includes a potential solution.

The parallel branch and bound algorithm is shown in Algorithm 1.

3 Results of Experiments

Various test functions ($n \geq 2$) for global minimization from [1, 3, 4] have been used in our experiments. Test functions with ($n = 2$ and $n = 3$) are numbered according

Algorithm 1 Parallel branch and bound algorithm.

Cover feasible region D by $I = \{I_k | D \subseteq \cup I_k, k = 1, \ldots, n!\ \}$ using face-to-face vertex triangulation [6].
$UB(D) = \infty$.
while (I is not empty: $I \neq \emptyset$) **do**
  $k =$ amount of simplices(I)
  omp_set_num_threads(p)
  #pragma omp parallel private(LB, m**)**
  #pragma omp for schedule(dynamic, 1) nowait
  for ($j = 0$; $j <= k$; j++) **do**
    Exclude I_j from I
    $LB(I_j) = \max_{x_v \in I}(f(x_v) - \min\{L_1 \max_{x \in I} \|x - x_v\|_\infty, L_2 \max_{x \in I} \|x - x_v\|_2, L_\infty \max_{x \in I} \|x - x_v\|_1\})$
    if ($LB(I_j) < UB(D) + \varepsilon$) **then**
      Branch I_j into 2 simplices: I_{j_1}, I_{j_2}
      for $m = 1, 2$ **do**
        # pragma omp critical $(UB(D))$
        $UB(D) = \min\{UB(D), \min_{x_v \in I_{jm}}\{f(x_v)\}\}$
        $LB(I_{jm}) = \max_{x_v \in I}(f(x_v) - \min\{L_1 \max_{x \in I} \|x - x_v\|_\infty, L_2 \max_{x \in I} \|x - x_v\|_2, L_\infty \max_{x \in I} \|x - x_v\|_1\})$
        if ($LB(I_{jm}) < UB(D) + \varepsilon$) **then**
          # pragma omp critical (I)
          $I = \{I, I_{jm}\}$
        end if
      end for
    end if
  end for
  end of parallel section
end while

to [1] and [4]. For $(n \geq 4)$, function names from [3] and [4] are used. Experiments were performed using the following hardware and software system:

- INTEL Core 2 Quad Q6600 (2.4 GHz), 2, 3 or 4 cores are used for the parallel version of the algorithm;
- 2GB RAM;
- Windows Vista Business;
- Visual Studio 2005 (SP1);
- Intel C++ Compiler 10.1.

The numbers of function evaluations and execution time of sequential program using Visual C++ and Intel C++ compilers are shown in Table 1. In the presented chapter, the default options of compilers are used.

The numbers of function evaluations do not depend on the compiler used. The numbers of function evaluations do not change when the parallel version is used for almost all test functions, although a very small difference exists for some functions. However, optimization time depends on the compiler used, and it is much smaller when the algorithm is compiled using Intel C++ Compiler.

Speedup s_m and efficiency e_m of parallelization solving test functions with different compilers are shown in Tables 2 and 3. For some low-dimensional test functions

Table 1 Number of function evaluation and execution time of sequential program.

No.	Function	n	ε	f. evaluation	VS2005 time(s)	INTEL time(s)
1.	1.[1]	2	0.355	987	0.0673	0.006
2.	2.[1]	2	0.0446	218	0.052	0.012
3.	3.[1]	2	11.9	7543	0.1147	0.0369
4.	4.[1]	2	0.0141	10	0.052	0.004
5.	5.[1]	2	0.1	63	0.052	0.004
6.	6.[1]	2	44.9	1753	0.0673	0.0085
7.	7.[1]	2	542.0	20202	0.2133	0.0958
8.	8.[1]	2	3.66	523	0.0573	0.0026
9.	9.[1]	2	62900	40316	0.3793	0.1916
10.	10.[1]	2	0.691	2377	0.0677	0.0116
11.	11.[1]	2	0.355	5656	0.099	0.0271
12.	12.[1]	2	0.804	29359	0.291	0.1403
13.	13.[1]	2	6.92	22247	0.2393	0.1119
14.	20.[1]	3	10.6	111520	2.0020	1.1825
15.	21.[1]	3	0.369	4149	0.1353	0.0635
16.	23.[1]	3	41.65	49154	0.9517	0.537
17.	24.[1]	3	3.36	62148	1.1750	0.652
18.	25.[1]	3	0.0506	27298	0.4783	0.243
19.	26.[1]	3	4.51	24750	0.4940	0.2585
20.	5.[4]	3	5000.0	138476	2.3767	1.3985
21.	Shekel 5 [3]	4	$2L_2$	403480	12.3005	7.939
22.	Shekel 7 [3]	4	$2L_2$	14043	0.4365	0.2345
23.	Shekel 10 [3]	4	$2L_2$	58480	1.8565	1.1245
24.	Levy 9 [3]	4	$2L_2$	58480	1.8175	1.1275
25.	Levy 15 [3]	4	$2L_2$	58480	1.8175	1.132
26.	Schwefel 1.2 [3]	4	$2L_2$	126226	3.2215	2.0725
27.	Powell [3]	4	$2L_2$	4326	0.195	0.1745
28.	Rosenbrock [4]	4	$2L_2$	56792	1.7235	1.0655
29.	Levy 16 [3]	5	$2L_2$	409093	20.233	13.36
30.	Rosenbrock [4]	5	$2L_2$	1198095	57.72	38.571
31.	Levy 10 [3]	5	$2L_2$	2373721	113.537	75.467
32.	Schwefel 3.7 [3]	5	$2L_2$	152	0.202	0.04
33.	Levy 17 [3]	6	$4L_2$	822664	59.889	42.868
34.	Rosenbrock [4]	6	$4L_2$	1925841	138.372	98.629

efficiency is quite small, because total numbers of calls of the objective function is small and parallelization is not effective. In Fig. 3, efficiency using 2, 3, and 4 processes is shown. In Fig. 4, functions are arranged in ascending order of the number of function calls of sequential algorithm. Efficiency increases when the numbers of calls of the objective function increase. Efficiency of parallelization is better when the test functions are more difficult. Efficiency decreases when the number of processors increases as non-parallelized parts of the algorithm take a larger proportion of total time and distribution of work becomes less balanced. However, the efficiency of parallelization for difficult test functions decreases less than for simple test functions.

Table 2 Speedup and efficiency for $n \geq 2$ (dimensional test) functions using Visual C++ Compiler.

No.	s_2	s_3	s_4	e_2	e_3	e_4
1.	0.99	1.29	1.08	0.50	0.43	0.27
2.	1.00	1.00	1.00	0.50	0.33	0.25
3.	1.22	1.34	1.57	0.61	0.46	0.39
4.	1.00	1.00	0.91	0.50	0.33	0.23
5.	1.00	1.00	1.00	0.50	0.33	0.25
6.	1.08	1.08	1.07	0.54	0.36	0.27
7.	1.37	1.71	2.05	0.68	0.57	0.51
8.	1.00	1.00	1.01	0.50	0.33	0.25
9.	1.46	1.92	2.28	0.73	0.64	0.57
10.	1.01	1.09	1.08	0.50	0.36	0.27
11.	1.19	1.27	1.47	0.59	0.42	0.37
12.	1.44	1.80	2.24	0.72	0.60	0.56
13.	1.39	1.84	2.09	0.70	0.61	0.52
14.	1.73	2.47	3.08	0.87	0.82	0.77
15.	1.30	1.44	1.37	0.65	0.48	0.34
16.	1.69	2.32	2.86	0.85	0.77	0.72
17.	1.74	2.46	2.97	0.87	0.82	0.74
18.	1.61	2.14	2.56	0.81	0.71	0.64
19.	1.58	2.02	2.32	0.79	0.67	0.58
20.	1.72	2.48	2.86	0.86	0.83	0.71
21.	1.81	2.68	3.39	0.90	0.89	0.85
22.	1.47	2.00	2.43	0.74	0.67	0.61
23.	1.76	2.59	3.17	0.88	0.86	0.79
24.	1.74	2.51	3.19	0.87	0.84	0.80
25.	1.69	2.48	3.15	0.84	0.83	0.79
26.	1.71	2.58	3.13	0.85	0.86	0.78
27.	1.32	1.47	1.78	0.66	0.49	0.45
28.	1.74	2.54	3.07	0.87	0.85	0.77
29.	1.79	2.65	3.39	0.90	0.88	0.85
30.	1.80	2.65	3.51	0.90	0.88	0.88
31.	1.81	2.64	3.51	0.90	0.88	0.88
32.	1.00	1.08	1.08	0.50	0.36	0.27
33.	1.89	2.71	3.66	0.95	0.90	0.91
34.	1.92	2.78	3.72	0.96	0.93	0.93

From Tables 2 and 3, we see that the results are very similar although different compilers are used: efficiency of parallelization is better when the test functions are more difficult and parallelization is not effective when functions are solved very fast—the number of function calls is small. For most test functions, better efficiency is achieved through using Intel C++ Compiler, but the difference is very small. Also implementation is faster when the algorithm is compiled using Intel C++ Compiler.

The ratio of optimization time using Visual C++ and Intel C++ compilers is shown in Fig. 5, where functions are arranged in ascending order of the number of function calls of sequential algorithm. When the test functions are difficult (the

Table 3 Speedup and efficiency for $n \geq 2$ (dimensional test) functions using Intel C++ Compiler.

No.	s_2	s_3	s_4	e_2	e_3	e_4
1.	1.25	1.46	1.20	0.63	0.49	0.30
2.	1.00	1.00	1.00	0.50	0.33	0.25
3.	1.51	1.68	2.00	0.76	0.56	0.50
4.	1.00	1.00	1.00	0.50	0.33	0.25
5.	1.00	1.00	1.00	0.50	0.33	0.25
6.	1.31	1.37	1.70	0.65	0.46	0.43
7.	1.47	1.81	2.21	0.73	0.60	0.55
8.	1.00	1.00	1.00	0.50	0.33	0.25
9.	1.71	2.08	2.29	0.85	0.69	0.57
10.	1.32	1.87	2.32	0.66	0.62	0.58
11.	1.40	1.44	1.73	0.70	0.48	0.43
12.	1.68	2.18	2.68	0.84	0.73	0.67
13.	1.45	1.89	2.38	0.72	0.63	0.60
14.	1.93	2.76	3.36	0.96	0.92	0.84
15.	1.37	1.65	1.94	0.68	0.55	0.49
16.	1.81	2.46	3.20	0.91	0.82	0.80
17.	1.81	2.39	2.82	0.90	0.80	0.71
18.	1.75	2.43	3.10	0.88	0.81	0.78
19.	1.60	2.05	2.35	0.80	0.68	0.59
20.	1.87	2.76	3.46	0.94	0.92	0.86
21.	2.00	2.92	3.77	1.00	0.97	0.94
22.	1.45	1.74	2.08	0.72	0.58	0.52
23.	1.85	2.48	3.00	0.92	0.83	0.75
24.	1.85	2.49	3.15	0.93	0.83	0.79
25.	1.86	2.42	3.15	0.93	0.81	0.79
26.	1.98	2.71	3.41	0.99	0.90	0.85
27.	1.47	1.59	1.73	0.73	0.53	0.43
29.	1.87	2.69	3.41	0.93	0.90	0.85
30.	1.94	2.83	3.57	0.97	0.94	0.89
31.	1.93	2.79	3.60	0.96	0.93	0.90
32.	1.00	1.00	1.00	0.50	0.33	0.25
33.	1.97	2.86	3.58	0.98	0.95	0.90
34.	1.98	2.87	3.76	0.99	0.96	0.94

number of function calls is larger than 10000), optimization time is by average 1.73 smaller when Intel C++ Compiler 10.1 is used. For simpler test functions, some fluctuations exist, but optimization is faster when Intel C++ Compiler is used.

4 Conclusions

This chapter presents the general parallel simplex based branch and bound algorithm for Lipschitz optimization. The algorithm has been implemented using C++ and

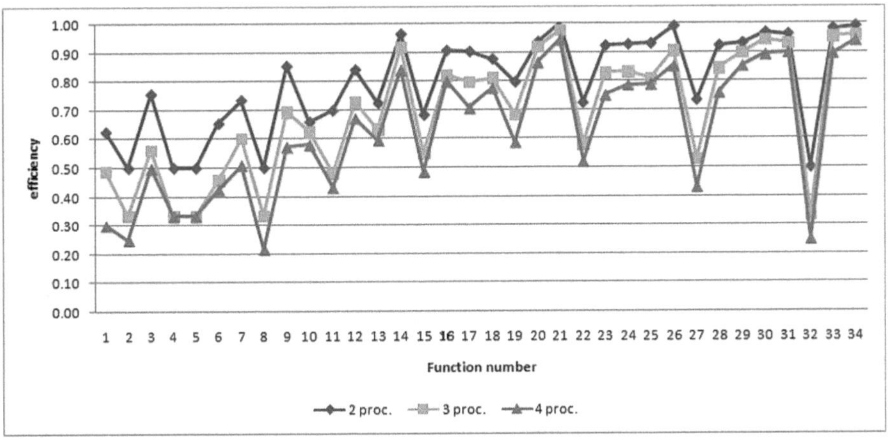

Fig. 3 Efficiency for various test functions.

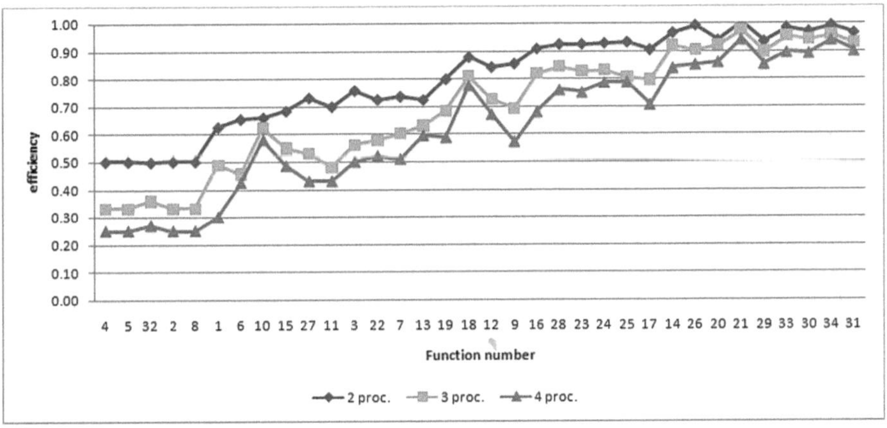

Fig. 4 Efficiency in ascending order of the number of function calls.

OpenMP. The criteria of parallelization, speedup and efficiency of the algorithm for test functions of various dimensionality $n \geq 2$, have been evaluated and compared.

The investigation showed that the efficiency of parallelization is better for difficult test functions—when the number of calls of the objective function is larger. Moreover, the efficiency of parallelization for difficult test functions decreases less when the number of processors increases compared with simple test functions.

Acknowledgments The research is partially supported by Lithuanian State Science and Studies Foundation within the project B-03/2007 "Global optimization of complex systems using high performance computing and GRID technologies".

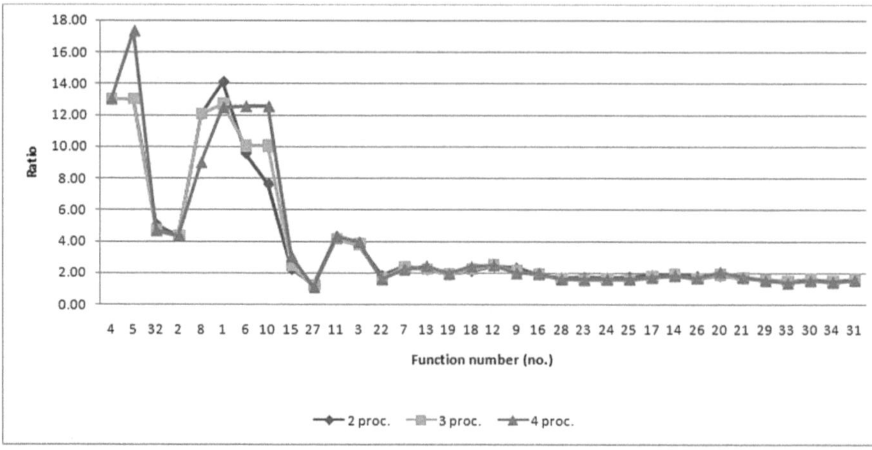

Fig. 5 Comparison of time of optimization depending on C++ compiler: ratio of time of implementations compiled using Visual C++ and Intel C++ compilers.

References

1. Hansen, P., Jaumard, B.: Lipshitz optimization. In: R. Horst, P.M. Pardalos (eds.) Handbook of Global Optimization, *Nonconvex Optimization and Its Applications*, vol. 2, pp. 404–493. Kluwer Academic Publishers (1995)
2. Horst, R., Pardalos, P.M., Thoai, N.V.: Introduction to Global Optimization, *Nonconvex Optimization and Its Applications*, vol. 3. Kluwer Academic Publishers (1995)
3. Jansson, C., Knüppel, O.: A global minimization method: The multi-dimensional case. Tech. rep., TU Hamburg-Harburg (1992)
4. Madsen, K., Žilinskas, J.: Testing branch-and-bound methods for global optimization. Tech. Rep. IMM-REP-2000-05, Technical University of Denmark (2000)
5. Paulavičius, R., Žilinskas, J.: Analysis of different norms and corresponding Lipschitz constants for global optimization. Technological and Economic Development of Economy **12**(4), 301–306 (2006)
6. Paulavičius, R., Žilinskas, J.: Analysis of different norms and corresponding Lipschitz constants for global optimization in multidimensional case. Information Technology and Control **36**(4), 383–387 (2007)
7. Žilinskas, J.: Optimization of Lipschitzian functions by simplex-based branch and bound. Information Technology and Control **14**(1), 45–50 (2000)
8. Žilinskas, J.: Branch and bound with simplicial partitions for global optimization. Mathematical Modelling and Analysis **13**(1), 145–159 (2008)

Experimental Investigation of Local Searches for Optimization of Grillage-Type Foundations

Sergėjus Ivanikovas, Ernestas Filatovas, and Julius Žilinskas

Abstract In grillage-type foundations, beams are supported by piles. The main goal of engineering design is to achieve the optimal pile placement scheme in which the minimal number of piles is used and all the reactive forces do not exceed the allowed values. This can be achieved by searching for the positions of piles where the difference between the maximal reactive forces and the limit magnitudes of reactions for the piles is minimal. In this study, the values of the objective function are given by a separate modeling package. Various algorithms for local optimization have been applied and their performance has been investigated and compared. Parallel computations have been used to speed-up experimental investigation.

1 Introduction

Many problems in engineering may be reduced to problems of global minimization. Mathematically, the problem of global optimization is formulated as

$$f^* = \min_{\mathbf{x} \in \mathbf{D}} f(\mathbf{x}),$$

where $f(\mathbf{x})$ is a nonlinear objective function of continuous variables $f : \mathbf{R}^n \to \mathbf{R}$, $\mathbf{D} \subset \mathbf{R}^n$ is a feasible region, and n is a number of variables. Besides the global minimum f^*, one or all global minimizers $\mathbf{x}^* : f(\mathbf{x}^*) = f^*$ should be found. No assumptions on unimodality are included into formulation of the problem—many local minima may exist.

Local optimization is often used in algorithms for global optimization to improve efficiency. In this chapter, various methods for local optimization of grillage-type foundations have been applied and the results have been compared.

Sergėjus Ivanikovas · Ernestas Filatovas · Julius Žilinskas
Institute of Mathematics and Informatics, Akademijos 4, LT-08663 Vilnius, Lithuania
e-mail: ivanikovas@gmail.com · ernest.filatov@gmail.com · zilinskasjulius@gmail.com

2 Optimization of Grillage-Type Foundations

The grillage-type foundations are the most conventional and effective scheme of foundations, especially in the case of weak grounds. Grillage consists of separate beams, which are supported by piles or reside on other beams. As piles may reach length of tens meters, reducing the number of piles will lead to substantial savings. The optimal scheme of grillage should possess the minimum possible number of piles. Theoretically, reactive forces in all piles should approach the limit magnitudes of reactions for those piles [3]. These goals can be achieved by choosing appropriate pile positions. The piles should be positioned minimizing the largest difference between reactive forces and limit magnitudes of reactions.

A designer may arrive at the acceptable pile placement scheme by engineering tests algorithms. However, obtaining optimal schemes is likely only in the case of simple geometries, simple loadings, and the limited number of design parameters. Practically, this is difficult to achieve for grillages of complex geometries. To be on the safe side, the number of piles in design schemes is usually overestimated.

The problems may be approached using global optimization [5]. These are "black box" optimization problems: the values of objective function are evaluated by an independent package that obtains the values of forces in the grillage using finite element method. Gradient may be estimated using sensitivity analysis implemented in the modeling package. The properties of the objective function are not known. The number of piles is n, usually $n \geq 10$. The position of a pile is defined by a real number, which is mapped to a two-dimensional position by the modeling package. Possible values are from zero to the sum of all beams, which is equal to 75 in the considered problem. The feasible region of the problems is $[0,75]^n$. The experiments have shown that the problems are difficult and parallel optimization is helpful [2, 6].

3 Methods for Local Optimization of Grillage-Type Foundations

MATLAB was selected as the main implementation for optimization of objective function. The Optimization Toolbox extends MATLAB technical computing environment with tools and widely used algorithms for optimization. There are two optimizations methods realized in MATLAB that can solve our problem, i.e., Nelder–Mead method and the Quasi-Newton methods.

The Nelder–Mead method is a numerical method for minimizing an objective function in a many-dimensional space [13]. This is a direct search method that does not use numerical or analytic gradients. The method uses the concept of a simplex, which is a polytope of $n+1$ vertices in n dimensions; a line segment on a line, a triangle on a plane, a tetrahedron in three-dimensional space, and so forth. The method approximately finds a locally optimal solution to a problem with n variables when the objective function varies smoothly. The Nelder–Mead method is realized in *fminsearch* MATLAB function. This is generally referred to as unconstrained nonlinear optimization.

Among the methods that use gradient information, the most favored are the Quasi-Newton methods. These methods are realized in *fminunc* MATLAB Optimization Toolbox function. *Fminunc* finds the minimum of unconstrained multivariable objective function starting at an initial estimate. A large number of Hessian updating methods have been developed. Three general methods of updating Hessian matrix and choosing the search direction are realized in *fminunc* function. The choices are BFGS, DFP, and STEEPDESC. The BFGS method developed by Broyden [4], Fletcher [9], Goldfarb [12], and Shanno [16] is thought to be the most effective. This method uses the BFGS [4, 9, 12, 16] formula for updating the approximation of the Hessian matrix. The other formula realized in DFP [7, 10, 11] can be selected, which approximates the inverse Hessian matrix. A Steepest Descent method for updating the approximation of the Hessian matrix can be selected also [1, 8, 14]. In MATLAB, optimization function *fminunc* is realized as STEEPDESC, although this method is not recommended.

Final experiments were carried out using minimization function based on the gradient method implemented in Fortran language. Variable metric method was used in these experiments [15]. Variable metric method also belongs to the Quasi-Newton method group. This method requires a function's gradient, or first partial derivatives, at arbitrary points. Variable metric methods come in two main flavors: DFP and BFGS. These schemes differ only in details of their round-off error, convergence tolerances, and some others. However, it has become generally recognized that, empirically, the BFGS scheme is superior in these details. So, BFGS scheme [15] was used in the algorithm.

4 Experimental Research

Experimental tests were performed using 10-dimensional problem (there are 10 decision variables corresponding to 10 piles or supports). The set of starting points (10-dimensional vectors) was generated before the experiments. A component of each vector is a real random number uniformly distributed in the interval [0,75]. MATLAB *rand* function was used to generate these vectors. The same starting vectors were used in all experiments so we were able to compare investigated algorithms and the results of minimization.

Experiments were performed using the following hardware and software:

- AMD Athlon(tm) 64 X2 Dual Core Processor 5000+ (2.6 GHz)
- 2GB RAM
- Windows Vista Ultimate
- MATLAB 7.4.0 (R 2007 a)
- Intel Fortran Compiler 9.0

The experimental research had two stages. At the first stage, the objective function was minimized using MATLAB methods. During these tests, two MATLAB functions were used: *fminsearch* and *fminunc*.

First tests were carried out using the simplex method (*fminsearch* MATLAB function), and 1006 local searches from the generated vectors were done. The experiment lasted for 152.8 hours. So, one minimization on the average ran on for about 9.1 minutes.

Working with gradient methods (*fminunc* function), several different modifications of minimization algorithm were tested. First test was performed supplying the gradient values to the MATLAB minimization function. This test did not give good results. The problem was the accuracy of the calculation of the gradient value. As the gradient value, estimated by the modeling package, is not precise, the MATLAB minimization function encourages problems with calculations. So, further tests were performed without supplying the gradient values. Gradient was calculated by the MATLAB minimization function using finite differences. Three modifications of the local search algorithm were investigated. Different methods for choosing the search direction in the Quasi-Newton algorithm were used: STEEPDESC, DFP [7, 10, 11], BFGS [4, 9, 12, 16]. A total of 1008 local searches were performed working with each modification of the algorithm. The experiments lasted for 78.9 hours using STEEPDESC method, 57.4 hours using DFP method, and 60 hours using BFGS method. One minimization on the average ran on for about 4.7 minutes for STEEPDESC, about 3.4 minutes for DFP, and about 3.6 minutes for BFGS. This means that Quasi-Newton methods run for about two times faster than do simplex. The fastest modification is DFP method.

Experiments showed that working with a given function the simplex method is slower, but more accurate.

The best values of the objective function ($f(\mathbf{x}) < 200$) were achieved using the simplex method. Gradient methods did not give such results. Among 1006 local searches carried out using the simplex method were 8 values where $f(\mathbf{x}) < 200$ (approximately 1% of all cases). These best function values are presented in Table 1.

Local search starting from the point where simplex method achieved the best result gave rather good result working with gradient methods [193.2449 and 237.4228 (Table 2) respectively]. The best function value obtained using gradient method is 221.8085; it was found using STEEPDESC modification.

The experimental results obtained working with the MATLAB system are presented in diagrams below. To compare the efficiency of the methods, the results achieved running different methods for the same time are presented in Fig. 1. The

Table 1 The best minimizers found using simplex method.

$\mathbf{x}$										$f(\mathbf{x})$
3.78	19.73	20.17	29.45	36.00	46.29	51.00	58.24	61.94	63.64	193.24
3.17	5.83	18.47	24.80	31.00	39.71	49.07	54.59	58.72	72.37	196.12
3.44	4.32	20.54	21.72	26.56	29.17	34.36	48.17	49.98	61.81	196.37
13.56	22.15	28.61	30.80	38.00	45.87	51.18	53.16	55.85	60.34	196.79
7.64	17.65	21.35	23.47	30.00	33.65	39.10	45.86	49.19	51.11	197.44
11.35	14.43	35.47	37.40	45.34	49.81	53.95	54.19	59.79	59.82	197.64
3.01	12.92	16.10	27.15	33.52	37.41	49.52	52.44	59.56	60.44	198.04
3.58	6.97	20.48	23.53	27.78	37.92	47.71	50.92	53.46	74.83	198.60

Table 2 The best minimizer found using gradient method.

					x					$f(\mathbf{x})$
3.54	18.98	19.49	29.51	34.60	47.07	49.01	58.89	62.09	63.59	237.42

time of the fastest method tested in the MATLAB system (57.4 hours) was selected as the time value. During this time, the simplex method performed 378, STEEPDESC 734, BFGS 963, and DFP 1008 local searches. In the diagram, only acceptable results with $f(\mathbf{x}) < 500$ are presented. Analyzing the diagram, we can be sure that the simplex method is more precise than gradient methods. Despite the fact that the simplex method is almost two times slower than gradient methods, the better results have been obtained even running tests for the same time. The accuracy of the simplex method can be explained by the complexity of the objective function, which has many "valleys" with almost no descent.

Best results obtained by the simplex method were analyzed additionally. A total of 100 new starting vectors were generated from the surrounding of those starting vectors that gave the best results. New local searches were performed starting from those points. The results are presented in Fig. 2. The distribution is only slightly different from that in Fig. 1, which shows that additional search in the surrounding of the best minimizers is not always worthwhile. Among 100 local searches performed, 3 results are obtained where the value of the objective function is less than 200. The best function value obtained in this test is 195.2995.

As the gradient methods performed more than two times faster than did simplex method, it was decided to run new experiments using minimization function based on the gradient method implemented in Fortran language. Variable metric method from [15] was used in this stage of our research. The gradient value supplied by the modeling package was modified rejecting very small and very big values:

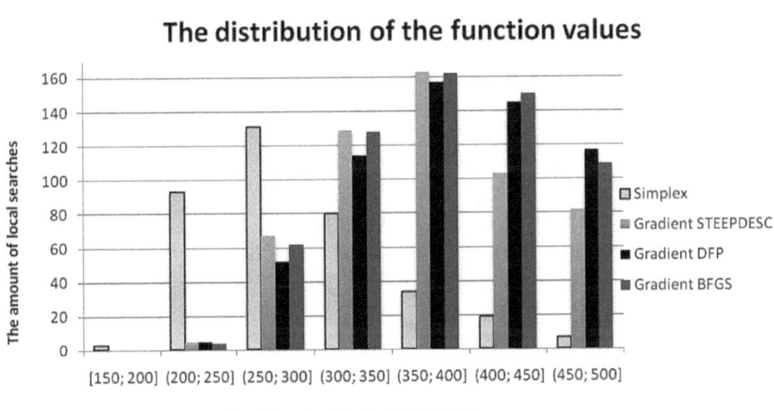

Fig. 1 Values of objective function achieved during the same time minimizing with gradient and simplex methods.

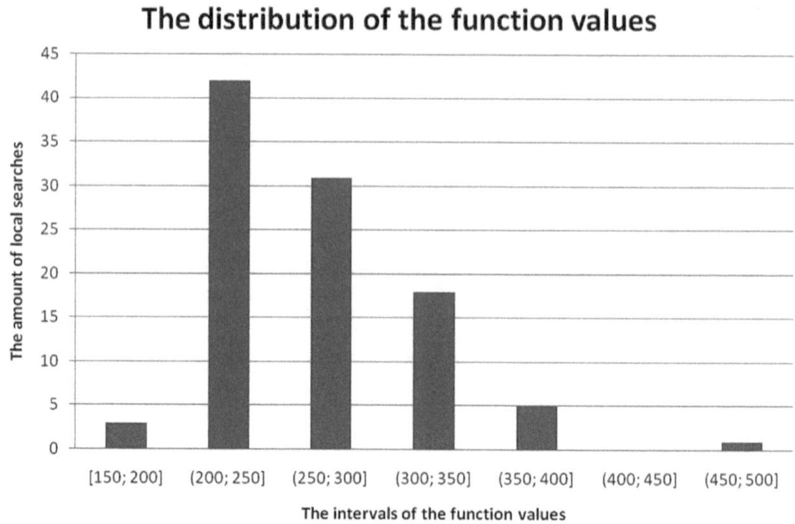

Fig. 2 The distribution of the function values in the surrounding of the best minimizers.

```
DO i = 1, n_xx
IF (ABS(RMAXD(IND(i))) .LT. 0.001) THEN
df(i) = 0.001
ELSEIF (RMAXD(IND(i)) .GT. 1000) THEN
df(i) = 1000
ELSEIF (RMAXD(IND(i)) .LT. –1000) THEN
df(i) = –1000
ELSE
df(i) = RMAXD(IND(i))
ENDIF
END DO
```

During the test, the same starting vectors were used. A total of 1008 local searches were performed. Experiment lasted for 17.38 hours. So, one minimization ran on for about 1.03 minutes. The best function value obtained during this test was 191.0907.

Variable metric algorithm implemented in Fortran runs much quicker than MATLAB functions. All the obtained results are presented in Fig. 3. In the diagram, the results of three investigated methods are presented. All the local searches with acceptable results ($f(\mathbf{x}) < 500$) are included.

As we see from the diagram, the simplex method finds minimal function values. The variable metric algorithm finds only two points where ($f(\mathbf{x}) < 200$), but one of those function values found is even better than the best results of the simplex method. These best minimizers are presented in Table 3. The variable metric algorithm implemented in Fortran is not as precise as the simplex method, but it is better than the gradient DFP method implemented in MATLAB. The best function value obtained with the variable metric algorithm was improved using the simplex method implemented in MATLAB. The minimizer is presented in Table 4; its function value is 190.15.

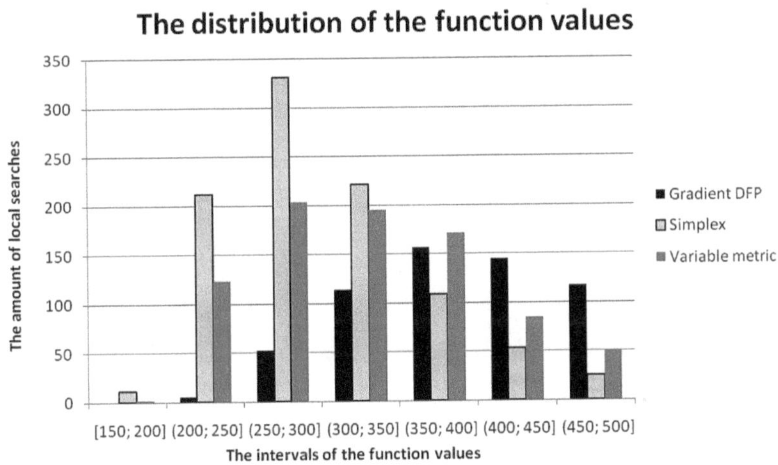

Fig. 3 The distribution of the function values working with gradient DFP, simplex, and variable metric methods.

The results obtained running DFP, simplex, and variable metric algorithms for the same time are presented in Fig. 4. All the algorithms ran on for 17.38 hours, as the fastest of them did. During this period of time, the variable metric algorithm performed 1008 local searches, gradient DFP 303, and simplex 113 local searches. The advantage of the variable metric algorithm implemented in Fortran can be noticed in Fig. 4.

Continuing investigation, it was analyzed how far the minimal point, found by the local search, is from the starting point. The covered distances obtained using simplex, gradient DFP, and variable metric methods are presented in Fig. 5. The distance was calculated using the following formula: $d = \sqrt{\sum_{i=1,\ldots,10} (x_i - x_i^*)^2}$, where x_i is a coordinate of the starting point, and x_i^* is the corresponding coordinate of the minimal point found in the local search. As it can be noticed in Fig. 5, the gradient DFP method is able to find the local minimum not far from the starting point. The variable metric method covers the longest distances searching for the minimum.

Table 3 The best minimizers found using the variable metric algorithm.

x										$f(\mathbf{x})$
2.92	10.65	18.47	24.03	30.95	36.33	44.00	46.01	54.66	59.18	191.09
6.58	16.61	18.02	24.40	34.37	50.41	51.30	61.17	67.05	69.24	196.52

Table 4 The best minimizer found using the variable metric algorithm improved with the simplex method.

x										$f(\mathbf{x})$
2.93	10.80	18.46	23.98	30.85	36.34	43.93	46.13	54.70	59.17	190.15

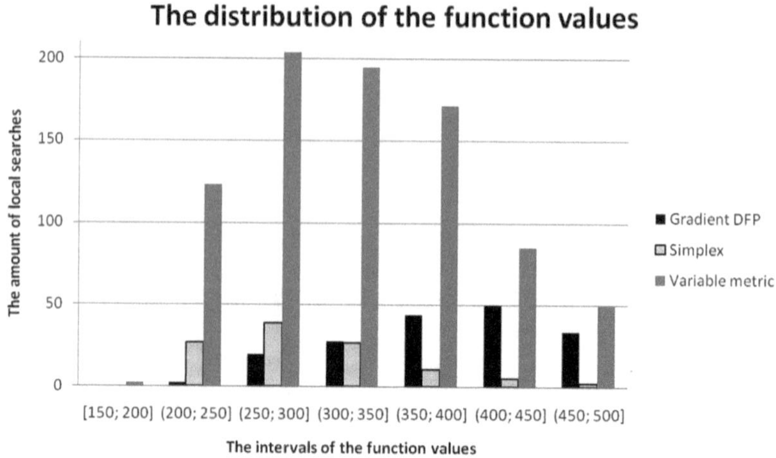

Fig. 4 The distribution of the function values working with gradient DFP, simplex, and variable metric methods during the same time.

In the presented experiments of optimization of grillage-type foundations, the absolute value of maximal reactive force for supports is minimized. Investigation of reactive forces for all supports may give additional information for analysis of foundation schemes. Reactive forces for supports of the best foundation schemes found using the simplex method in MATLAB and the variable metric algorithm implemented in Fortran are given in Table 5. Reactive forces for most of the supports of the best foundation scheme found using the simplex method are very similar, there are five supports with equal maximal reactive forces, but reactive force for one of the supports is quite different. In the best foundation scheme found using the

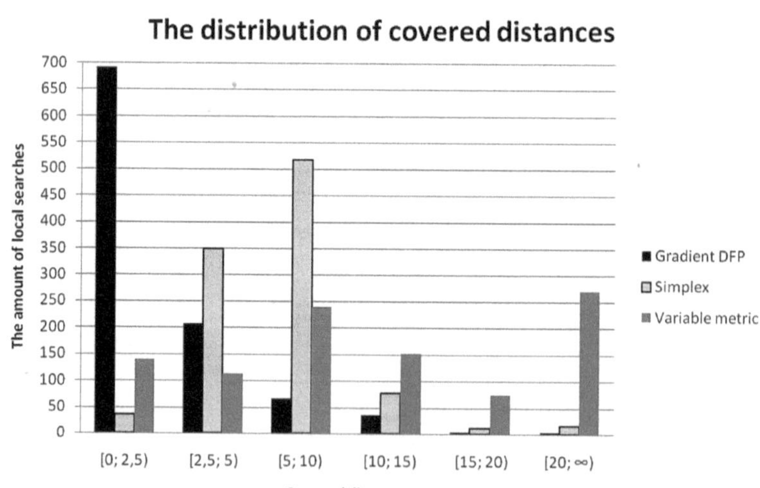

Fig. 5 The distribution of the covered distances working with gradient DFP, simplex, and variable metric methods.

Table 5 Reactive forces for supports of best foundation schemes.

Reactive forces										$f(\mathbf{x})$
−193.2	−193.2	−193.2	−193.2	−193.2	−191.6	−189.3	−185.8	−183.8	−121.0	193.24
−191.1	−191.0	−190.9	−190.7	−190.1	−189.8	−183.6	−183.0	−173.8	−153.7	191.09
−190.2	−190.2	−190.2	−190.1	−190.0	−189.9	−189.6	−182.7	−177.1	−147.8	190.15

variable metric algorithm, reactive forces for all supports are quite similar, but only one is maximal. In the best foundation scheme found using a combination of the two methods (the best scheme found using the variable metric algorithm and improved with the simplex method), reactive forces for all supports are quite similar, and there are three supports with equal maximal reactive forces. The ideal foundation scheme for this problem would be with all supports equal to 183.77. The found schemes are not far from ideal.

Continuing the experiments, the parallel version of the algorithm was created. Data parallelism was used in this parallel algorithm. The starting array was divided into two parts, and parallel processing of those parts was organized. Creating the parallel version of the algorithm, OpenMP standard was used. All the tests were performed using PC with AMD Athlon(tm) 64 X2 Dual Core Processor. Parallel program was also executed on this PC. The working time of the program was 9.01 hours. A total of 1008 local searches were performed. So, one local search ran on for about 0.54 minutes. This means that the parallel algorithm is almost two times faster than the serial one. The speed-up of the algorithm is 1.93, and the efficiency is 0.97.

Further test were performed using quad-core system (Intel Core2 Quad Q6600 (2.4GHz); 2GB RAM; Windows Vista Business). Such results were obtained while working with this system: the working time of the serial program was 15.97 hours (one local search on the average ran on for about 52.25 seconds), the parallel program with 2 threads ran on for 9.43 hours (one local search on the average lasted for about 30.87 seconds), and the four-threaded parallel program ran on for 4.65 hours (one local search on the average ran on for about 15.22 seconds). The speed-up of the algorithm while using 2 threads is 1.69. Working with 4 threads, the speed-up of the algorithm is 3.43. The efficiency is 0.85 while working with 2 threads and 0.86 working with 4 threads.

5 Conclusions

This chapter presents the evaluation of local minimization methods applied to the considered engineering problem, modeled using the finite element method. The experiments were carried out using different realizations of the methods. Local search methods presented in MATLAB and algorithms implemented in Fortran were analyzed. The tests showed that working with the MATLAB system, the best results are achieved using the simplex method. For the considered objective function, gradient methods in MATLAB are not precise enough. All gradient methods in MATLAB do not accept the gradient value supplied by the modeling package.

The investigation showed that the objective function is rather complicated. The best results were achieved working with the simplex method implemented in the MATLAB system and with the variable metric method implemented in Fortran.

The local minimization method implemented in Fortran enabled us to use the gradient value supplied by the modeling package and to speed up the minimization. The accuracy of this method is not as good as that of simplex, but this method is about 9 times faster.

The parallel realization of the experiment with the Fortran local minimization method was created using data parallelism. The parallel program enabled us to speed up the calculation almost twice working with dual processor system and more than three times working with quad-core system. Working with cluster or multi-processor system, we can speed up the program even more.

Acknowledgments The research is partially supported by Lithuanian State Science and Studies Foundation within the project B-03/2007 "Global optimization of complex systems using high performance computing and GRID technologies".

References

1. Arfken, G., Weber, H.: Mathematical Methods for Physicists, 6th edn., chap. 7.3, pp. 489–496. Academic Press (2005)
2. Baravykaitė, M., Belevičius, R., Čiegis, R.: One application of the parallelization tool of master-slave algorithms. Informatica **13**(4), 393–404 (2002)
3. Belevičius, R., Valentinavičius, S., Michnevič, E.: Multilevel optimization of grillages. Journal of Civil Engineering and Management **8**(2), 98–103 (2002)
4. Broyden, C.G.: The convergence of a class of double-rank minimization algorithms 1. General considerations. IMA Journal of Applied Mathematics **6**(1), 76–90 (1970)
5. Čiegis, R.: On global minimization in mathematical modelling of engineering applications. In: A. Törn, J. Žilinskas (eds.) Models and Algorithms for Global Optimization, *Springer Optimization and Its Applications*, vol. 4, pp. 299–310. Springer (2007)
6. Čiegis, R., Baravykaitė, M., Belevičius, R.: Parallel global optimization of foundation schemes in civil engineering. Lecture Notes in Computer Science **3732**, 305–312 (2006)
7. Davidon, W.C.: Variable metric method for minimization. SIAM Journal on Optimization **1**(1), 1–17 (1991)
8. Erdelyi, A.: Asymptotic Expansions. Dover (1956)
9. Fletcher, R.: A new approach to variable metric algorithms. Computer Journal **13**(3), 317–322 (1970)
10. Fletcher, R.: Practical Methods of Optimization, 2nd edn. John Wiley & Sons (2000)
11. Fletcher, R., Powell, M.J.D.: A rapidly convergent descent method for minimization. Computer Journal **6**(1), 163–168 (1963)
12. Goldfarb, D.: A family of variable-metric methods derived by variational means. Mathematics of Computation **24**(109), 23–26 (1970)
13. Lagarias, J.C., Reeds, J.A., Wright, M.H., Wright, P.E.: Convergence properties of the Nelder-Mead simplex method in low dimensions. SIAM Journal of Optimization **9**(1), 112–147 (1998)
14. Morse, P.M., Feshbach, H.: Methods of Theoretical Physics, chap. 4.6, pp. 434–443. McGraw-Hill, New York (1953)
15. Press, W.H., Teukolsky, S.A., Vetterling, W.T., Flannery, B.P.: Numerical Recipes in Fortran 77: The Art of Scientific Computing, 2nd edn. Cambridge University Press (1992)
16. Shanno, D.F.: Conditioning of Quasi-Newton methods for function minimization. Mathematics of Computation **24**(111), 647–656 (1970)

Part III
Management of Parallel Programming Models and Data

Comparison of the UK National Supercomputer Services: HPCx and HECToR

David Henty and Alan Gray

Abstract We give an overview of the two current UK national HPC services: HPCx and HECToR. HPCx was the flagship UK academic supercomputer from 2002 to 2007, at which point the larger HECToR system was installed. It is particularly interesting to compare the two machines as they will now be operating together for some time, so users have a choice as to which machine best suits their requirements. Results of extensive experiments are presented.

1 Introduction

In this chapter, we will give an overview of the two current UK national HPC services: HPCx [7] and HECToR [5]. HPCx was the flagship UK academic supercomputer from 2002 to 2007, at which point the larger HECToR system was installed. It is particularly interesting to compare the two machines as they will now be operating together for some time, so users have a choice as to which machine best suits their requirements.

In terms of this specific workshop, the machines are interesting for two reasons. First, many of the UK speakers will have used these systems for their research. Second, they are both accessible under the HPC-Europa Transnational Access program [6] if visits are made to EPCC in Edinburgh. HPC-Europa was identified at this workshop as a key mechanism for Lithuanian researchers to access world-leading supercomputers, so the capabilities of HPCx and HECToR are of interest.

The HPCx system is based on IBM Power technology, whereas HECToR is supplied by Cray. An enormous amount of research has been done worldwide studying IBM and Cray architectures over the past years, but it is not our aim to cover these studies here. We will focus on studies specifically done of HPCx and HECToR,

David Henty · Alan Gray
Edinburgh Parallel Computing Centre, The University of Edinburgh, UK
e-mail: d.henty@epcc.ed.ac.uk · a.gray@epcc.ed.ac.uk

mostly comprising technical reports produced by the support staff of these two systems. For further sources, please consult the reference sections of these technical reports.

2 System Overview

The UK's major academic supercomputer services are funded by the Engineering and Physical Sciences Research Council (EPSRC), available to all UK researchers with access granted after peer review. Although the systems are mainly used for EPSRC science, they are also utilized by Natural Environment (NERC) and Biological Science (BBSRC) users.

These flagship supercomputers have always been targeted at "Capability Computing", that is enabling scientists to perform simulations that are simply not possible on smaller systems. In parallel computing terms, this usually means running on many hundreds or even thousands of processors. This is in contrast with "Capacity Computing", which involves running a large number of small jobs. Although capacity computing applications can also produce world-leading science, it is simply not cost-effective to run such jobs on these large national supercomputers. One of the major costs of a supercomputer is the high-performance interconnect, which enables parallel jobs to run efficiently on large numbers of processors. Capacity computing does not require this level of interconnect performance so is better suited to smaller clusters where money is spent on raw processing power as opposed to the network.

The national supercomputer systems are normally scheduled to run for around six years, and so to keep up with Moore's law, there are always a number of planned hardware upgrades during the lifetime of the service. These are typically every two years, so a single service will usually comprise at least three different machines over the years. Although this keeps the machine at the leading edge of technology, it presents a significant challenge to the systems team to provide a seamless service to users across these potentially major upgrades.

2.1 HPCx

The HPCx project was originally scheduled to operate from December 2002 until December 2008, although this may yet be extended. The service is provided by a consortium of EPCC at the University of Edinburgh, STFC Daresbury Laboratory, and IBM. The machine is physically located at Daresbury in the north of England, with distributed systems and science support teams located at both Daresbury and Edinburgh.

The budget for the entire HPCx service was £53M (around €80M) in 2002. This represents the total cost of the service for six years and includes all staff and running

Table 1 History of HPCx systems.

	Phase 1	Phase 2	Phase 2a	Phase 3
Year of introduction	2002	2004	2005	2006
Technology	Power4	Power4+	Power5	Power5
Clock frequency (GHz)	1.3	1.7	1.5	1.5
SMP width	8	32	16	16
Interconnect	SP Switch2	HPS	HPS	HPS
Total cores	1280	1600	1536	2560
TFlop Linpack	3.2	6.2	7.4	12.9
Position in top 500	9	18	45	43

costs (such as power) as well as the hardware. The IBM systems provided by HPCx have been based on IBM Power processors, which are packaged into shared-memory nodes with several processors under the control of a single AIX operating system. All the HPCx systems have therefore been shared-memory clusters, connected by networks also built by IBM. These networks are based on multi-way switches, with more than one level of switch required for the largest systems.

The service has so far offered four different systems (or "phases"), which are summarized in Table 1. Note that the SP Switch2 and High Performance Switch (HPS) networks are often referred to as "Colony" and "Federation", respectively.

The various upgrades achieved the aim of doubling the system performance every two years. However, although this is in line with Moore's law, it is not enough to maintain the same position in the top 500 list [11]. This is because the number of processors in a single supercomputer is also increasing rapidly. The top machine in the 2002 list had just over 5000 cores; in 2006, this had increased to well over 100,000. However, the planned upgrade path for HPCx was very successful in maintaining a world-class system (one of the top 50 in the world) for over four years.

Of course, it is worth noting that Linpack performance is not necessarily a good indicator of performance for real science applications. For HPCx, the Phase 1 to Phase 2 upgrade practically doubled the Linpack rating by increasing the clock frequency by 30% and the number of cores 25%. The associated switch upgrade to HPS only contributed a further 20% to the Linpack figure. However, this had a much more significant impact on applications performance due to the higher communications bandwidth and lower latency of HPS. For example, certain applications could now scale to much higher processor counts [1].

Similarly, although the Phase 2a upgrade from Power4 to Power5 represented a 20% increase in Linpack (despite the drop in clock frequency), many applications saw a much larger increase in sustained performance due to an improved memory bandwidth [2]. A slight increase in memory latency did adversely affect some areas such as Molecular Dynamics simulations, but the latency-hiding Simultaneous Multi-Threading (SMT) capabilities of the Power5 architecture were often able to solve this problem [4].

Table 2 Current and planned HECToR systems (all Phase 2 data is estimated).

	Phase 1	Phase 2
Year of introduction	2007	2009
Technology	AMD Opteron	AMD Opteron
Clock frequency (GHz)	2.8	??
Cores per chip	2	4
Interconnect	Seastar2	Gemini
Total cores	11328	22656
TFlop Linpack	55	200
Position in top 500	17	??

2.2 HECToR

The HECToR service is scheduled to run for six years from October 2007, with a total budget of £113M (around €160M). Similar to HPCx, HECToR is operated by EPCC at the University of Edinburgh. The systems support is provided in collaboration with STFC Daresbury Laboratory, with scientific support provided by NAG Ltd. The machine, a Cray XT4, is physically located at the University of Edinburgh. It is housed within the Advanced Computing Facility, which was purpose-built to accommodate the university's supercomputer facilities.

HECToR has a very different architecture to HPCx. There are few computing companies large enough to supply their own hardware, operating system, compilers, and system software as IBM does for HPCx. Cray has followed the major recent trend of using commodity hardware and software whenever possible, concentrating on only designing those particular components, such as the interconnect, which are specific to HPC. Cray also provides a special-purpose stripped-down operating system for the compute nodes. However, even this is now based on standard Linux. The Cray interconnect is the major component that differentiates the XT4 from a commodity cluster. Cray's Seastar2 chip is used to construct a 3D toroidal network, designed specifically to enable applications to scale to tens of thousands of cores.

Like HPCx, HECToR has two planned upgrades during its lifetime, which will substantially increase the performance. The first two phases of the HECToR hardware are summarized in Table 2. Note that the precise configuration of Phase 2 is subject to change, although it is expected to be around four times as powerful as the Phase 1 system. The additional factor of two over the increased core count comes from a doubling of the flops per cycle of each Opteron core.

3 System Comparison

Before looking at achieved applications performance on the two systems, it is worth doing a slightly more detailed comparison between the HPCx Phase 3 and HECToR Phase 1 machines.

3.1 Processors

Despite having very different processor technologies (16-way SMP IBM Power5 vs dual-core AMD Opteron), the peak per-core performance of the two machines is very similar—see Table 3. Measurements using the STREAMS benchmark also indicate that the sustained per-core memory bandwidths of the two architectures are also very similar [3].

To allow for a more direct comparison, we have compared one core of a dual-core Power5 with one core of a dual-core Opteron. In practice, the fact that the Power5 is a dual-core processor is not particularly important on HPCx; what is more relevant to users is that there are eight Power5 processors (sixteen cores) per SMP node. On the XT4, there is only one Opteron processor (two cores) per node.

Despite having almost half the frequency, the Power5 has virtually the same peak performance as the Opteron as it can perform twice as many flops per cycle. The Floating Point Unit (FPU) in both architectures can issue two instructions per cycle. For the Opteron, this can be one multiplication and one addition, whereas for the Power5 it can be two Fused Multiplication-Addition (FMA) operations.

It is very clear from Table 3 that the fourfold peak performance increase from HPCx to HECToR is almost entirely due to a fourfold increase in the number of cores. As noted above in respect to the number of cores in the leading machine in the top 500, supercomputers are getting faster simply because they contain more processor cores. The days of Moore's law translating into an increased per-core performance appear, unfortunately, to be over; today it just means more cores per processor.

3.2 Interconnect

Although it may seem quite straightforward to increase the number of cores in a supercomputer, the most difficult engineering challenge is building a network that enables scientific applications to scale to these new levels. As a result, the network performance of HECToR will be critical in maintaining the capability focus of the national services.

Table 3 Comparison of HPCx and HECToR processor-cores.

	HPCx	HECToR
Processor	IBM Power5	AMD Opteron
Cores per chip	2	2
Frequency (GHz)	1.5	2.8
FPUs	2 FMA	1M, 1A
Peak Gflops	6.0	5.6
Total cores	2560	11328
Tflops Linpack	12.9	54.6

Table 4 Comparison of HPCx and HECToR networks.

	HPCx	HECToR
Inter-node latency (μs)	5.5	5.8
Inter-node bandwidth (MB/s)	141	513
1024-core bisection bandwidth (GB/s)	20	27
8192-core bisection bandwidth (GB/s)	—	133

It is difficult to compare the communications performance based on low-level network parameters numbers due to the very different architectures of the two networks (hierarchical switch vs 3D torus). It is much more instructive to run actual communications benchmarks. However, one design feature of the Cray Seastar2 network is worth mentioning in terms of scalability. The connection between two nodes can sustain substantially more bandwidth than can be injected by a single node. This leaves additional capacity for routing of through-traffic between non-neighboring nodes, which should increase the scalability especially for bandwidth-dominated operations.

Figures derived from the standard IMB communications benchmark are given in Table 4. When performing bandwidth tests, we ensured that all cores in a node were active at the same time, which is normally the situation in a real application. If only a single core is used in each node, that core has exclusive access to the network and the performance will be much higher. The inter-node latency was measured using ping-pong and the inter-node bandwidth from multi ping-ping. The effective bisection bandwidth was estimated from the asymptotic all-to-all performance [3].

The fact that the latencies of the two networks are so close is very interesting and leads to almost identical performance for messages below 1KB. However, the HPCx network saturates much earlier than does HECToR, giving HECToR a performance advantage of between three and four for large messages.

The effective bisection bandwidth, relevant for global communications patterns such as required by parallel FFTs, is roughly comparable on 1024 cores. The results for 8192 cores (only measurable on HECToR) are very encouraging as they show that this continues to scale up to almost the full system size. Interestingly, it appears that the limiting factor in both cases is the rate at which processes can inject data into the network and not the internal network capacity itself.

The performance figures above would suggest that the two systems should be roughly equivalent on a per-core basis, with HECToR perhaps scaling slightly better due to the increased inter-node bandwidth. We would also expect that applications that scale to large processor counts on HPCx will continue to scale to the much larger counts available on HECToR. However, an application that cannot currently scale to fill HPCx will see little benefit on HECToR.

4 Applications Performance

The performance of a large number of scientific applications has been studied on HPCx and HECToR [3], but here we choose to concentrate on two representative codes that employ substantially different simulation techniques.

The first is PDNS3D [9], a finite difference formulation for Direct Numerical Simulation (DNS) of turbulence from the University of Southampton, UK. The relatively simple geometry of the simulation and the fact that the dominant communications pattern is nearest-neighbor mean that this code typically scales very well. However, like many CFD codes, it places heavy requirements on the local memory system.

NAMD [8, 10] is a widely used Molecular Dynamics (MD) package that can simulate very large biomolecular systems. Achieving good scaling for MD codes is very challenging, especially if global communications are required for the computation of long-range electrostatic forces. NAMD was specifically designed for large parallel systems and uses a novel approach to load-balancing via Charm++ parallel objects.

4.1 PDNS3D

The benchmark chosen for this study used the large T3 data set, which uses a global grid of size $360 \times 360 \times 360$. When initially ported, the performance on HECToR was poorer than expected. Each timestep took roughly 50% longer per-core than on HPCx. However, the code was rewritten to be more cache-friendly and therefore to make less demands on the memory system. This improved the performance significantly on HECToR, although the performance on HPCx was unchanged. After this work, each timestep took about the same time on the two systems.

The scaling of PDNS3D is shown in Fig. 1 where the two graphs differ only in scale.

The results confirm that the per-core performance of PDNS3D on the two machines is roughly the same, with similar results on 500 cores or less. On HPCx,

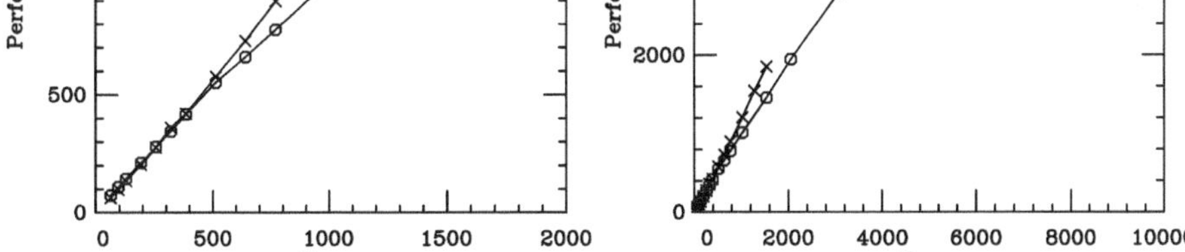

Fig. 1 Performance of PDNS3D on HPCx (crosses) and Cray XT4 (circles). All data is normalized to the HPCx performance on 64 cores (taken as 64.0); the two graphs differ only in their scales.

scaling is linear up to around 500 cores and then becomes super-linear. This is presumably due to cache effects rather than to any feature of the network: the local problem size is reduced as the number of processors is increased, and at some point the problem starts to fit into cache. On HECToR, linear scaling gives parallel efficiencies of around 100% on up to 2000 cores, and the performance continues to improve all the way to in excess of 8000 cores. Although the scaling is not perfect, HECToR achieves efficiencies of around 80% at 4000 cores and 55% at 8000. These figures could of course be increased if a larger data set were used, which is what would probably be done in practice for studies on supercomputers of the size of HECToR.

4.2 NAMD

The benchmark chosen for this study used the standard F1-ATP data set, which contains around 327000 atoms. As noted previously, the IBM Power5 has a moderately high memory access latency. This leads to HECToR having a per-core performance advantage of around 35% on small core counts. The performance scaling of NAMD is shown in Fig. 2 where the two graphs differ only in scale.

Although HECToR has an absolute performance advantage over HPCx, the scaling of the two machines is remarkably similar with both achieving a parallel efficiency of around 75% at 1000 cores. Although the performance on HECToR continues to increase, the parallel efficiency is, not surprisingly for an MD simulation, significantly lower than for PDNS3D. At 4000 cores, it has dropped to around 33%. Again, we would expect that this would improve if larger data sets were used.

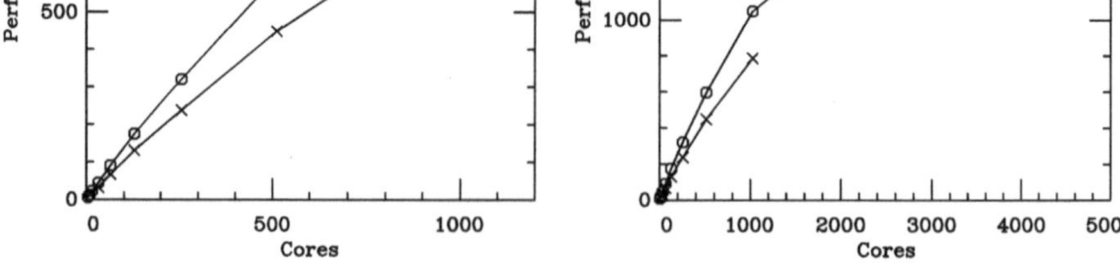

Fig. 2 Performance of NAMD on HPCx (crosses) and HECToR (circles). All data is normalized to the HPCx performance on 32 cores (taken as 32.0); the two graphs differ only in their scales.

5 Conclusions

With HECToR and the long-standing HPCx services, the UK continues to be one of the leading countries in the world for supercomputing. The two systems are very different in terms of their architectures but exhibit very similar performance characteristics in terms of low-level benchmarks and sustained performance on real applications. The only significant difference is in the network performance, enabling applications to scale to the much larger core counts available on HECToR.

The change from HPCx to HECToR is representative of the evolution of supercomputing worldwide, with core speeds remaining relatively static but core counts increasing rapidly. This has a fundamental impact on the way that scientfic research is carried out on these machines: applications codes must be made to scale to many thousands (if not tens of thousands) of cores to achieve any increase in capability. It is simply not cost effective to use these enormous facilities for capacity computing. This scaling problem is the key challenge facing computational science today. Although some classes of applications such as finite difference codes (for example, PDNS3D) may readily be capable of such scaling, other applications such as molecular dynamics (for example, NAMD) face much bigger challenges. It will be very interesting to see if and how these problems are solved in the coming years.

Acknowledgments We would like to thank Dr. Joachim Hein of EPCC for discussions regarding quantifiying the network performance of HPCx and HECToR. We would also like to thank all the members of the HPCx Science Support and HECToR User Support teams who have peformed the in-depth studies of the two systems and the IBM and Cray support staff who have provided their expertise.

References

1. Ashworth, M., Bush, I.J., Guest, M.F., Plummer, M., Sutherland, A.G., Hein, J.: HPCxTR0417: Application Performance on the High Performance Switch. URL http://www.hpcx.ac.uk/research/hpc/
2. Gray, A., Ashworth, M., Booth, S., Bull, J.M., Bush, I.J., Guest, M.F., Hein, J., Henty, D., Plummer, M., Reid, F., Sutherland, A.G., Trew, A.: HPCxTR0602: A Performance Comparison of HPCx Phase 2a to Phase 2. URL http://www.hpcx.ac.uk/research/hpc/
3. Gray, A., Ashworth, M., Hein, J., Reid, F., Weiland, M., Edwards, T., Knight, P., Johnstone, R.: HPCxTR0801: A Performance Comparison of HPCx and HECToR. URL http://www.hpcx.ac.uk/research/hpc/
4. Gray, A., Hein, J., Plummer, M., Sutherland, A.G., Smith, L., Simpson, A., Trew, A.: HPCxTR0604: An investigation of Simultaneous Multithreading on HPCx. URL http://www.hpcx.ac.uk/research/hpc/
5. HECToR Website. URL http://www.hector.ac.uk
6. HPC-Europa Website. URL http://www.hpc-europa.org
7. HPCx Website. URL http://www.hpcx.ac.uk
8. NAMD Website. URL http://www.ks.uiuc.edu/Research/namd/

9. PDNS3D Website. URL http://www.cse.scitech.ac.uk/arc/pdns3d.shtml
10. Phillips, J.C., Zheng, G., Kumar, S., Kalé, L.V.: NAMD: biomolecular simulation on thousands of processors. In: Proceedings of the 2002 ACM/IEEE conference on Supercomputing, pp. 1–18. IEEE Computer Society Press (2002)
11. Top500 Website. URL http://www.top500.org

DL_POLY_3 I/O: Analysis, Alternatives, and Future Strategies

Ilian T. Todorov, Ian J. Bush, and Andrew R. Porter

Abstract The molecular dynamics (MD) method is the only tool to provide detailed information on the time evolution of a molecular system on an atomistic scale. Although novel numerical algorithms and data reorganization approches can speed up the numerical calculations, the actual science of a simulation is contained in the captured frames of the system's state and simulation data during the evolution. Therefore, an important bottleneck in the scalability and efficiency of any MD software is the I/O speed and reliabilty as data has to be dumped and stored for postmortem analysis. This becomes increasingly more important when simulations scale to many thousands of processors and system sizes increase to many millions of particles. This study outlines the problems associated with I/O when performing large classic MD runs and shows that it is necessary to use parallel I/O methods when studying large systems.

1 Introduction

DL_POLY_3 [1] is a general purpose molecular dynamics (MD) package developed by I.T. Todorov and W. Smith at STFC Daresbury Laboratory to support researchers in the UK academic community [11]. This software is designed to address the demand for large-scale MD simulations on multi-processor platforms, although it is also available in serial mode. DL_POLY_3 is fully self-contained and written in Fortran 95 in a modularized manner with communications handled by MPI. The standards conformance of the code has been very rigorously checked using the NAGWare95 and FORCHECK95 analysis tools, so guaranteeing exceptional portability. Parallelization is achieved by equi-spatial domain decomposition distribution, which guarantees excellent load balancing and full memory distribution

Ilian T. Todorov · Ian J. Bush · Andrew R. Porter
STFC Daresbury Laboratory, Warrington WA4 4AD, UK
e-mail: i.t.todorov@dl.ac.uk

provided the particle density of the system is fairly uniform across space [7]. This parallelization strategy results in mostly point to point communication with very few global operations [12] and excellent scaling [10]; one might compare it with the halo exchange algorithms employed in computational fluid dynamics.

However, parallelization of the computation is only part of the story. For DL_POLY_3 to be an effective tool for researchers, all parts of the simulation must scale, and this includes the input and output (I/O) stages of the code. Historically, this has most often been performed in an essentially serial manner, not only in DL_POLY_3 but also in many other large-scale packages. However, with the scale of calculations now possible, it is clear that this approach will not scale to the next generation of machines as the time taken by I/O is becoming prohibitive. In this short study, we shall describe how DL_POLY_3 has performed I/O in the past, what problems this has resulted in, and first efforts to address the problems.

2 I/O in DL_POLY_3

The main I/O in DL_POLY_3 is, as is the case for all classic MD codes, reading and writing *configurations*. These are simply lists of the coordinates, velocities, and forces acting on the particles that comprise the system. In DL_POLY_3, this has traditionally been performed using formatted I/O for portability; whereas the MD run itself may be done on the supercomputer, the analysis of the results is often done on a workstation at home. Although this is very portable, there are a couple of potential problems:

1. Formatted I/O is not very efficient.
2. For large systems, the files can get very big and expensive to write.

It is the second point that is causing us to re-evaluate our I/O strategy. With top-end machines like CRAY XT3/4, IBM P575, and BG/L/P now capable of performing classic MD simulations with millions (or in some cases even billions) of atoms, the time taken to write these large files is now beginning to impact the amount of science that users of DL_POLY_3 can perform.

As a matter of fact, it is not the reading of configurations that is the issue as that is typically done only once to define the initial state of the system. It is the writing of configurations. Not only need this be done for the final state of the system but also many times during the simulation of the system so that the time evolution can be studied. Thus it is the writing of configurations that needs to be parallelized, at least initially. In the remainder of this section, we shall describe how the writing of configurations has been performed historically and also our new parallel methods.

2.1 Serial Direct Access I/O

Historically, the writing of configurations has been done very simply in DL_POLY_3 using a master slave method. One processor, the master, receives in turn the coordinates, velocities, and forces held by the other processors and writes them to file. While simple, portable, and robust, this strategy has one obvious drawback, it is inherently serial and so may impact scalability through simple Amdahls law effects. For instance, on a large BG/L system, we have observed one run, of a 14.6 million atom system on 16,384 compute cores, where while a timestep in the MD took approximately 0.5 seconds to perform, the dumping of a configuration took 450 seconds. As a configuration is typically dumped every 100–10,000 timesteps, this is not a very satisfactory situation! We shall call this the SDAW (Serial Direct Access Write) algorithm.

Reading of the initial configuration is the reverse of the above. The master reads chunks of the initial configuration and sends them in turn to the slave processors. As input is not so much of an issue at present, all results given here for reading will use this algorithm.

2.2 Parallel Direct Access I/O

The obvious problem with the SDAW algorithm is that only one processor ever performs the I/O, and thus all the other processors must wait on it. As an alternative we have developed a very simple PDAW (Parallel Direct Access Write) algorithm where all processors participate in writing to the file. Given the simple and regular format of the configuration file, it is very simple to calculate where the data for a given atom needs to be written; it depends solely on the global index of that atom. Imposing a fixed-length record, as in databases, it is very simple to use Fortran direct access files and have each processor writing to the appropriate records for the atoms it holds. So in our PDAW algorithm, all the processors open and write to the same file at once, as each processor can calculate where in the direct access file the data need be written.

While this algorithm is simple to implement, it is not robust and will not work on some platforms, for instance the Cray XT4 series. The reason for this is that it does not conform to the Fortran standard, which is *only* applicable if one process will ever be accessing a file at once. The problem of many processes accessing the same file simultaneously is in fact related to whether the I/O buffers of different processes holding the data to be written and the file system are "cash coherent". If they are not, the I/O buffers contents are pasted in the file without being smoothly appended and the contiguity of the data file is compromised. However, if cash coherency is guaranteed, all processes can open the same file, and if they write in an ordered manner, i.e., do not write over each other's data (data coherency), then the contiguity of the data file is intact. In our experience, this appears to be the case for most platforms relying on GPFS.

2.3 MPI-I/O

The main drawback of the PDAW algorithm, its lack of portability, is easily removed when using MPI-I/O. The implementation of this strategy, MPIW (Message Passing

Interface Write), is similar in spirit to PDAW as all the processors write to the file at once. However, the standardization of MPI I/O ensures portable behavior.

2.4 Serial I/O Using NetCDF

Another possible approach to improving I/O performance while retaining portability of the resulting files is to use NetCDF (network Common Data Form) [6]. This provides a set of software libraries and machine-independent data formats for array-based, scientific data and is widely used. The current stable release of this software does not support parallel I/O but the next release (currently in alpha) will, through the use of MPI-I/O.

The reason that NetCDF potentially provides performance benefits, even in serial form, is that its data formats are binary rather than ASCII-based. This can significantly reduce the quantity of data that must be written to disk without compromising precision. (The data itself is encoded using XDR to ensure portability.) We have recently implemented serial, NetCDF-based I/O within DL_POLY_3 and shall refer to this as the SNCW (Serial NetCDF Write) algorithm. The resulting trajectory files conform to the format used by the Molecular Modeling Tool Kit [3].

3 Results and Discussion

To test the I/O, the system we used was an oxygen-deficient pyrochlore $Gd_2Zr_2O_7$ (zirconite) with a size of 3,773,000 particles, which requires a 1.1GB configuration dump file. It is worth mentioning that no machine was available for exclusive use while benchmarking, which could have contributed to the fluctuations of the observed times at low processor counts. The parallel timings presented are from runs undertaken on the following platforms:

- The UK National Supercomputing facility HPCx [4], sited at STFC [9], comprising 160 IBM p5-575 nodes, totalling 2560 POWER5 1.5GHz compute cores.
- The BG/L, 1024 PowerPC 440 700 MHz compute cores, and BG/P, 4096 PowerPC 450 850MHz compute cores, clusters [5] sited at STFC [9].
- The UK National Supercomputing facility HECToR [2], sited at the University of Edinburgh, with 60 Cray XT4 cabinets totaling 11,328 AMD 2.8GHz Opteron compute cores.
- The Swiss Supercomputing Centre CSCS [8], comprising 18 Cray XT3 cabinets, totaling 3328 AMD 2.6GHz Opteron compute cores.

Tables 1 and 2 show the results for three IBM systems; BG/L, BG/P, and P5-575, respectively. The times for writing a single configuration are given, as is the time to perform a single timestep (in seconds). The values in bold are estimates of the I/O bandwidth in MBytes per second. All systems have a GPFS file system.

Table 1 BG/L and BG/P timings.

CPUs	SDAW		PDAW		MPIW		Timestep
			BG/L				
32	228.7	**4.9**	21.5	**52.5**	5443.0	**0.2**	28.22
64	244.5	**4.6**	20.8	**54.2**	2938.7	**0.4**	13.69
128	242.1	**4.7**	16.6	**67.9**	1534.0	**0.7**	7.23
256	238.9	**4.7**	18.4	**61.2**	1091.0	**0.1**	3.86
512	248.9	**4.5**	22.5	**50.1**	10800.0	**0.01**	2.03
1024	252.2	**4.5**	39.5	**28.5**	N/A	**N/A**	1.16
2048	253.0	**4.5**	58.8	**19.2**	N/A	**N/A**	0.77
			BG/P				
32	239.4	**4.6**	58.3	**18.9**	N/A	**N/A**	28.22
64	257.2	**4.3**	40.8	**27.0**	N/A	**N/A**	12.44
128	281.4	**3.9**	29.2	**37.6**	N/A	**N/A**	8.32
256	314.2	**3.5**	52.0	**21.1**	N/A	**N/A**	4.15
512	354.5	**3.1**	115.2	**9.6**	N/A	**N/A**	2.02
1024	384.9	**2.9**	131.6	**8.4**	N/A	**N/A**	1.17
2048	419.3	**2.6**	229.0	**4.8**	N/A	**N/A**	0.65

Table 2 P5-575 timings.

CPUs					P5-575				
	SDAW		SNCW		PDAW		MPIW		Timestep
32	98.3	**11.2**	71.9	**15.3**	11.3	**97.3**	1158.5	**0.9**	12.02
64	118.3	**9.3**	74.9	**14.7**	9.7	**113.4**	822.6	**1.3**	5.95
128	153.5	**7.2**	76.6	**14.4**	9.1	**120.9**	729.2	**1.5**	3.03
256	120.0	**9.2**	84.9	**13.0**	13.1	**84.0**	262.4	**4.2**	1.58
512	134.1	**8.2**	88.6	**12.4**	13.3	**82.7**	320.6	**3.4**	0.87
1024	148.3	**7.4**	89.0	**12.4**	17.1	**64.3**	N/A	**N/A**	0.58

Tables 3 and 4 show the results for two Cray systems; XT3 and XT4, respectively, in single-core and dual-core configured runs. The times for writing a single configuration are given, as is the time to perform a single timestep (in seconds). The values in bold are estimates of the I/O bandwidth in MBytes per second. Though both systems have a LUSTRE filesystem, they use different versions.

Table cells marked with **N/A** signify failure of the writing algorithm in question to deliver on the particular platform. We note that the failure took different forms:

1. The failure of the MPIW on high processor counts on BG/L is because after 5 hours the job had still not completed. This behavior was also found on *all* processor counts on BG/P.
2. The failure of MPIW on 1024 processors of the IBM p575 cluster was due to the implementaton of MPI-I/O being unable to run on this number of processors.
3. On the XT3, the SDAW method took over 9 hours to complete. This might have been a consequence of the use of CRAY XT3 Catamount microkernel.
4. As noted, the non-standard PDAW method gave incorrect results on the XT4.

Table 3 XT3 timings.

CPUs	SDAW		PDAW		MPIW		Timestep
			XT3 Single Core				
64	N/A	**N/A**	16.5	**66.7**	21.4	**51.4**	2.93
128	N/A	**N/A**	16.3	**67.5**	27.1	**40.6**	1.61
256	N/A	**N/A**	14.9	**73.8**	93.3	**11.8**	0.87
512	N/A	**N/A**	17.3	**63.6**	62.6	**17.6**	0.48
			XT3 Dual Core				
64	N/A	**N/A**	15.0	**73.3**	20.1	**54.6**	3.62
128	N/A	**N/A**	12.5	**88.0**	27.2	**40.5**	2.30
256	N/A	**N/A**	14.8	**74.3**	42.3	**26.0**	1.26
512	N/A	**N/A**	19.6	**56.1**	104.2	**10.6**	0.64
1024	N/A	**N/A**	11.45	**96.1**	131.0	**8.4**	0.44

It can be seen that the PDAW algorithm was performing markedly superiorly to the SDAW or MPIW methods when supported by the platform (HECToR's Cray XT4 platform does not) for this particular size of messages (73 Bytes ASCII per message). Improvements by an order of magnitude can be obtained, and it is clear that if regular writing of configurations is required when studying a system, that this parallel strategy will markedly improve the scaling of the whole code, *even though the I/O is not scaling especially well itself.*

It is also worth noting that for all the benchmarked platforms, it was *only* the Cray XT3/4 where the MPI-I/O write performed consistently well and much better than the standard serial direct access write. However, as seen on the Cray XT3, this

Table 4 XT4 timings.

CPUs	SDAW		PDAW		MPIW		Timestep
			XT4 Single Core				
32	**874.8**	**1.3**	N/A	N/A	20.7	**53.2**	4.07
64	**1182.4**	**0.9**	N/A	N/A	21.4	**51.4**	2.15
128	**1455.3**	**0.8**	N/A	N/A	27.1	**40.6**	1.15
256	**1819.1**	**0.6**	N/A	N/A	93.3	**11.8**	0.59
512	**2204.3**	**0.5**	N/A	N/A	62.6	**17.6**	0.31
1024	**2621.6**	**0.4**	N/A	N/A	92.8	**11.9**	0.19
2048	**3285.0**	**0.3**	N/A	N/A	164.0	**6.7**	0.12
			XT4 Dual Core				
32	**857.8**	**1.3**	N/A	N/A	25.1	**43.9**	4.25
64	**1120.9**	**1.0**	N/A	N/A	20.1	**54.6**	2.27
128	**1419.0**	**0.8**	N/A	N/A	27.2	**40.5**	1.27
256	**1706.7**	**0.6**	N/A	N/A	42.3	**26.0**	0.66
512	**2134.4**	**0.5**	N/A	N/A	104.2	**10.6**	0.36
1024	**2488.1**	**0.4**	N/A	N/A	131.0	**8.4**	0.23
2048	**3145.8**	**0.3**	N/A	N/A	141.9	**7.8**	0.15

was still not the fastest I/O write algorithm but the parallel direct access one. This suggests that there is considerable room for improvement in the MPI-I/O implementation. Furthermore, the failure of the XT3 to deliver with the traditional SDAW and the XT4 with the PDAW (whereas both performed the same with the MPI-I/O strategy) probably reflects different problems in the different LUSTRE I/O systems the platforms had.

When writing was forced to native binary, a speed-up by a factor of 2.5 to 3 was seen in all benchmarks on all machines. This relates very well to the ratio of the size of the messages sent per record in either case: 73 ASCII characters (73 Bytes) to three double precision reals (3×8 Bytes $= 24$ Bytes). Unfortunately, binary files are not portable between platforms, and as analysis of DL_POLY_3 runs is often performed on a different machine to that on which the code originally executed, this is not acceptable.

As mentioned in Sect. 2.4, NetCDF solves this portability issue by using XDR-encoded binary data. It does, indeed, considerably reduce the quantity of data involved. This was confirmed by the fact that a trajectory file containing a single frame for the pyrochlore system was 329MB compared with 1.03GB in ASCII form. This also accounts for the low bandwidth figures for the SNCW method relative to the SDAW algorithm in Table 2. It is clear that there is a further saving (3–7% storage-wise) and the scaling will improve when more frames are stored because the MMTK format, in contrast with the ASCII, includes the particles' specifications (chemical type, global index, mass, charge) just once. Including such sort of information will not significantly affect the I/O performance of the SNCW strategy as this data need only be written to the trajectory file once, at the beginning of a simulation.

Comparing the fastest write strategies on all platforms, we see that those from IBM (BG/P/L and P5-575) deliver the best I/O performance when running on 128 compute cores, whereas the Cray XT3/4 performs best on 32 compute cores in single-core mode or 64 in dual-core mode.

4 Conclusions

We have demonstrated that the use of a parallel I/O strategy in writing MD configuration files during simulations can bring a dramatic improvement in the overall performance, *even though the I/O is not scaling especially well itself.* It is clear from the discussion above that for DL_POLY_3 to be able to address larger systems, we must use a parallel I/O strategy. However, it is also clear that neither the PDAW nor MPIW strategy is optimal as either it

1. is not portable or
2. does not scale well or
3. does not perform as well as one might hope.

Although the SNCW strategy provides performance and storage benefits over the SDAW algorithm, it will ultimately be limited due to its lack of parallelism. We therefore intend to look at other possible solutions, such as a parallel version of the SNCW strategy, to solve the problems outlined in this study. We hope that this will also address the MPI-I/O performance problem as, ultimately, parallel NetCDF relies on it.

References

1. The DL_POLY webpage, URL http://www.ccp5.ac.uk/DL_POLY/
2. HECToR – the UK supercomputing service, URL http://www.hector.ac.uk/
3. Hinsen, K.: The molecular modeling toolkit: a new approach to molecular simulations. J. Comp. Chem. **21**, 79–85 (2000)
4. The HPCx supercomputing facility, URL http://www.hpcx.ac.uk/
5. The IBM BlueGene, URL http://www.research.ibm.com/bluegene/
6. NetCDF, URL http://www.unidata.ucar.edu/software/netcdf/
7. Pinches, M.R.S., Tildesley, D., Smith, W.: Large scale molecular dynamics on parallel computers using the link-cell algorithm. Mol. Simulation **6**, 51–87 (1991)
8. The Swiss National Supercomputing Centre, URL http://www-users.cscs.ch/
9. The Science and Technology Facilities Council, URL http://www.stfc.ac.uk/
10. Todorov, I.T., Allan, N.L., Purton, J.A., Dove, M.T., Smith, W.: Use of massively parallel molecular dynamics simulations for radiation damage in pyrochlores. J. Mater. Sci. **42**(6), 1920–1930 (2007)
11. Todorov, I.T., Smith, W.: DL POLY 3: the CCP5 national UK code for molecular-dynamics. simulations. Phil. Trans. R. Soc. Lond., A **362**, 1835–1852 (2004)
12. Todorov, I.T., Smith, W., Trachenko, K., Dove, M.T.: *DL_POLY_3*: new dimensions in molecular dynamics simulations via massive parallelism. J. Mater. Chem. **16**, 1911–1918 (2006)

Mixed Mode Programming on HPCx

Michał Piotrowski

Abstract Clusters of shared memory nodes have become a system of choice for many research and enterprise projects. Mixed mode programming is a combination of shared and distributed programming models and naturally matches the SMP cluster architecture. It can potentially exploit features of the system by replacing the message exchanges within a node with faster direct reads and writes from memory, using message passing only to exchange information between the nodes.

Several benchmark codes, based on a simple Jacobi relaxation algorithm, were developed; a pure MPI (Message Passing Interface) version and three mixed mode versions: Master-Only, Funneled, and Multiple. None of the mixed mode versions managed to outperform the pure MPI, mainly due to longer MPI point-to-point communication times. Results will be presented and reasons behind the performance losses discussed.

1 Introduction

Clusters of shared memory nodes have become a very popular HPC solution in recent years. Although they may never reach the peak performance of the large Massive Parallel Processors (MPP) systems, their relatively low cost, reliability, and availability make them systems of choice for many research and enterprise projects.

The HPCx system, a large SMP cluster, combines distributed memory parallelism between the nodes with shared memory parallelism inside each node. The vast majority of codes written for SMP clusters are pure MPI [2]. Memory is treated as fully distributed, ignoring the specific machine architecture. The hybrid programming model could potentially exploit features of the SMP cluster by replacing the message exchanges within the node with much faster direct reads and writes from

Michał Piotrowski
Edinburgh Parallel Computing Centre, University of Edinburgh, UK
e-mail: mpiotrow@epcc.ed.ac.uk

memory, using message passing only to exchange information between the nodes. This could possibly combine advantages of both parallelization paradigms and improve performance.

Three possible mixed mode programming styles are compared, that is, Master-Only, Funneled, and Multiple. In the Master-Only style, MPI communication is handled outside OpenMP [3] parallel regions by the master thread. In the Funneled style, inter-node communication is still performed by a single thread, however MPI routines are called within parallel regions. In the most sophisticated Multiple style, MPI communication routines may be called simultaneous by more than one thread. Appropriate benchmark codes based on a simple Jacobi algorithm were developed. Computation and communication performance of the codes was compared with the corresponding results of the pure MPI version.

2 Benchmark Codes

The MPI processes are arranged in a 3D cuboid (as shown in Fig. 1) by using the `MPI_Cart_create()` method. All communications between processors will be performed in the Cartesian communication space. Processes arranged in this way allow for 3D data decomposition that matches the rank of the distributed array. This decomposition scenario is the most popular solution in a wide range of scientific codes, e.g., fluid dynamics simulations.

All benchmark codes are based on the Jacobi relaxation method, an iterative algorithm that finds discretized solutions to set of differential equations. The algorithm that is used is a modified version of a reverse edge-detection algorithm, which operates on a 3D data set. All versions of the algorithm use three static 3D arrays of doubles: `edge`, `new`, and `old`. With given `edge` data, original data set can be determined iteratively using a Jacobi algorithm:

$$\text{new}_{ijk} = \frac{\text{old}_{i+1jk} + \text{old}_{i-1jk} + \text{old}_{ij+1k} + \text{old}_{ij-1k} + \text{old}_{ijk+1} + \text{old}_{ijk-1} - \text{edge}_{ijk}}{6}$$

where `old` is data calculated in previous iteration and *edge* is the input data.

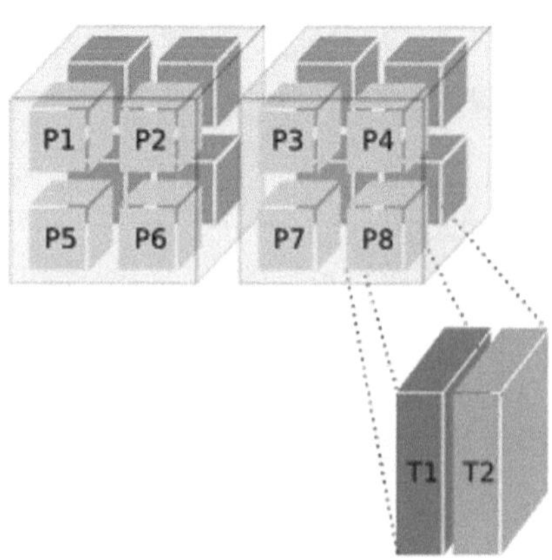

Fig. 1 MPI and OpenMP decomposition.

Algorithm 1 Simple iterative pure MPI algorithm.

```
while Δ is greater than tolerance
    exchange halos
    perform Jacobi relaxation
    calculate global Δ
    update 'old' array
end while
```

The algorithm satisfies the main requirements for a hybrid model benchmark code. It is computationally expensive, works on a regular data set that can be easily distributed between processors, and neighboring nodes have to exchange their halo areas in each iteration. The simple iterative pure MPI algorithm is shown in Algorithm 1. Each iteration starts with the `delta` check. Iterative image reconstruction is performed until the result has a satisfactory precision level (`delta<tolerance`) or the number of iterations reaches a predefined maximum. If neither condition is met, halo regions are exchanged between processors. When this is done, program calculates `new` value for each element in the 3D matrix using Jacobi relaxation method. At the end of every fifth iteration, Δ value is calculated.

$$\Delta^2 = \frac{1}{MNK} \sum_{i=1;j=1;k=1}^{i=M;j=N;k=K} (\mathrm{new}_{i,j,k} - \mathrm{old}_{i,j,k})^2$$

Δ needs to be computed locally on each process, then we perform a global sum before taking the final square root.

Part of the experiment was to determine how different MPI communication techniques will influence mixed mode code performance. In addition to testing how blocking and non-blocking MPI calls perform, we tried to move the non-blocking communication to the background so the processors could use the former waiting time for computation.

3 Mixed Mode

Mixed mode is a combination of message passing and shared memory programming. All three mixed mode versions implemented for purposes of this experiment follow the hierarchical model as shown in Fig. 1. The larger cuboid represents sub-arrays of data distributed between MPI processes organized in 3D Cartesian topology; these sub-arrays are then divided between OpenMP threads using a 1D decomposition. Having distributed data in that way, each thread operates on contiguous address space. Experiments show that 1D thread decomposition is almost always the best solution. Operating on large, contiguous chunks of data minimizes the probability of cache misses and prevents threads from false sharing.

Algorithm 2 Master-Only mixed mode algorithm.

```
distribute data between threads
while Δ is greater than tolerance
    if my thread id == 0 then
        exchange halos
    end if
    perform Jacobi relaxation
    OpenMP Barrier
    calculate global Δ
    update 'old' array
end while
```

This results in a reduced number of messages exchanged within a node, which are replaced by faster direct reads and writes to memory. The number of inter-node messages is also reduced. However, the average message size will increase. This may also have an impact on the performance, depending on how the interconnect deals with large messages.

Master-Only

The first of the mixed mode versions, Master-Only, is the most straightforward version to implement. It differs from other models in the way the MPI communication is handled. All messages are exchanged by the master thread outside parallel regions. This means that only one thread is handling MPI communication, and while this is done all other processors are idle. This approach does not promise the highest efficiency, as we are wasting some of the computing resources. However, this style has its advantages. It is easy to implement, and it is very straightforward to modify existing MPI code. Modifications made to the pure MPI version are limited to adding several OpenMP directives. Thread parallelism is achieved by inserting `omp parallel for` directive whenever program operates on the `edge`, `old`, and `new` arrays. The values of the iterator of the outermost loop is then distributed between threads, so each of them is working on its own 'slice' of the 3D array. The last modification made concerns residual calculation. Delta is first calculated locally, summed up first by an OpenMP reduction clause and finally by `MPI_Allreduce`. The Master-Only mixed mode algorithm is shown in Algorithm 2.

Funneled

The Funneled version also uses only one thread to perform MPI communication, however MPI routines may be called within OpenMP parallel regions. The Funneled algorithm model is shown in Algorithm 3. Instead of using an OpenMP `parallel for` directive, loop bounds are specified for each thread and are passed as a parameter to the `jacobistep()` function. This allows us to employ other threads in the

Algorithm 3 Funneled mixed mode algorithm.

```
distribute data between threads (load balance)
while Δ is greater than tolerance do
    if my thread id == 0 then
        exchange halos
        perform Jacobi relaxation
    else
        perform Jacobi relaxation ((non-boundary space))
    end if
    OpenMP Barrier
    calculate global Δ
    update 'old' array
end while
```

Jacobi computation while the master thread is exchanging data with neighboring processes. All threads are then synchronized using OpenMP barrier.

Loop boundaries are determined by a `ThreadDistribute()` function based on the thread ID value. By modifying the amount of work assigned to the master thread, we can balance the amount of work equally between all threads. This style of mixed mode programming is more difficult to implement, as more modifications have to be made to the pure MPI version than for the Master-Only version. However, in return we are able to employ all threads when MPI communication is handled, which may result in better performance.

Multiple

The Multiple version of the algorithm uses yet a different way of handling MPI communication inside the parallel regions. Here, all threads participate in the exchange of the halo regions. Each thread sends and receives the part of the boundary data that it is working on, independently of other threads. Threads that are located on the sides have to additionally exchange 'side' halo blocks. The Multiple version of the algorithm is shown in Algorithm 4. Compared with the Funneled version, no synchronization is needed before threads enter their halo regions, as each thread waits individually for the relevant MPI communication to complete.

As we can see in Fig. 2, multiple messages are exchanged between neighboring processes in one algorithm iteration. In order to organize which message is sent by which thread, we tag each message with an integer value that is based on the thread ID and the direction in which the message is being sent.

4 Hardware

The HPCx Phase 3 features 160 IBM p575 compute nodes, each containing 16 IBM Power5 processors, and 8 IBM eServer 575 LPARs for login and disk I/O. The Power5 processor is a 1.5GHz 64-bit RISC processor. The service offers a total of 2560 processors for computations [1].

Algorithm 4 Multiple version mixed mode algorithm.

```
distribute data between threads
while Δ is greater than tolerance
    exchange 'top', 'bottom', 'front' and 'back' halos
    if my thread id == 0 then
       exchange 'left' halo
    else if my thread id == max thread id
       exchange 'right' halo
    end if
    perform jacobi relaxation
    OpenMP Barrier
    calculate global Δ
    update 'old' array
end while
```

The 64-bit Power5 contains two microprocessor cores. It is equipped with a 64KB level-1 instruction cache and a 32KB level-1 data cache per core. The level-2 combined data and instruction cache of 1.9MB and level-3 cache directory and controls are shared between 2 processors. It also has a build-in fabric controller that controls the flow of information and control data between the L2 and L3 and between chips.

Each core contains two fixed-point execution units (FXU; FX Exec Unit), two floating-point execution units (FPU; FP Exec Unit), two load/store execution units (LDU; LD Exec Unit), one branch execution unit (BRU; BR Exec Unit), and one execution unit to perform logical operations on the condition (CRU; CR Exec Unit) [4]. It also supports speculative, out-of-order execution.

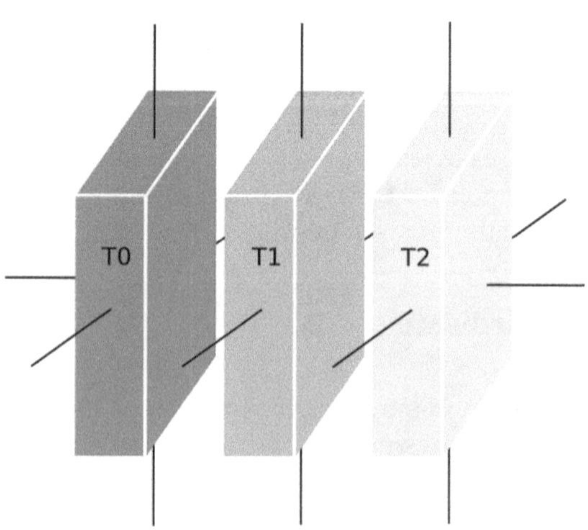

Fig. 2 Comms: Multiple version.

5 Experimental Results

The number of processors used in all experiments was fixed to 256, that is 16 SMP nodes. Problem size is fixed per processor, it is $192 \times 192 \times 192$ array of doubles, independent of the balance between the number of threads and processes. Distributing the problem this way excludes factors connected with domain decomposition shape.

The HPCx system is a shared resource; this means that resources (interconnect and processors) may be shared with system demons or other users. It is important to request exclusive access to all the nodes that will be used by our program. This is done by setting `node_usage = not_shared` in a job command file. In order to get accurate timings, the `MPI_Wtime` function was used. It returns a double precision floating point number of seconds, representing elapsed wall-clock time since some time in the past. High-resolution timings are obtained without adding significant overhead to the code. Each version of the code was set to run for 11 iterations. However, only the last 10 iterations were taken into consideration. The first run through the algorithm usually takes more time, as data has to be loaded into memory and program suffers from a greater number of cache misses. Because of the small number of iterations in total, the first iteration would distort the final result with greater effect.

Results of the first experiment are shown in Fig. 3. Here we compare performance of the pure MPI code with all mixed mode versions. In addition, we experiment

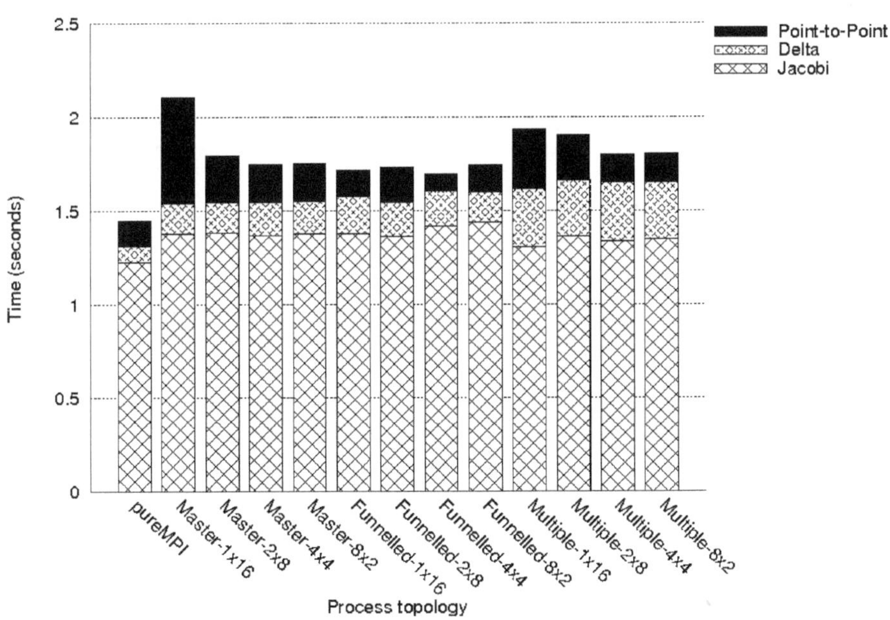

Fig. 3 Mixed Mode vs. pure MPI (blocking comms).

with different processes/threads decomposition schemes (the first number represents number of processes used).

None of the mixed mode codes was able to outperform pure MPI version. MPI is about 20% faster than the Funneled version, which turned out to be the most efficient mixed mode implementation. Looking at results for different topologies, we can see that decomposing problem over the maximum number of threads, that is 16 per node, significantly lowers program performance. The best efficiency is achieved for 4 threads / 4 processes per node, where the best balance between longer MPI communication and faster collectives is achieved.

The main performance loss in Master-Only and Funneled versions is observed during the message assembly and disassembly process, which is done using MPI derived data types. This is performed by a single thread that is responsible for MPI communication, and hence all other threads are idle at this time. In addition, data recently accessed by other threads has to be fetched from the caches of other processors. Remote memory access may result in series of cache misses during master's thread computational part of the program, which impacts overall program performance. Other reasons behind poor performance might be the cost of the explicit OpenMP barrier and the actual message size. When 16 threads are spawned, instead of sending 16 small messages like in pure MPI version, only one message is built. On HPCx, interconnect bandwidth is better utilized by a large number of small messages sent in parallel rather by a few very large messages.

In the Multiple version, the size of messages is exactly the same as in the pure MPI, except the case where side halo regions are exchanged. In addition, each thread is assembling and disassembling its own messages. However, we still cannot outperform pure MPI. More analysis (using the Paraver diagnostic tool) showed that MPI implementation couldn't cope with simultaneous calls to the MPI library. Instead of sending messages in parallel, messages were sent sequentially. This results not only in the longer MPI communication time but also in longer time spent for the delta calculation, as this includes the time needed for threads to synchronize before performing a global reduction.

Figure 4 shows the performance for different communication schemes used in the benchmark codes. There is not much difference between blocking and non-blocking communication, except in the Multiple version, where we observe a significant performance loss, which may be the result of the MPI sequentialization mentioned above. However, this cannot be confirmed without complete access to IBM's MPI implementation details. We tried to move the communication process to the background, so in theory processors could use the former waiting time for computation. However, this clearly doesn't work for either pure MPI or mixed mode.

In Fig. 5, we compare results of a different work balance attempts in the Funneled version of the code. The number on the bottom scale represents the amount of computation scheduled for the master thread, the thread that is also responsible for handling the MPI communication. In the unbalanced distribution, each thread is working on a $192 \times 192 \times 192$ slice. Results show that we were able to gain a little performance boost by assigning a smaller chunk of data to the master thread.

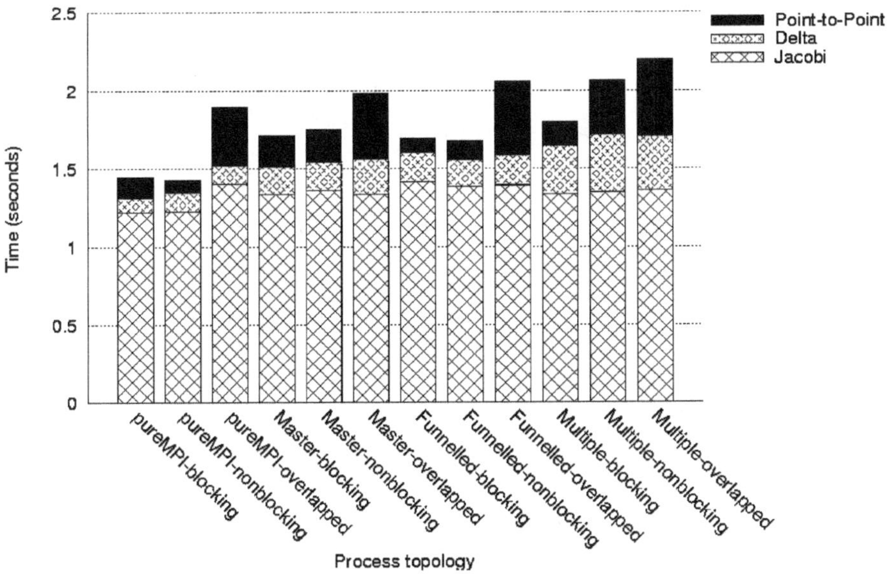

Fig. 4 Mixed mode vs. pure MPI: Communication.

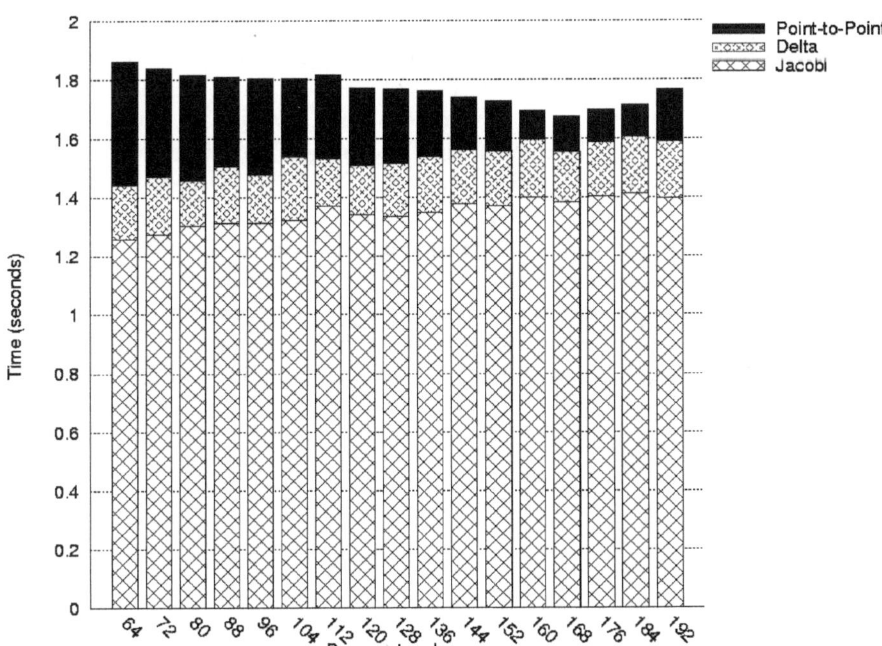

Fig. 5 Funneled version: load balance.

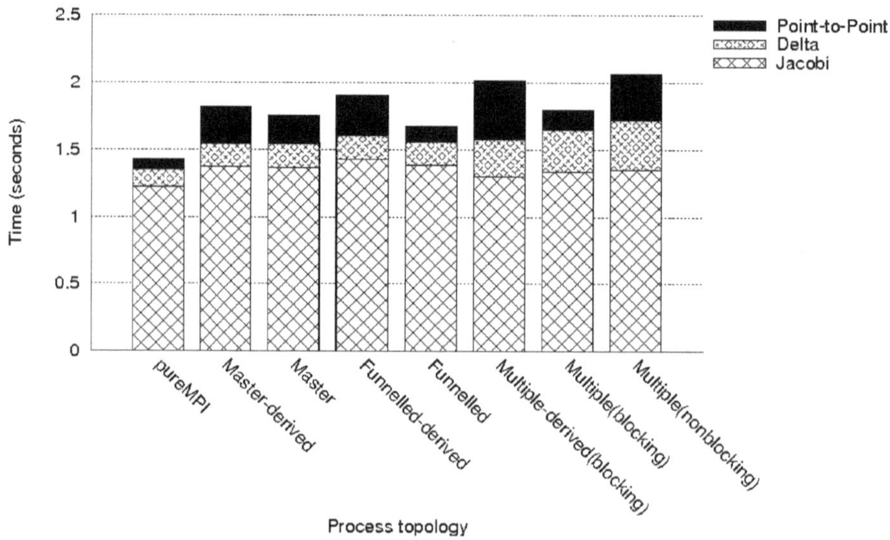

Fig. 6 Manual assembly vs. derived datatypes.

Figure 6 shows the performance for each model where different methods of message assembly and disassembly were used. In the derived version, in order to build a message that will be sent to another process, we use the MPI_Type_vector data type. It produces a new datatype by making copies of an existing data type and allows for regular stride in the displacements. However, using a predefined MPI structure doesn't allow us to distribute the assembly process between the threads (for Master-Only and Funneled) and creates additional multi-threaded calls to the MPI library for the Multiple version, which we have already seen to be problematic.

In the second version, we assembly the message by hand and distribute this task between threads, which also ensures more local memory accesses. We observe a significant improvement of the point-to-point communication time (which includes message assembly) for the Funneled and Multiple versions of the code.

6 Conclusions

The comparison of the pure MPI and mixed mode implementation shows the superiority of the pure MPI in terms of performance. All three mixed mode versions were outperformed in almost all sections of the code. The most efficient Funneled mixed mode implementation turned out to be 15% slower than the pure MPI.

The main reasons behind poor performance are ineffective point-to-point communication and limited multi-thread support provided by the MPI library (IBM MPI). Another issue might be a non-uniform structure of HPCx memory, thus memory access time may very depending on processor and memory location.

In addition, the intra-node MPI communication is well optimized on HPCx, i.e., it uses shared memory within the node instead of explicit exchange of messages, thus can be hard to outperform by a handwritten code.

Although currently slower than pure MPI, there are several scenarios where mixed mode could show its potential. An algorithm that forces data replication between the threads might be a good example where we could overcome memory limitations by using mixed mode. It is also much easier to balance the amount of work between threads rather than processes, which was demonstrated by our Funneled example. This could be beneficial when working on unbalanced problems.

References

1. HPCx Capability Computing (2006). URL http://www.hpcx.ac.uk
2. Message Passing Interface Forum (2007). URL http://www.mpi-forum.org
3. OpenMP Application Programming Interface (2007). URL http://www.openmp.org
4. Sinharoy, B., Kalla, R.N., Tendler, J.M., Eickemeyer, R.J., Joyner, J.B.: Power5 system microarchitecture. IBM Journal of Research and Development **49**(4/5), 505–521 (2005). URL http://researchweb.watson.ibm.com/journal/rd/494/sinharoy.html

A Structure Conveying Parallelizable Modeling Language for Mathematical Programming

Andreas Grothey, Jonathan Hogg, Kristian Woodsend, Marco Colombo, and Jacek Gondzio

Abstract Modeling languages are an important tool for the formulation of mathematical programming problems. Many real-life mathematical programming problems are of sizes that make their solution by parallel techniques the only viable option. Increasingly, even their generation by a modeling language cannot be achieved on a single processor. Surprisingly, however, there has been no effort so far at the development of a parallelizable modeling language. We present a modeling language that enables the modular formulation of optimization problems. Apart from often being more natural for the modeler, this enables the parallelization of the problem generation process making the modeling and solution of truly large problems feasible. The proposed structured modeling language is based on the popular modeling language AMPL and implemented as a pre-/postprocessor to AMPL. Unlike traditional modeling languages, it does not scramble the block-structure of the problem but passes this on to the solver if wished. Solvers such as block linear algebra exploiting interior point solvers and decomposition solvers can therefore directly exploit the structure of the problem.

1 Introduction

Algebraic modeling languages are recognized as an important tool in the formulation of mathematical programming problems. They facilitate the easy construction of models through an intuitive notation and features such as automatic differentiation, abolishing the need for tedious and error-prone coding work.

Many of today's large-scale optimization problems are not only sparse but structured. By structured we mean that the constraint matrix displays a special structure such as being primal- or dual-block angular. Other possible structures include

Andreas Grothey · Jonathan Hogg · Kristian Woodsend · Marco Colombo · Jacek Gondzio
School of Mathematics, University of Edinburgh, UK
e-mail: A.Grothey@ed.ac.uk

network matrices, projection matrices, or matrices obtained by a low-rank correction to other structures. These structures usually appear in the problem (and the problem matrices) due to some underlying generating process: They may reflect a discretization process (over time as in optimal control problems), over space (for PDE-constrained optimization), or over a probability space (as in stochastic programming). Further, they might reflect a structure inherent in the concept being modeled (e.g., a company consisting of several divisions). In many cases, these structures are nested. The structure (or at least the process generating the structure) is usually known to the modeler. Because many promising solution approaches for large-scale optimization problems, such as decomposition and interior point methods, can efficiently exploit this structure, it seems natural to pass the knowledge about the structure from the modeler to the solver.

Truly large-scale problems may not only require parallel processing for the solution of the problem but also for the model generation stage. The Asset and Liability Management problem solved in [17] requires 60GB of memory just to store the matrices and vectors describing the problem. Even the raw data from which the matrices are generated requires 2GB of storage. Clearly for problems of these sizes, the model generation should be done in parallel.

Current modeling languages do not in general offer this possibility. Whereas some languages offer the possibility to explicitly define special problem types such as network constraints or stochastic programming, none of the currently available modeling languages offer the flexibility to model nested structures and pass this information to the solver. No current modeling language attempts to use structure information to parallelize the model generation.

In addition, the memory and time requirements for the generation of large-scale structured mathematical programming problems frequently grow at faster than linear rate making the generation of very-large-scale problems impractical [19].

In fact, most really large problems designed to test the limits of state-of-the-art solvers have to be modeled by dedicated model generators, greatly increasing development effort. A structure exploiting model generation would enable the scalability of models to larger dimensions.

In the following section, we will give more details on the concepts of mathematical programming, structured problems, and modeling languages. Section 3 reviews efficient parallelizable solution techniques for structured optimization problems. In Sect. 4, we describe the design and implementation of our structured modeling language SML. Finally in Sect. 5, we draw our conclusions.

2 Background

2.1 Mathematical Programming

A mathematical programming problem can be written in the form

$$\min_x f(x) \text{ s.t. } g(x) \leq 0, x \geq 0, x \in X, \tag{1}$$

where $X \subset \mathbb{R}^n, f : \mathbb{R}^n \to R, g : \mathbb{R}^n \to \mathbb{R}^m$. Functions f, g are usually assumed to be sufficiently smooth: $f, g \in C^2$ is common. The vector-valued constraint function g can be thought of as vector of scalar constraints $g(x) = (g_1(x), \dots, g_m(x))^T$. Frequently, mathematical programming can reach very large sizes: n, m being of the order of millions is not uncommon. For problems of these sizes, the constraints $g_i(x) \le 0$ are usually generated by repeating a limited number of constraint prototypes over large index sets.

Consider as an example problem a multicommodity network flow [10] (MCNF) problem: Given is a network with nodes $i \in \mathscr{V}$ and (directed) arcs $j \in \mathscr{E}$. Further, we are given a set of commodities $k \in \mathscr{C}$ where each commodity is a triple (s_k, t_k, d_k) consisting of start node $s_k \in \mathscr{V}$, end node $t_k \in \mathscr{V}$ and amount to be shipped $d_k \in \mathbb{R}$. All arcs have a capacity Cap_j and a usage cost c_j. In a multicommodity network flow problem, we wish to find the minimum cost routing of all commodities such that the capacity of the arcs is not exceeded. The decision variables $x_{k,j}$ represent the flow on the jth arc induced by routing the kth commodity. The MCNF problem can be written mathematically as (2).

$$
\begin{aligned}
\min & \sum_{j\in\mathscr{E}} \sum_{k\in\mathscr{C}} c_j x_{k,j} \\
\text{s.t. } & Ax_k = b_k, \quad \forall k \in \mathscr{C} \\
& \textstyle\sum_{k\in\mathscr{C}} x_{k,j} \le \text{Cap}_j, \quad \forall j \in \mathscr{E} \\
& x_{k,j} \ge 0, \quad \forall k \in \mathscr{C}, j \in \mathscr{E}
\end{aligned}
\tag{2}
$$

Here $A \in \mathbb{R}^{|\mathscr{V}|\times|\mathscr{E}|}$ is the node-arc incidence matrix, and $b_k \in \mathbb{R}^{|\mathscr{V}|}$ is the demand vector for the kth commodity:

$$
A_{i,j} = \begin{cases} -1 & \text{node } i \text{ is source of arc } j, \\ 1 & \text{node } i \text{ is target of arc } j, \\ 0 & \text{otherwise.} \end{cases}
$$

$$
b_{k,i} = \begin{cases} -d_k & \text{node } i \text{ is source of demand } k, \\ d_k & \text{node } i \text{ is target of demand } k, \\ 0 & \text{otherwise.} \end{cases}
$$

Further, $x_k = (x_{k,1}, \dots, x_{k,|\mathscr{V}|}), b_k = (b_{k,1}, \dots, b_{k,|\mathscr{V}|})$. The MCNF problem displays a block structure (Fig. 1): for each commodity, there is a network block (light blocks) on the diagonal corresponding with matrices A, with the capacity constraints (dark blocks) linking all blocks.

2.2 Modeling Languages

A solution method for a mathematical program will typically need to evaluate $f(x), g_i(x), \nabla f(x), \nabla g_i(x), \nabla^2 f(x), \nabla^2 g_i(x)$ at various points $x \in X$. Although it would be possible to write routines to evaluate the functions f, g_i and its derivatives

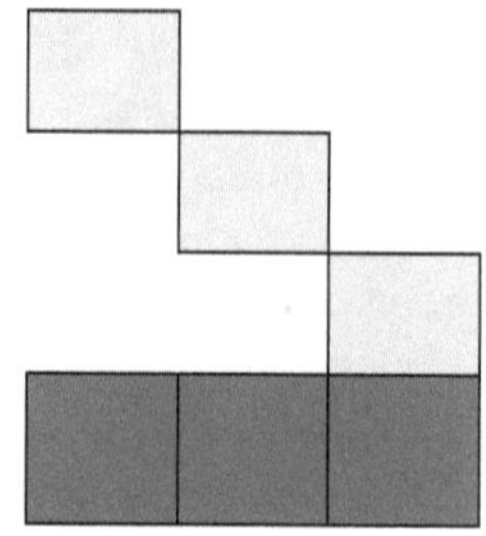

Fig. 1 Block structure of MCNF problem.

in a high-level programming language, this is tedious and error-prone work. Algebraic Modeling Languages (AML) provide facilities to express a mathematical program in a form that is both close to the mathematical formulation and easy to parse by an automated tool that will then provide the solver with the required information about the constraint and objective functions. One of the key paradigms of AMLs is the separation of model and data, so that different instances of the same model can be solved by using different data only. In practice, this is achieved by separate model and data files.

Modeling languages allow the mathematical programming problem to be expressed in terms of variables, constraints, objectives, parameters, and (indexing) sets. They typically use an internal tree representation of the entities defining the problem. This will then be used to calculate the value or the derivatives of constraint and objective functions at a given point when needed by the solver. Use is made of facilities such as automatic differentiation.

Traditional modeling languages such as AMPL [12], GAMS [5], AIMMS [3], or Xpress-Mosel [7] generate a problem in a monolithic, unstructured way. They will analyze the sparsity pattern of the resulting system matrices and pass these to the solver but scramble any further structure that might be present in the model.

AMPL is a widely used algebraic modeling language. Apart from ease of use, one of its main advantages is its well-defined user interface that makes linking to new solvers and language extension by pre-/postprocessing straightforward. AMPL communicates with the solver through `*.nl`-files. These are a binary representation of AMPL's internal model description, holding all the information necessary to evaluate objective and constraint functions. Using AMPL as a front end to a solver involves two phases: in the generation phase, AMPL reads in the model and the corresponding data file, parses them, expands all indexing sets to calculate the problem characteristics such as numbers of variables and constraints and sparsity patterns, and writes out the `.nl` file. In the solution phase, the solver can get problem information, such as problem dimensions or the values of problem functions and its derivatives at specified points through callback routines. These routines are provided in the `amplsolver` library [15] that serves as an interface to read AMPL's `.nl`-file model representation.

```
set NODES, ARCS, COMM;
param cost{ARCS}, Cap{ARCS}, arc_src{ARCS}, arc_trg{ARCS};
param comm_src{COMM}, comm_trg{COMM}, comm_demand{COMM};

var Flow{COMM, ARCS} >=0;

subject to FlowBalance{k in COMM, l in NODES}: %flow into = flow out of node
  sum{j in ARCS:arc_trg[j]==l} Flow[k,j]
                              + if (comm_src[k]==l) then comm_demand[k]
  = sum{j in ARCS:arc_src[j]==l} Flow[k,j]
                              + if (comm_trg[k]==l) then comm_demand[k];

subject to Capacity{j in ARCS}: sum{k in COMM} Flow[k, j] <= Cap[j];

minimize obj: sum{j in ARCS, k in COMM}  Flow[k,j]*cost[j];
```

Fig. 2 AMPL model for the MCNF problem.

Figure 2 gives the AMPL formulation of the MCNF problem (2). As can be seen, the model description is close to the mathematical notation. However, the block-structure of the problem is not immediately apparent.

3 Solution Approaches to Structured Problems

Several efficient solution approaches exist that exploit the block structure in problems like (2). They can be classified into decomposition approaches and interior point method (IPM) based approaches. Both classes have been demonstrated to parallelize very well. One possible advantage of IPM is their broader applicability to linear as well as nonlinear problem formulations.

3.1 Decomposition

In decomposition (e.g., Benders decomposition [1] or Dantzig–Wolfe decomposition [8]), the problem is decomposed into two (or more) levels: In the bottom level, primal values for complicating variables (linking columns) or the dual values for complicating constraints (linking rows) are temporarily fixed, resulting in a separation of the problem into many smaller problems (corresponding with the diagonal blocks in Fig. 1). The top level optimizes these complicating primal/dual variables by building up value surfaces for the optimal subproblem solutions.

3.2 Interior Point Methods

In interior point methods, the main work of the algorithm is to solve systems of equations with the augmented system matrix $\Phi = \begin{bmatrix} -\Theta^{-1} & A^T \\ A & 0 \end{bmatrix}$ where $A = \nabla c(x)$ is the Jacobian of the constraints as pictured in Fig. 1 and Θ is a diagonal ma-

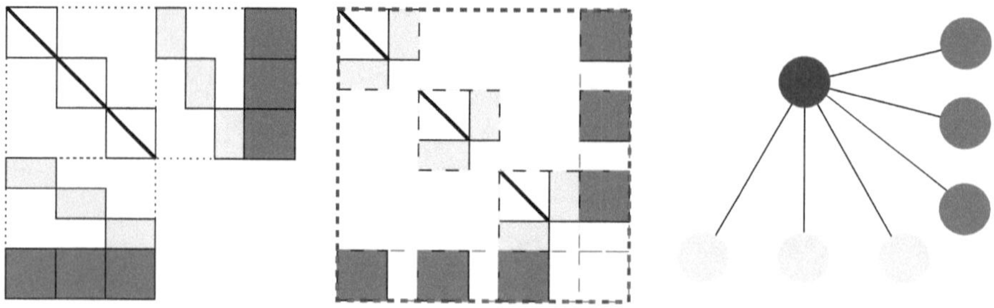

Fig. 3 OOPS: Structured matrices, structured augmented system, and the resulting algebra tree for the MCNF problem.

trix. For problems with a block structured A matrix, Φ itself can be reordered into a block-structured matrix (see Fig. 3). The block-structure can be exploited in the linear algebra of the IPM, using Schur-complement techniques and the Sherman-Morrison-Woodbury formula. The object-oriented parallel interior point solver OOPS [16] does this by building a tree of submatrices, with the top node corresponding with the whole problem and the nodes further down the tree corresponding with the constituent parts of the problem matrices. All linear algebra operations needed for the interior point algorithm (such as factorizations, backsolves, and matrix-vector products) can be broken down into operations on the tree: i.e., a factorization on the top node of the tree can be broken down into factorizations, solves, and matrix-vector products involving nodes further down the matrix tree. This approach generalizes easily to more complicated nested structures and facilitates an efficient parallelization of the IPM. The power of this approach has been demonstrated by solving an Asset and Liability Management problem with over 10^9 variables on 1280 processors in under 2 hours [17]. A similar approach is followed by Steinbach [22] and Blomvall and Lindberg [4].

4 Structure Conveying Modeling Languages

To apply any of these structure exploiting solution approaches, however, information about the block structure of problems must be passed to the solution algorithm. This information is not immediately apparent from the model (Fig. 2). Moreover AMLs will order rows and column of the generated system matrices by constraint and variable types rather than by structural blocks and therefore lose information about the problem structure. There have been attempts to automatically recover such a structure from the sparsity pattern of the problem matrices [9]. However, these methods are generally computationally expensive, are not parallelizable themselves, and obtain far from perfect results. Moreover in most cases the structure of the problem (or at least the process generating the structure) is known to the modeler: it was lost in the model generation process. It would be desirable to enable the modeler to express the structure directly in the modeling language.

4.1 Other Structured Modeling Approaches

The need for a structure conveying modeling language in mentioned in various places in the recent literature. However, there are very few actual implementations and no universally accepted approach exists.

Most structure conveying modeling languages that have been described deal specifically with the case of stochastic programming. Examples are SPInE [23], StAMPL [13], sMAGIC [6], MusMod [18], and the proposed stochastic programming extension to AMPL. Their main concern is with easing the work for the modeler, rather than conveying additional structure to the solver or exploiting parallelism in the implementation. sMAGIC uses a recursive model formulation that can be exploited in nested Benders Decomposition. Unfortunately, it is limited to linear programming (LP) problems. StAMPL and MusMod go one step further in that they separate not only the data from the model but also the scenario tree used in stochastic programming. This can be interpreted as separating the *structure* of the problem from the model, which makes their approach similar in spirit to ours.

There have been very few attempts at a general structure conveying modeling languages: SET [14] offers facilities to declare the structure of a model as a postscript to the model: An additional structure file declares the blocks making up the structure of the problem and a list of row and column names that make up each block. This approach requires that the full (unstructured) problem be processed first by the modeling language and is hence not suited for parallel model generation. Further, considerable (and in our opinion needless) effort goes into unscrambling the problem following the structure definition file. The authors of AMPL argue in [11] that AMPL's suffix facility can be used to the same effect (but with the same restrictions). Finally, MILANO [21] is an interesting concept of an object-oriented modeling language. However, its available documentation is scarce and it seems to have been abandoned. None of these approaches however attempt to parallelize the generation process itself.

Also worth mentioning is SMPS [2], which is not strictly speaking a modeling language. It can be most adequately described as a matrix description language, targeted at the special matrices encountered in stochastic programming. It is an extension of the standard MPS format [20] used to describe LP problems. However, it possesses important features that have inspired our modeling extension, notably the separation of the problem data into a *core matrix*, divisions of the core matrix, and a *scenario tree* that indicates how the various divisions of the core matrix are to be repeated. This corresponds respectively with our use of a *flat model tree*, indexing sets, and the *expanded model tree*.

4.2 Design

The dependency of the blocks that make up the structure of the MCNF problem structure in Fig. 1 can be described by a tree: the root node of the tree is the whole

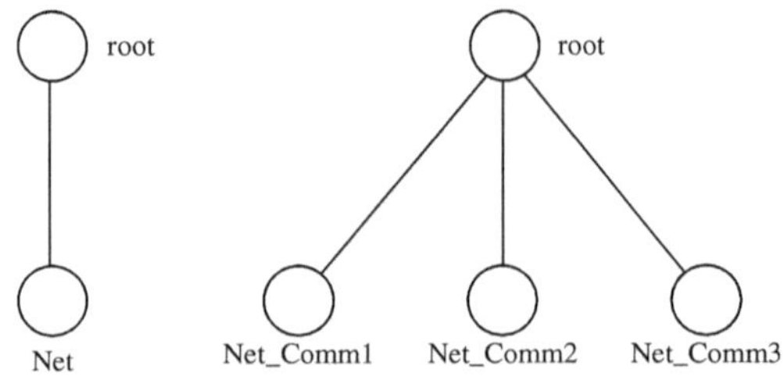

Fig. 4 Flat vs expanded model tree for the MCNF problem.

problem. Its children are the network routing problems: one each for every commodity. One can think easily of situations where submodels are nested deeper, resulting in a tree of dependencies. This model tree is presented in the right-hand side of Fig. 4 and corresponds to the matrix tree used by OOPS (Fig. 3). We call this the *expanded model tree*.

On the other hand, the relations between different parts of the model can also be described by a tree (see left-hand side in Fig. 4). This is the *flat model tree*. It has one node for each *type* of block rather than for each actual block present in the expanded problem. The expanded tree (and therefore the full problem) can be obtained from the flat tree by repeating nodes (in the case of MCNF for all commodities).

The main contribution of our modeling language is the `block` command to define a submodel. Its power comes from the fact that any `block` can be repeated by indexing it. This is a natural way of writing the flat model tree in the model file. A submodel definition using the `block` command takes the following form

```
block nameofblock{i in indexingset}:
  ...
end block
```

Within the scope delimited by `block`/`end block`, any number of `set`, `param`, `subject to`, `var`, `minimize`, or indeed further `block` commands can be placed. The understanding is that they are all repeated over the indexing expression used for the `block` command. Clearly, the nesting of such `blocks` creates a tree structure of blocks. A block command creates a variable *scope*: within it entities (sets, parameters, variables, constraints) defined in this block or any of its parents can be used. Entities defined in one of its children can be referred to by using the form

```
nameofblock[i].nameofentity
```

Entities defined in nodes that are not direct children or ancestors of the current block cannot be used.

Using the `block` command, the MCNF model of Fig. 2 can be written more concise as in Fig. 5.

There are several advantages of the SML formulation over the plain AMPL formulation: first the nested structure of the complete model (Fig. 1) is immediately apparent from the model file (and therefore to the solver). Further, much of the indexing in the AMPL model, in particular the double indexing of `Flow`, is not needed anymore, making the problem easier to read.

```
set NODES, ARCS, COMM;
param cost{ARCS}, Cap{ARCS}, arc_src{ARCS}, arc_trg{ARCS};

block Net{k in COMM}:
  param comm_src, comm_trg, comm_demand;
  var Flow{ARCS} >=0;

  % flow into node equals flow out of node
  subject to FlowBalance{l in NODES}:
    sum{j in ARCS:arc_trg[j]==l} Flow[j] + if (comm_trg==l) then comm_demand
      = sum{j in ARCS:arc_src[j]==l} Flow[j]
                                    + if (comm_src==l) then comm_demand;
  minimize obj: sum{j in ARCS}: Flow[j]*cost[j];
end block

subject to Capacity{j in ARCS}: sum{k in COMM} Net[k].Flow[j] <= Cap[j];

minimize obj: sum{k in COMM}  Net[k].obj;
```

Fig. 5 SML model for the MCNF problem.

4.3 Implementation

In our implementation, the SML model as in Fig. 5 is used to create AMPL model files for all the nodes of the flat model tree. The SML file is parsed to extract the flat model-tree, the list of entities associated with each flat-tree node, and the dependency graph of the entities. For every node in the flat model-tree, one AMPL model file is created. This file includes definitions for all the entities belonging to the flat-tree node plus all their dependencies. Figure 6 shows these model files for the MCNF problem. References to entities are changed to a global naming and indexing scheme. These are easily generated from the SML file by generic text substitutions. Note that the variable and constraint definitions in the two submodel files are very similar to the ones in the original unstructured AMPL model. By means of the SML description, we have managed to separate the AMPL model into the appropriate submodels for every node of the model trees.

The nodes of the flat and the expanded tree are represented in our implementation as C++ classes. The flat and expanded trees themselves are hence a tree of C++ objects. The `ExpandedTreeNode` class corresponds to one block of the problem as needed by a parallel, structure exploiting solver. It corresponds to a node in the OOPS matrix tree or a sub problem/master problem in a decomposition scheme. For every node of the expanded tree, there will be a separate `*.nl` file that coveys all the information on this node to the solver via the `amplsolver` library. The `*.nl` file for this node is produced locally on the appropriate processor from the corresponding submodel model file. Note that the underlying submodel is the same for all expanded tree nodes that correspond with the same flat tree node. However, they are produced with different data instances: namely, different choices of elements from the `block` indexing sets.

In our implementation, therefore, the `*.nl` files are associated with the `ExpandedTreeNode` objects, whereas the submodel files are associated with the `FlatTreeNode`.

```
%------------------------------ root.mod -------------------------------
set ARCS, COMM;
param Cap{ARCS};
var Net_Flow{COMM, ARCS};

subject to Capacity{j in ARCS}: sum{k in COMM} Net_Flow[k, j] <= Cap[j];

minimize obj: sum{k in COMM, j in ARCS}: Net_Flow[k, j]*cost[j];

% ------------------------- root_Net.mod ---------------------------
set NODES, ARCS, COMM;
param cost{ARCS}, Cap{ARCS}, arc_src{ARCS}, arc_trg{ARCS};

set COMM_SUB within COMM
param comm_src{COMM_SUB}, comm_trg{COMM_SUB}, comm_demand{COMM_SUB};

var Net_Flow{COMM_SUB, ARCS} >=0;

% flow into node equals flow out of node
subject to Net_FlowBalance{k in COMM_SUB, l in NODES}:
sum{j in ARCS:arc_trg[j]==l} Net_Flow[k, j]
                             + if (comm_trg[k]==l) then comm_demand[k]
= sum{j in ARCS:arc_src[j]==l} Net_Flow[k, j]
                             + if (comm_src[k]==l) then comm_demand[k];
```

Fig. 6 Generated AMPL model files for the MCNF submodels.

In a parallel implementation, it is the expanded tree nodes that are allocated to processors (Fig. 7). However, producing of the expanded tree from the flat tree requires the expansion of indexing sets, which should be done only on the processors allocated to this particular branch of the expanded model tree. To circumvent this bootstrapping problem, we will expand the flat tree by working down recursively through the expanded tree starting from the root node, splitting the computation among the processors when appropriate. These thoughts lead to the bootstrapping scheme explained below. We use the following notation: With every expanded tree node $n_E \in \mathscr{T}_E$ we associate: a flat tree node $t_F(n_E) \in \mathscr{T}_F$, a set of processors $\wp(n_E)$, a submodel file $\texttt{submod}(n_E)$:=$\texttt{submod}(t_F(n_E))$, and a data instance $\texttt{dat}(n_E)$. $\{1,\ldots,P\}$ is the set of processors available, and p is the current processor.

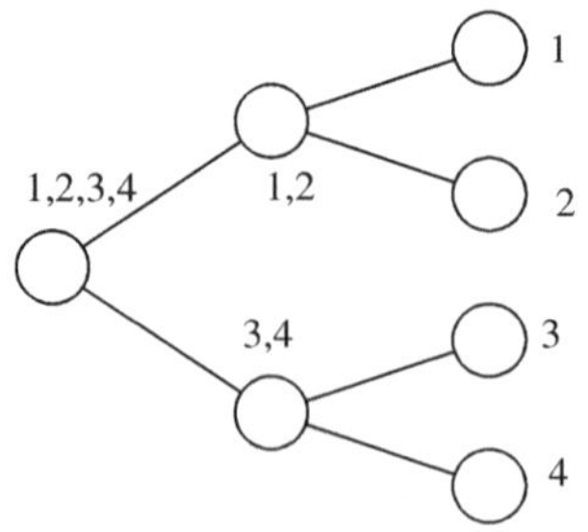

Fig. 7 Recursive processor allocation for an expanded model tree.

1. Parse the SML model file and set up the flat-model tree $\mathscr{T}_F$ made up of `FlatTreeNode` objects on all processors.
2. Generate the submodel files $\texttt{submod}(n), \forall n \in \mathscr{T}_F$ on all processors.
3. Create the expanded model tree $\mathscr{T}_E$ recursively: set $\mathscr{T}_E = \{n_0\}$ where n_0 is the root node and set $\wp(n_0) = \{1, \ldots, P\}$.
   `for` all nodes $\{n_E \in \mathscr{T}_E : p \in \wp(n_E)\}$ `do:`
   a. Expand the indexing sets of the children of node $t_F(n_E)$ in the flat-tree. This gives the set $C(n_E)$ of children in the expanded tree.
   b. Allocate a set of processors $\wp(n) \subseteq \wp(n_E)$ to each $n \in C(n_E)$ (as in Fig. 7).
   c. Add $C(n_E)$ to $\mathscr{T}_E$, if $C(n_E) \neq \emptyset$.

   `end do`
4. Generate the $*$`.nl` files: For every $n_E \in \mathscr{T}_E : p \in \wp(n_E)$: create the corresponding `.nl` file for this node by processing submodel $\texttt{submod}(t_f(n_E))$ and data instance $\texttt{dat}(n_E)$ by AMPL.

Two remarks about the algorithm are in order: first, the allocation of processors $\wp(n_E)$ to the child nodes of n_E in step 3(b) is done by distributing the processors in $\wp(n_E)$ equally among its children. This may of course raise a load balancing problem if the children are not of equal size. Secondly, for leaf nodes $t_F(n_E)$ in the flat-tree the set $C(n_E)$ will be empty, thus guaranteeing that the algorithm will finish.

5 Conclusions

We have presented a modeling language based on AMPL that allows the modeling of block-structured problems. The strength of the approach lies in the fact that this block structure is passed on to a structure exploiting solver and is used to parallelize the model generation process itself. We believe that this feature will become of major importance in the future as optimization models keep growing in size and although solvable by modern parallel solvers are now frequently beyond the reach of current modeling languages.

Acknowledgments This research was supported by Intel Corporation under project "Conveying Structure from Modelling Language to a Solver".

References

1. Benders, J.F.: Partitioning procedures for solving mixed-variables programming problems. Numerische Mathematik **4**(1), 238–252 (1962)
2. Birge, J.R., Dempster, M.A.H., Gassmann, H.I., Gunn, E.A., King, A.J., Wallace, S.W.: A standard input format for multiperiod stochastic linear programs. Committee on Algorithms Newsletter **17**, 1–19 (1987)

3. Bisschop, J., Entriken, R.: AIMMS The Modeling System. Paragon Decision Technology DG Haarlem, The Netherlands (2001)
4. Blomvall, J., Lindberg, P.O.: A Riccati-based primal interior point solver for multistage stochastic programming. European Journal of Operational Research **143**(2), 452–461 (2002)
5. Brooke, A., Kendrick, D., Meeraus, A.: GAMS: A User's Guide. The Scientific Press, Redwood City, California (1992)
6. Buchanan, C.S., McKinnon, K.I.M., Skondras, G.K.: The recursive definition of stochastic linear programming problems within an algebraic modeling language. Annals of Operations Research **104**(1-4), 15–32 (2001)
7. Colombani, Y., Heipcke, S.: Mosel: an extensible environment for modeling and programming solutions. In: Proceedings of the Fourth International Workshop on Integration of AI and OR Techniques in Constraint Programming for Combinatorial Optimisation Problems (CP-AI-OR'02), pp. 277–290. Le Croisic, France, Mar. 25–27 (2002)
8. Dantzig, G.B., Wolfe, P.: The decomposition algorithm for linear programming. Econometrica **29**(4), 767–778 (1961) Published online at http://www.emn.fr/x-info/cpaior/Proceedings/CPAIOR.pdf
9. Ferris, M.C., Horn, J.D.: Partitioning mathematical programs for parallel solution. Mathematical Programming **80**(1), 35–61 (1998)
10. Ford, L.R., Fulkerson, D.R.: Suggested computation for maximal multi-commodity network flows. Management Science **5**(1), 97–101 (1958)
11. Fourer, R., Gay, D.M.: Conveying problem structure from an algebraic modeling language to optimization algorithms. In: M. Laguna, J.L. Gonzáles-Velarde (eds.) Computing Tools for Modeling, Optimization and Simulation, pp. 75–89. Kluwer Academic Publishers, Dordrecht (2000)
12. Fourer, R., Gay, D.M., Kernighan, B.W.: AMPL: A Modeling Language for Mathematical Programming. The Scientific Press, San Francisco, California (1993)
13. Fourer, R., Lopez, L.: Stampl: A filtration-oriented modeling tool for stochastic programming. Tech. rep., University of Arizona and Northwestern University (2006)
14. Fragnière, E., Gondzio, J., Sarkissian, R., Vial, J.P.: A structure-exploiting tool in algebraic modeling languages. Management Science **46**(8), 1145–1158 (2000)
15. Gay, D.M.: Hooking your solver to AMPL. Tech. rep., Bell Laboratories, Murray Hill, New Jersey (1993; revised 1994, 1997)
16. Gondzio, J., Grothey, A.: Exploiting structure in parallel implementation of interior point methods for optimization. Tech. Rep. MS-04-004, School of Mathematics, University of Edinburgh, Edinburgh, Scotland, UK (2004)
17. Gondzio, J., Grothey, A.: Direct solution of linear systems of size 10^9 arising in optimization with interior point methods. In: R. Wyrzykowski, J. Dongarra, N. Meyer, J. Wasniewski (eds.) Parallel Processing and Applied Mathematics, *Lecture Notes in Computer Science*, vol. 3911, pp. 513–525. Springer-Verlag, Berlin (2006)
18. Hochreiter, R.: Simplified stage-based modeling of multi-stage stochastic programming problems. In: Conference Talk: SPXI. Vienna, Austria (2007)
19. Hogg, J.: Model generation for large scale structured problems. Master's thesis, School of Mathematics, University of Edinburgh (2005)
20. Murtagh, B.A.: Advanced Linear Programming: Computation and Practice. McGraw-Hill, New York (1981)
21. Sengupta, P.: MILANO, An Object-Oriented Algebraic Modeling System. Simulation Sciences Inc. URL http://www.simsci-esscor.com/NR/rdonlyres/C0061CFF-7F7A-4526-8710-D4362C6DCD36/0/102797.pdf
22. Steinbach, M.: Hierarchical sparsity in multistage convex stochastic programs. In: S. Uryasev, P.M. Pardalos (eds.) Stochastic Optimization: Algorithms and Applications, pp. 385–410. Kluwer Academic Publishers, New York (2001)
23. Valente, P., Mitra, G., Poojari, C., Kyriakis, T.: Software tools for stochastic programming: A stochastic programming integrated environment (SPInE). Tech. Rep. TR/10/2001, Brunel University (2001)

Computational Requirements for Pulsar Searches with the Square Kilometer Array

Roy Smits, Michael Kramer, Ben Stappers, and Andrew Faulkner

Abstract One of the key science projects of the Square Kilometer Array (SKA) is to provide a strong-field test of gravitational physics by finding and timing pulsars in extreme binary systems, such as a pulsar-black hole binary. We studied the computational requirements for beam forming and data analysis, assuming the SKADS (SKA Design Studies) design for the SKA, which consists of 15-meter dishes and an aperture array (AA). Beam forming of the 1-km core of the SKA requires no more than $2 \cdot 10^{15}$ ops. This number can be reduced when the dishes are placed in identical sub-arrays. Limiting the total field of view (FoV) of the AA to 3 deg^2, the maximum data rate from a pulsar survey using the 1-km core becomes about $2.7 \cdot 10^{13}$ bytes per second and requires a computation power of about $2.6 \cdot 10^{17}$ ops for a deep real-time analysis.

1 Introduction

Pulsars are rapidly rotating neutron stars that emit radio waves with extremely regular periods. We want to perform an all-sky pulsar survey for a number of reasons [2, 5].

- Finding rare objects that provide the greatest opportunities as physics laboratories. This includes:
  - Binary pulsars with black hole companions.
  - Pulsar in the Galactic Center will probe conditions near a $3 \times 10^6 \,\mathrm{M}_\odot$ black hole.
  - Millisecond pulsars (MSPs) spinning faster than 1.5 ms can probe the equation-of-state of pulsars.

Roy Smits · Michael Kramer · Ben Stappers · Andrew Faulkner
Jodrell Bank Centre for Astrophysics, University of Manchester, UK
e-mail: Roy.Smits@manchester.ac.uk

  - Pulsars with translational speeds in excess of $10^3\,\mathrm{km\,s^{-1}}$ probe both core-collapse physics and the gravitational potential of our Galaxy.
  - Rotating radio transients (RRATs).
  - Intermittent pulsars.

- To provide understanding of the advanced stages of stellar evolution.
- MSPs can be used as detectors of cosmological gravitational waves.
- To probe the interstellar medium in the Galaxy.
- To understand pulsars themselves.

The handling of the data rates for finding pulsars with the SKA might be compared with the ATLAS (A Toroidal LHC ApparatuS) experiment at the LHC (Large Hadron Collider) accelerator at CERN, for which the raw data rate is estimated to be 64 Tb/s (e.g., [4, 8]). Paper [6] describes in great detail a search algorithm for finding pulsar binaries that is suitable for their survey and computational resources. Here we are only concerned with the required data rates and computational power to achieve a pulsar and pulsar binary survey with the SKA. These requirements need to be considered in the design of the SKA.

2 SKA Configuration

We will express different sensitivities as a fraction of one 'SKA unit' defined as $2\times10^4\,\mathrm{m^2\,K^{-1}}$. Following the preliminary SKADS specifications for the SKA [7], we assume that the SKA will consist of three basic elements:

- A low-frequency array operating up to 500MHz.
- A high-frequency AA operating from 300MHz to 1GHz with a sensitivity of 0.5 SKA units and a maximum FoV of $250\,\mathrm{deg}^2$.
- Single pixel 15-meter dishes operating above 1GHz with a bandwidth of 500MHz and a collective sensitivity of 0.5 SKA units.

The concentration of both the AA and the dishes are assumed to be 20% within a 1-km radius and 50% within a 5-km radius. The low-frequency array will be ignored, as we will not consider using it for pulsar surveys in this article.

Technical Considerations

The FoV of a survey has a maximum that depends on the characteristics of the used elements. In the case of circular dishes with only one receiver, the FoV can be approximated as $(\lambda/D_{\mathrm{dish}})^2$, where λ is the wavelength of the observed radiation and D_{dish} is the dish diameter. In the case of the AA, the FoV of the elements is about half the sky. The actual size of the FoV that can be obtained will be limited by the available computational resources. This is because the signals from the elements will be combined coherently, resulting in 'pencil beams', the size of which scales with $1/D_{\rm tel}^2$, where $D_{\rm tel}$ is the telescope diameter. In order to obtain a large FoV, many of these pencil beams need to be formed. The computation time for the beam forming can be reduced by using only the elements in the core of the telescope, which will, however, reduce the sensitivity of the survey.

3 Computational Requirements

A pulsar survey requires the coherent addition of the signals from the elements in the core of the array to form sufficient pencil beams to create a sufficiently large FoV. These pencil beams will produce a large amount of data that needs to be transported either in the memory of a computer or to disk. This data then needs to be analyzed. To estimate the required computation power and data rates, we assume the SKA to have 2400 15-meter dishes, a bandwidth of 500MHz and 2 polarizations, and an AA with a frequency range from 300MHz to 1GHz and 2 polarizations.

3.1 Beam Forming

Following [1], we estimate the number of operations to fill the entire FoV with pencil beams as:

$$N_{\rm osb} = F_c N_{\rm dish} N_{\rm pol} B \left(\frac{D_{\rm core}}{D_{\rm dish}}\right)^2, \tag{1}$$

where F_c is the fraction of dishes inside the core, $N_{\rm dish}$ is the total number of dishes, $N_{\rm pol}$ is the number of polarizations, B is the bandwidth, $D_{\rm dish}$ is the diameter of the dishes, and $D_{\rm core}$ is the diameter of the core. The number of operations per second to beam-form the 1-km core of the 15-meter dishes then becomes $2 \cdot 10^{15}$. This is much less than the specifications for the correlator, which should be able to achieve 10^{17} operations per second. Still, it is possible to reduce this number by considering beam forming in two stages, where we assume that the dishes in the core are positioned such that they form identical sub-arrays. In the first stage, beam forming of the full FoV is performed for each sub-array. In the second stage, the final pencil beams are formed by coherently adding the corresponding beams of each sub-array formed in the first stage. This leads to the following expression for the number of operations required for beam forming in two stages:

$$N_{\rm osb2} = N_{\rm sa} N_{\rm dishsa} N_{\rm pol} B \left(\frac{D_{\rm sa}}{D_{\rm dish}}\right)^2 + \left(\frac{D_{\rm sa}}{D_{\rm dish}}\right)^2 N_{\rm sa} N_{\rm pol} B \left(\frac{D_{\rm core}}{D_{\rm sa}}\right)^2, \tag{2}$$

where $N_{\rm sa}$ is the number of sub-arrays in the core, $N_{\rm dishsa}$ is the number of dishes in one sub-array, and $D_{\rm sa}$ is the diameter of the sub-arrays. Substituting $N_{\rm dishsa} = F_c N_{\rm dish}/N_{\rm sa}$ and $N_{\rm sa} = (D_{\rm core}/D_{\rm sa})^2$ yields

$$N_{\text{osb2}} = F_c N_{\text{dish}} N_{\text{pol}} B \left(\frac{D_{\text{sa}}}{D_{\text{dish}}}\right)^2 + N_{\text{pol}} B \left(\frac{D_{\text{core}}}{D_{\text{dish}}}\right)^2 \left(\frac{D_{\text{core}}}{D_{\text{sa}}}\right)^2, \tag{3}$$

which has a minimum at $D_{\text{sa}} = D_{\text{core}}/\sqrt[4]{F_c N_{\text{dish}}}$. At this minimum, the number of operations for beam forming in two stages is

$$N_{\text{osb2}}^{\text{min}} = 2\sqrt{F_c N_{\text{dish}}} N_{\text{pol}} B \left(\frac{D_{\text{core}}}{D_{\text{dish}}}\right)^2. \tag{4}$$

For the same parameters as above, the number of operations per second becomes $2 \cdot 10^{14}$.

Figure 1 shows the number of operations per second for the beam forming as a function of core diameter. To obtain the benchmark concentration as a continuous function of core diameter between 0 and 10 kilometers, we used the following expression:

$$F_c(D_{\text{core}}) = a[1 - \exp(-bD_{\text{core}})], \tag{5}$$

where a and b were tuned to the specifications of 20% and 50% of the dishes within a 1-km and 5-km core, respectively, which leads to $a = 0.56$ and $b = 0.45 \cdot 10^{-3}\,\text{m}^{-1}$.

As an alternative to beam forming by coherently adding the signals from the dishes up to a certain core size, it is also possible to perform the beam forming by incoherently adding the signals from sub-arrays. This process is similar to beam forming in two stages as mentioned above, except that in the second stage the beams

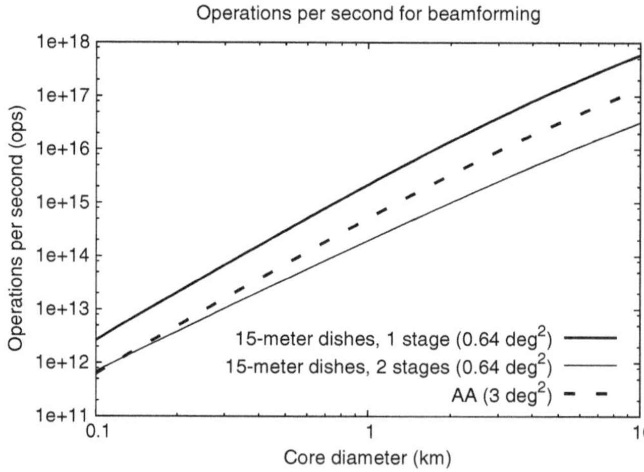

Fig. 1 The number of operations per second required to perform the beam forming for the 15-meter dishes and 3 deg^2 of FoV using the AA. It is assumed that there are 2400 15-meter dishes with a bandwidth of 500MHz. For the AA it is assumed that the total collective area is 500,000 m^2 and the frequency range is 0.3 to 1 GHz. In all cases the number of polarizations is 2. The thick black line corresponds to the beam forming of the 15-meter dishes in one stage. The thin black line corresponds to the beam forming of the 15-meter dishes in two stages. The striped/dotted line corresponds to the beam forming of the AA.

that were formed in the first stage are added incoherently. This leads to a much larger beam size, which reduces the required computation power for beam forming significantly. It also reduces the total data rate and the required computation power for the data analysis. The drawback is that it reduces the sensitivity of the telescope by a factor of $\sqrt{N_{sa}}$, where N_{sa} is the number of sub-arrays. However, this can be partially compensated as this method allows utilizing the full collective area of the telescope.

For the beam forming of the AA, the calculations are slightly different. Initial beam forming will be performed at the stations themselves, leading to a beam equivalent to those of a 60-meter dish for each station. Demanding a total FoV of 3 deg^2 over the frequency range of 0.3 to 1 GHz and assuming a total collective area of 500,000 m^2, the number of operations becomes:

$$N_{osb} = F_c N_{dish} N_{pol} B \cdot 3 \left(\frac{\pi}{180c}\right)^2 D_{core}^2 \nu_{max}^2. \tag{6}$$

As Fig. 1 shows, the required computation power for beam forming the AA is less than that required for beam forming the 15-meter dishes in 1 stage.

3.2 Data Analysis

The amount of data that will need to be analyzed will be large due to two reasons. Firstly, the FoV will be split up into many pencil beams, each of which will need to be searched for pulsar signals. Secondly, the SKA will be able to see the great majority of the total sky, all of which we want to search for pulsars and pulsar binaries. There are two ways to achieve this. The first option is to analyze the data as it is received, immediately dispensing with the raw data after analysis. This requires the analysis to take place in real time. The second option is to store all the data and analyze them at any pace that we see fit. Both approaches pose serious technical challenges, which we will discuss here.

First we consider the dishes for which we estimate the data rate of one pencil beam as

$$DR_{dish} = \frac{1}{T_{samp}} \frac{B}{\Delta\nu} N_{pol} \frac{N_{bits}}{8} \quad \text{bytes per second}, \tag{7}$$

where T_{samp} is the sampling time, B is the bandwidth, $\Delta\nu$ is the frequency channel width, N_{pol} is the number of polarizations, and N_{bits} is the number of bits used in the digitization. $\Delta\nu$ can be estimated by demanding that the dispersion smearing within the frequency channel does not exceed the sampling time. Thus:

$$\Delta\nu(\text{GHz}) = \frac{T_{samp}(\mu\text{s})\nu_{min}^3(\text{GHz})}{8.4 \cdot 10^3 DM_{max}}, \tag{8}$$

where ν_{min} is the minimum (lowest) frequency in the observation frequency band and DM_{max} is the maximum expected dispersion smearing. For the dishes, the number of pencil beams to fill up the FoV can be estimated as $(D_{\rm core}/D_{\rm dish})^2$. For the AA, the number of pencil beams becomes frequency dependent and is given by

$$N_{\rm pbAA} = 3\left(\frac{\pi}{180c}\right)^2 D_{\rm core}^2 \nu^2, \tag{9}$$

where c is the speed of light. However, each pencil beam needs to be centered on the same point on the sky for each frequency. Thus, the number of pencil beams will be constant over frequency and is determined by the highest frequency. The total data rate from the AA can then be estimated as

$$DR_{\rm AA} = \frac{1}{T_{\rm samp}} \frac{B}{\Delta\nu} N_{\rm pol} \frac{N_{\rm bits}}{8} 3\left(\frac{\pi}{180c}\right)^2 D_{\rm core}^2 \nu_{\rm max}^2. \tag{10}$$

Figure 2 shows the data rate from a pulsar survey for the 15-meter dishes and the AA as a function of core diameter, assuming $T_{\rm samp}$=100 μs, $DM_{\rm max}$=2000 for the dishes, $DM_{\rm max}$=500 for the AA, $N_{\rm pol}$=1, and 2 bits digitization. The AA was assumed to operate on the full frequency range of 0.3 to 1 GHz. The frequency range for the dishes was assumed to be 1 to 1.5 GHz. We can estimate the number of operations to search one 'pencil beam' for accelerated periodic sources as one Fourier transform of all the samples in the observation, for each trial DM-value and for each trial acceleration:

$$N_{\rm oa} = N_{\rm DM} N_{\rm acc} \cdot 5N_{\rm samp} \log_2(N_{\rm samp}), \tag{11}$$

where N_{DM} is the number of trial DM-values, which is equal to the number of frequency channels given by $B/\Delta\nu$, $N_{\rm acc}$ is the number of trial accelerations which scales with the square of $N_{\rm samp}$, which is the number of samples in one observa-

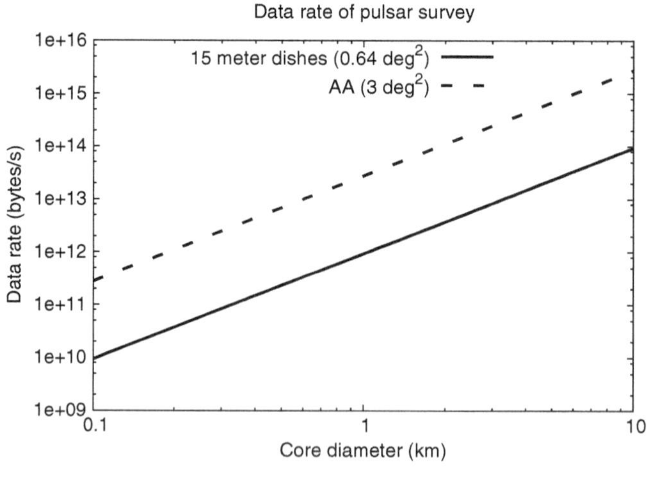

Fig. 2 Data rate from pulsar surveys using the 15-meter dishes or the AA, as a function of core diameter.

tion. Thus, the number of operations for the analysis scales as $N_{\text{samp}}^3 \log(N_{\text{samp}})$. This means that increasing the observation length is computationally very expensive. Once again, for the dishes the number of pencil beams to fill up the FoV can be estimated as $(D_{\text{core}}/D_{\text{dish}})^2$ and for the AA the number of pencil beams is given by (9), with $\nu=\nu_{\text{max}}=1$ GHz.

Figure 3 shows the number of operations per second required to perform a real-time analysis of a pulsar survey with the 15-meter dishes and the AA as a function of core diameter, assuming 100 trial accelerations, $\text{DM}_{\text{max}}=2000$ for the dishes, $\text{DM}_{\text{max}}=500$ for the AA, a sampling time of 100 μs, and an observation time of 1800 s.

In both Fig. 2 and Fig. 3, the values for the AA are much larger due to the large FoV and the low frequencies used, which leads to small frequency channels for the dedispersion.

An estimate of the computational power by 2015 is given by [3] to be 10 to 100 Pflop for \$100 M. This suggests that the required computation power for a real-time analysis of a pulsar survey from the AA needs to be reduced. This can be achieved by reducing the FoV, the bandwidth, the number of trial DM-values, the number of trial accelerations, or by increasing the sampling time.

Alternatively, all the data can be stored and analyzed at a much slower rate. From (7) we can estimate the total amount of data from a survey. Figure 4 shows the total amount of data from an all-sky survey and a survey of the galactic plane as a function of core diameter, assuming a frequency range of 1 to 1.5 GHz, $T_{\text{samp}}=100\,\mu$s, $DM_{\text{max}}=2000$, $N_{\text{pol}}=1$, an observation time of 1800 s, and 2 bits digitization.

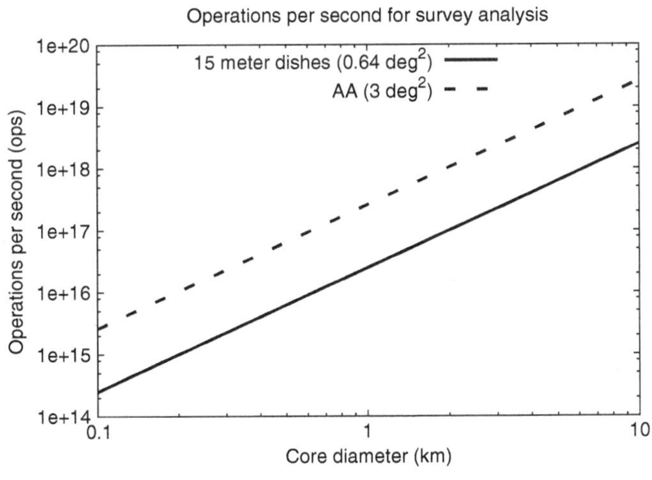

Fig. 3 Operations per second required to perform a real-time analysis of pulsar surveys using the 15-meter dishes or the AA, as a function of core diameter.

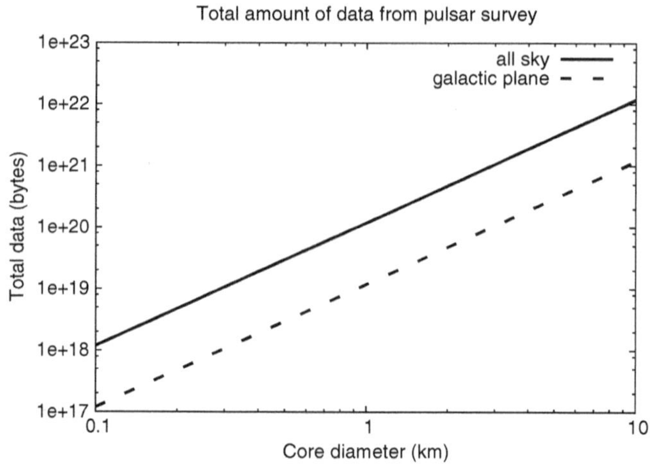

Fig. 4 Total amount of data from an all-sky survey and a survey of the galactic plane as a function of core diameter.

4 Conclusions

Performing a pulsar survey with the SKA requires the coherent addition of the signals of the individual dishes, forming sufficient pencil beams to fill the entire FoV of a single dish. Because of the extreme computational requirements that arise due to the large baselines, it is not possible to combine the signals of all of the dishes in the SKA. Rather, only a core of the SKA can be used for a pulsar survey. We have derived the computational requirements to perform such beam forming as a function of core-diameter for the 15-meter dishes and the AA. When the dishes are placed such that they form identical sub-arrays, the computational requirement for beam forming goes down significantly. We have also calculated the data rates and the computational requirements for applying a search-algorithm to the data to find binary pulsars. Both limit the usage of the SKA to a core of about 1 kilometer.

Acknowledgments The authors would like to thank Jim Cordes, Duncan Lorimer, Simon Johnston, Scott Ransom, and an anonymous referee for their help and useful suggestions. This effort/activity is supported by the European Community Framework Programme 6, Square Kilometre Array Design Studies (SKADS), contract no. 011938.

References

1. Cordes, J.M.: The SKA as a radio synoptic survey telescope: Widefield surveys for transients, pulsars and ETI (2007). URL http://www.skatelescope.org/PDF/memos/97_memo_Cordes.pdf
2. Cordes, J.M., Kramer, M., Lazio, T.J.W., Stappers, B.W., Backer, D.C., Johnston, S.: Pulsars as tools for fundamental physics & astrophysics. New Astronomy Review **48**, 1413–1438 (2004). DOI 10.1016/j.newar.2004.09.040

3. Cornwell, T.J.: SKA computing costs for a generic telescope model. Memo 64, SKA (2005). URL http://www.skatelescope.org/PDF/memos/64_Cornwell.pdf
4. Hörtnagl, C.: Evaluation of technologies of parallel computers' communication networks for a real-time triggering application in a high-energy physics experiment at CERN. Ph.D. thesis, CERN, Genève (1997). URL http://atlas.web.cern.ch/Atlas/documentation/thesis/hortnagl/writeup.ps.gz
5. Kramer, M., Backer, D.C., Cordes, J.M., Lazio, T.J.W., Stappers, B.W., Johnston, S.: Strong-field tests of gravity using pulsars and black holes. New Astronomy Review **48**, 993–1002 (2004). DOI 10.1016/j.newar.2004.09.020
6. Manchester, R.N., Lyne, A.G., Camilo, F., Bell, J.F., Kaspi, V.M., D'Amico, N., McKay, N.P.F., Crawford, F., Stairs, I.H., Possenti, A., Kramer, M., Sheppard, D.C.: The Parkes multi-beam pulsar survey - I. Observing and data analysis systems, discovery and timing of 100 pulsars. MNRAS **328**, 17–35 (2001). DOI 10.1046/j.1365-8711.2001.04751.x
7. Schilizzi, R.T., Alexander, P., Cordes, J.M., Dewdney, P.E., Ekers, R.D., Faulkner, A.J., Gaensler, B.M., Hall, P.J., Jonas, J.L., Kellermann, K.I.: Preliminary specifications for the Square Kilometre Array, v2.7.1. SKA draft (2007). URL http://www.skatelescope.org/PDF/Preliminary Specifications of the Square Kilometre Array_v2.7.1.pdf
8. Stancu, S.N.: Networks for the ATLAS LHC detector: Requirements, design and validation. Ph.D. thesis, Universitatea POLITEHNICA Bucuresti, Bucharest (2005). URL http://documents.cern.ch/cgi-bin/setlink?base=preprint&categ=cern&id=cern-thesis-2005-054

Part IV
Parallel Scientific Computing in Industrial Applications

Parallel Multiblock Multigrid Algorithms for Poroelastic Models

Raimondas Čiegis, Francisco Gaspar, and Carmen Rodrigo

Abstract The application of parallel multigrid for two-dimensional poroelastic model is investigated. First, a special stabilized finite difference scheme is proposed, which allows one to get a monotone approximation of the differential problem. The obtained systems of linear algebraic equations are solved by a multigrid method, when a domain is partitioned into structured blocks. This geometrical structure is used to develop a parallel version of the multigrid algorithm. The convergence for different smoothers is investigated, it is shown that the box Gauss–Seidel smoother is robust and efficient. Finally, the parallel multigrid method is tested for the Poisson problem.

1 Introduction

The fast and efficient solution of linear systems of equations is an important task for the computational simulation of many real-world problems. The different variants of the multigrid (MG) method are among the most popular tools, due to optimal complexity of such algorithms [1, 8, 17]. The MG can be used as an iterative method or as preconditioner for Krylov subspace methods.

MG methods are motivated by the fact that many iterative methods have a smoothing effect on the error between the exact solution and a numerical approximation. They also take into account that a smooth error can be well represented on a coarser grid where its approximation is substantially less expensive. Thus, MG

Raimondas Čiegis
Vilnius Gediminas Technical University, Saulėtekio al. 11, LT–10223, Vilnius, Lithuania
e-mail: rc@fm.vgtu.lt

Francisco Gaspar · Carmen Rodrigo
Departamento de Matematica Aplicada, Universidad de Zaragoza, 50009 Zaragoza, Spain
e-mail: fjgaspar@unizar.es · carmen.rodrigo.cardiel@gmail.com

methods use several grids in order to eliminate the different components of the error in a more efficient way. The design of efficient smoothers in multigrid methods for the iterative solution of systems of partial differential equations, however, often requires special attention.

Geometric MG methods are very popular for solving linear systems of equations arising after approximation of differential problems on structured grids. They possess the asymptotic optimality property of h-independent convergence. The application of standard smoothers in MG algorithms can be not efficient for problems having singularities such as strongly coupled unknowns in one direction or for systems of partial differential equations. Therefore, the design of nonstandard smoothers that are robust for special types of problems often requires special attention [15]. The poroelastic model is an example of such problem, and the investigation of nonstandard smoothers is one of our goals (see also [2, 6]).

This chapter also refers to the parallelization of the developed geometric MG algorithms. The parallelization of MG (both, geometric and algebraic) is actively investigated in many papers (see, e.g., [10, 12]). In order to achieve the high parallel efficiency, it is important to stay as close as possible to the sequential algorithm. The parallelization of geometric MG based on the domain decomposition method has been very efficient [9, 11]. In this chapter, we implement a block-structured sequential version of MG algorithm. Grids are composed of non-overlapped subgrids each of which is logically rectangular. This sequential algorithm is targeted at the efficient parallel implementation on the distributed memory computers. The block-structure provides a natural basis for the parallelization as each block can be assigned to a different processor. Because many robust smoothers loose their coupling over interior block boundaries, the number of iterations can grow up, and a typical MG property, the h-independent convergence rate, can be lost for problems in which anisotropies occur. Thus we investigate the dependence of the convergence rate of the proposed blocked version of MG algorithm on the number of sub-blocks used to decompose the grid.

Special attention is paid to an object-oriented implementation of the parallel code. The performance of the parallel method is separated from the main sequential code. This strategy gives a possibility to tune the communication part of the parallel algorithm to different types of the architecture of parallel computers, e.g., clusters of PCs or multicore computers.

The rest of the paper is organized as follows. In Sect. 2, the mathematical model of an elastic porous material and the fluid flow inside is presented. The special stabilized finite-difference scheme is developed for the approximation of this problem. Nonstandard MG methods are discussed in Sect. 3. They are based on the box-point and alternating box-line Gauss–Seidel smoothers. Results of numerical experiments with the standard alternating Gauss–Seidel and the alternating box-line Gauss–Seidel smoothers are presented in Sect. 4. The parallel version of the multigrid algorithm is presented and investigated in Sect. 5, and some final conclusions are given in Sect. 6.

2 Mathematical Model and Stabilized Difference Scheme

Poroelasticity theory addresses the time-dependent coupling between the deformation of an elastic porous material and the fluid flow inside. The mathematical model for a general situation was first proposed and analyzed by Biot, studying the consolidation of soils. Nowadays, poroelastic models are used to study problems in geomechanics, hydrogeology, and petrol engineering. Also, these equations have recently been applied in biomechanics to the study of soft tissue compression, to model the deformation and permeability of biological tissues, such as cartilage, skin, lungs, and arterial or myocardial tissues. The poroelastic model can be formulated as a system of partial differential equations for the unknowns displacements $\mathbf{u}$ and pressure of the fluid p. Here, we consider the case of a homogeneous, isotropic, and incompressible medium Ω, so the governing equations are given by:

$$\begin{cases} -\mu\tilde{\Delta}\mathbf{u} - (\lambda+\mu)\nabla(\nabla\cdot\mathbf{u}) + \nabla p = \mathbf{g}(\mathbf{x},t), & \mathbf{x}\in\Omega, \\ \dfrac{\partial}{\partial t}(\nabla\cdot\mathbf{u}) - \dfrac{\kappa}{\eta}\Delta p = f(\mathbf{x},t), & 0<t\le T, \end{cases} \tag{1}$$

where $\tilde{\Delta}$ represents the vector Laplace operator, λ and μ are the Lamé coefficients, κ the permeability of the porous medium, η the viscosity of the fluid, $\mathbf{g}$ is the density of applied forces, and f represents an injection or extraction process. Before fluid starts to flow and due to the incompressibility of the solid and fluid phases, the initial state satisfies the equation

$$\nabla\cdot\mathbf{u}(\mathbf{x},0) = 0, \quad \mathbf{x}\in\Omega.$$

When a load is applied on an elastic and saturated porous medium, the pressure suddenly increases and a sharp boundary layer can appear in the early stages of the time-dependent process. A spatial discretization with central finite differences and a temporal discretization with an implicit Euler algorithm leads to unstable solutions in the pressure field. In order to avoid this unstable behavior, we consider the corresponding discretization of the incompressible poroelasticity equations, in which a stabilization term $\varepsilon\partial\Delta p/\partial t$ with $\varepsilon = h^2/4(\lambda+2\mu)$ is added [5, 6]:

$$\begin{cases} -\mu\tilde{\Delta}\mathbf{u} - (\lambda+\mu)\nabla(\nabla\cdot\mathbf{u}) + \nabla p = \mathbf{g}(\mathbf{x},t), & \mathbf{x}\in\Omega, \\ \dfrac{\partial}{\partial t}(\nabla\cdot\mathbf{u}) - \dfrac{\kappa}{\eta}\Delta p - \varepsilon\dfrac{\partial}{\partial t}\Delta p = f(\mathbf{x},t), & 0<t\le T, \\ \nabla\cdot\mathbf{u}(\mathbf{x},0) - \varepsilon\Delta p(\mathbf{x},0) = 0. \end{cases} \tag{2}$$

This problem, i.e., the original system of equations in which a stabilization term is added, can be discretized on a collocated grid with central differences. We approximate the elasticity part by the difference operator (here the standard notation of the finite-difference method is used [16]):

$$A_h = \begin{pmatrix} A_{11} & A_{12} \\ A_{21} & A_{22} \end{pmatrix}, \quad A_{11} = -\mu\Delta_h - (\lambda+\mu)\,\partial_{h,xx}, \tag{3}$$

$$A_{12} = A_{21} = -\frac{\lambda+\mu}{4h^2}\begin{bmatrix} -1 & & 1 \\ & * & \\ 1 & & -1 \end{bmatrix}_h, \quad A_{22} = -\mu\Delta_h - (\lambda+\mu)\,\partial_{h,yy}.$$

Here the usual five-point stencil approximation of the Laplace operator is used. To approximate the coupling terms, ∇p and $\nabla \cdot \mathbf{u}$, we also use the classic second-order central difference approximations.

After the spatial approximation of (2), we get the following Cauchy problem for the system of differential-difference equations

$$\begin{cases} A\mathbf{u}_h(t) + Gp_h(t) = \mathbf{g}_h(t), \quad 0 < t \le T, \\ \dfrac{d}{dt}\Big(D\mathbf{u}_h(t) - \dfrac{h^2}{4(\lambda+2\mu)}\Delta p_h(t)\Big) - \dfrac{\kappa}{\eta}\Delta p_h(t) = f_h(t), \\ D\mathbf{u}_h(0) - \dfrac{h^2}{4(\lambda+2\mu)}\Delta p_h(0) = 0. \end{cases} \tag{4}$$

The implicit Euler scheme is applied for time discretization. We use a uniform grid for time discretization with step-size $\tau > 0$. Let $y^m(\mathbf{x}) = y(\mathbf{x}, t_m)$, where $t_m = m\tau$, $m = 0, 1, \ldots, M$, $M\tau = T$. The fully discrete scheme is given by

$$A_h\mathbf{u}_h^{m+1} + Gp_h^{m+1} = g_h^{m+1}, \tag{5}$$

$$\frac{D\mathbf{u}_h^{m+1} - D\mathbf{u}_h^m}{\tau} - \frac{h^2}{4(\lambda+2\mu)}\frac{\Delta_h p_h^{m+1} - \Delta_h p_h^m}{\tau} - \frac{\kappa}{\eta}\Delta_h p_h^{m+1} = f_h^{m+1}.$$

The convergence of the stabilized scheme in an energy norm is given in [5].

3 Multigrid Methods

Multigrid (MG) methods are well-known fast solvers for algebraic systems arising in the discretization of PDE problems. In order to reach a higher efficiency in the numerical resolution of our problem, we are going to apply an MG algorithm. Nevertheless, some research must be done to define an efficient MG solver for the poroelasticity system with the artificial pressure term, because of the fact that the performance of the method depends critically on the choices of the MG components.

In this section, we define a sequential version of block MG algorithm. The grids are composed of subgrids each of which is logically rectangular. Along the arisen artificial interior block boundaries, an overlap region is placed. This block-structure provides a natural basis for the parallelization of the MG algorithm.

There are several points in a MG algorithm, such as in the relaxation and residual calculation, where information from neighbor subgrids must be transferred. For this communication, each subdomain is augmented by an overlap-layer of so-called ghost nodes that surround it. The width of this overlap region is mainly determined by the extent of the stencil operators involved.

All the MG components that have to be specified should be as efficient as possible for the concrete problem. In addition, we will require that these components allow an efficient parallel implementation. In practice, for the overall parallel performance of multigrid, it is the most important to find a good parallel smoother, as the other multigrid components usually do not develop problems. The convergence rate of the sequential block version of the MG algorithm should be as close as possible to the convergence rate of the standard MG algorithm. We note that the selection of a good smoother is a challenging problem even in this case.

When discretizing the incompressible poroelasticity equations with the standard second-order central differences and an artificial pressure term, the development of multigrid smoothing methods is not straightforward. The Fourier smoothing analysis shows that the smoothing factors of standard collective point-wise relaxations are not satisfactory. A possibility to overcome this problem is to extend to collocated grids the idea of box relaxation, which has been proved to be a suitable smoother in the case of staggered grids [13]. A systematic comparison of various smoothing strategies in MG for the poroelasticity system is done in [7].

With respect to the coarse grid correction, we choose geometric grid coarsening on the Cartesian grids, i.e., the sequence of coarse grids is obtained by doubling the mesh size in each spatial direction, and the multigrid transfer operators, from fine-to-coarse and from coarse-to-fine, can easily be adopted from scalar geometric multigrid.

3.1 Box Relaxation

Firstly, we consider a pointwise box Gauss–Seidel. This smoother updates all unknowns appearing in the discrete divergence operator in the second equation of the system simultaneously. This means that five unknowns ($p_{i,j}$, $u_{i+1,j}$, $u_{i-1,j}$, $v_{i,j+1}$, $v_{i,j-1}$) centered around a pressure point are relaxed simultaneously, see Fig. 1a. Therefore, for each box, a small 5×5 system must be solved.

This smoother works satisfactorily but its alternating line version gives better results. Therefore, we have considered the alternating box-line Gauss–Seidel version. It consists of a first step of relaxing at the same time all the points appearing in Fig. 1b, this means for each line of the domain we relax the unknowns corresponding with p and u in the nodes of this line and the unknowns corresponding with v associated with the nodes in the upper and lower lines. After x-direction relaxation, we perform an analogous y-direction relaxation.

In a standard geometric MG processor, p needs access only to such data of a neighboring processor that correspond with variables inside the overlap region,

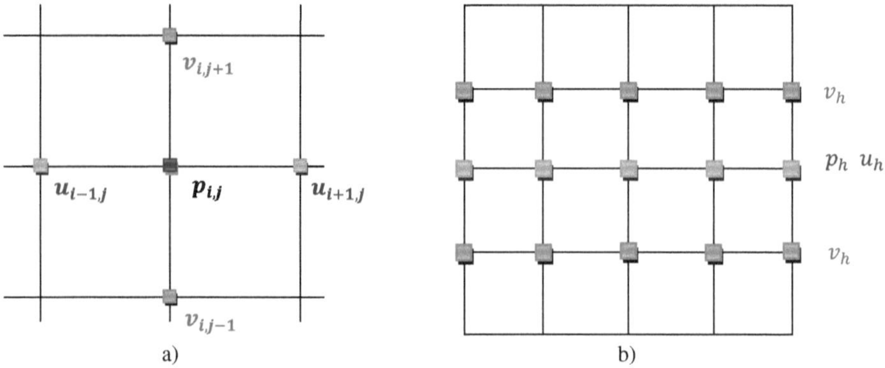

Fig. 1 Different smoothers: (a) box smoother with five unknowns centered around a pressure point; (b) alternating box-line smoother, unknowns updated simultaneously by the box x-line relaxation.

defined by the stencil of the finite-difference scheme. In our case, a geometric overlap of width 2 is required. The presented smoothers require exchanging the residual and the solution at some stages of the algorithm. The right-hand-side vectors are exchanged after the application of the restriction operator. These updates of the variables in overlap regions are implemented also in the sequential version of the code.

4 Numerical Experiments

In this section, we solve numerically a true 2D footing problem [5]. The computational domain is a block of porous soil $\Omega = (-60,60) \times (0,120)$. We assume that the bottom and the vertical walls are rigid, and on the central part of the top wall a load of intensity σ_0 is applied. The whole boundary is free to drain. More precisely, the boundary data is given as follows:

$$p = 0, \quad \text{on } \partial\Omega, \quad \sigma_{xy} = 0, \ \sigma_{yy} = -\sigma_0, \quad \text{on } \Gamma_1,$$

$$\sigma_{xy} = 0, \ \sigma_{yy} = 0, \quad \text{on } \Gamma_2, \ \mathbf{u} = \mathbf{0}, \quad \text{on } \partial\Omega \setminus (\Gamma_1 \cup \Gamma_2),$$

where

$$\sigma_{xy} = \mu \left(\frac{\partial u}{\partial y} + \frac{\partial v}{\partial x} \right), \quad \sigma_{yy} = \lambda \frac{\partial u}{\partial x} + (\lambda + 2\mu) \frac{\partial v}{\partial y}$$

and

$$\Gamma_1 = \{(x,y) \in \partial\Omega,\ |x| \leq 20,\ y = 120\},$$

$$\Gamma_2 = \{(x,y) \in \partial\Omega,\ |x| > 20,\ y = 120\}.$$

The material properties of the porous medium are given by $E = 3 \cdot 10^4$ $[N/m^2]$, $\nu = 0.2$, where λ and μ are related to the Young's modulus E and the Poisson's ratio ν by $\lambda = \dfrac{\nu E}{(1+\nu)(1-2\nu)}$, $\mu = \dfrac{E}{2(1+\nu)}$.

This problem is solved iteratively by MG with the smoothing method proposed above. A systematic parameter study is performed by varying the quantity κ/η. We compare the results obtained with the new alternating box-line Gauss–Seidel and the standard alternating line Gauss–Seidel smoothers. The latter is a straightforward generalization of the point-wise collective smoother. The stopping criterion per time step is that the absolute residual should be less than 10^{-8}. The $F(2,1)$-cycle, meaning two pre-smoothing and one post-smoothing steps, is used at each time step.

Results in Table 1 show the convergence of the alternating line Gauss–Seidel for different values of κ/η. It can be observed that this smoother is sensitive to the size of the diffusion coefficients. For small values of κ/η, the convergence is unsatisfactory.

Table 2 shows the corresponding convergence of the MG iterative algorithm with the alternating box-line Gauss–Seidel smoother. For all values of κ/η, a very satisfactory convergence is obtained. We observe a fast and h-independent behavior with an average of 9 cycles per time step. From this experiment, the box-line smoother is going to be preferred, as it results in a robust convergence.

Table 1 Results for the alternating line Gauss–Seidel smoother: $F(2,1)$ convergence factors and the average number of iterations per time step for different values of κ/η.

κ/η	48×48	96×96	192×192	384×384
10^{-2}	0.13 (8)	0.07 (7)	0.07 (7)	0.08 (8)
10^{-4}	0.30 (14)	0.28 (14)	0.38 (18)	>1
10^{-6}	0.30 (14)	0.24 (13)	0.41 (20)	>1
10^{-8}	0.30 (14)	0.24 (13)	0.41 (20)	>1

Table 2 Results for the alternating box-line Gauss–Seidel smoother.

κ/η	48×48	96×96	192×192	384×384
10^{-2}	0.07 (7)	0.09 (8)	0.09 (8)	0.09 (8)
10^{-4}	0.07 (7)	0.09 (8)	0.07 (8)	0.07 (8)
10^{-6}	0.07 (7)	0.09 (8)	0.09 (9)	0.09 (9)
10^{-8}	0.07 (7)	0.09 (8)	0.09 (9)	0.09 (9)

5 Parallel Multigrid

We note that even a sequential version of the proposed MG algorithm has a block structure, when grids are composed of non-overlapped structured subgrids. The block-structure provides a natural basis for the parallelization as each block can be assigned to a different processor. Each processor p needs access only to such data of a neighboring processor that correspond with variables inside the overlap region, defined by the stencil of the finite-difference scheme.

5.1 Code Implementation

The communication algorithms can be implemented very efficiently taking into account the structured geometry of subgrids (and, as a consequence, the overlapping area of grids). An optimized communication layer has been built over the MPI library, in order to support code portability issues. The domain distribution (achieving the load balancing and minimizing the amount of overlapping data), communication routines, and logics of the parallelization are hidden as much as possible in the implementation of the code at the special parallelization level. This can be done very efficiently due to the block-structure of the sequential algorithm, which also uses ghost-cells that must be exchanged at the special synchronization points of the MG algorithm.

All the multigrid components that have to be specified should be parallel or as parallel as possible, but at the same time as efficient as possible for the concrete problem. However, in practice, to find a good parallel smoother is crucial for the overall parallel performance of MG, as the other MG components usually do not develop problems. When discretizing the incompressible poroelasticity equations with standard second-order central differences and an artificial pressure term, the development of MG smoothing methods is not straightforward. In this section, we investigate the efficiency of the proposed alternating box-line Gauss–Seidel smoother. Due to the block version of the algorithm, the implementation of this smoother is fully parallel. The amount of computations on the hierarchy of coarser levels develops in a priori defined way due to the structured nature of the geometrical domain decomposition on the Cartesian grids, i.e., the sequence of coarse grids is obtained by doubling the mesh size in each spatial direction, and the multigrid transfer operators, from fine-to-coarse and from coarse-to-fine, can easily be adopted from scalar geometric multigrid. The data exchange among processors also require predictable and uniform for all levels communication patterns. Thus the total efficiency of the parallel implementation of the block MG algorithm depends only on the load balancing of the domain distribution algorithm.

But the problem of defining efficient parallel smoothers is much more complicated, as above we considered only the parallel implementation of the MG algorithm for a fixed number of blocks. When a grid is split into many blocks that are all smoothed simultaneously, the alternating line smoothers loose their coupling over

Table 3 Results for the alternating box-line Gauss–Seidel smoother for different numbers of blocks.

Blocks	Grid per block	Conv. factor	Iter.
1 (1×1)	576x576	0.065	8
9 (3×3)	192x192	0.09	9
36 (6×6)	96x96	0.09	9
144 (12×12)	48x48	0.09	9

interior block boundaries and the convergence rate of multiblock MG algorithm can be much slower than the convergence rate of one block MG [15]. In fact, with an increasing number of blocks, the multiblock MG method comes closer to a point relaxation method.

We have investigated the robustness of the proposed alternating box-line Gauss–Seidel smoother varying the number of blocks in which we split the domain. Here we fix the value of $\kappa/\eta = 10^{-4}$. We observe in Table 3 a fast and robust multigrid convergence even when the number of blocks is large. Thus there is no need to apply extra update of the overlap region or to use one additional boundary relaxation before the alternating sweep of the line smoother, as was done in [14, 15].

5.2 Critical Issues Regarding Parallel MG

Important issue of the parallel MG deals with the treatment of sub-domains of the coarsest subgrid-levels. For large number of subdomains (blocks), at a certain coarse level it becomes inefficient to continue parallel coarsening involving all processors. Although the number of variables per processors will be small, the total number of variables will be very large, and the convergence rate of the smoother will be not acceptable.

There are two possibilities how to improve the efficiency of the parallel algorithm. In both cases, the agglomeration of subgrids on the coarsest level is done. First, neighboring subdomains can be joined and treated by fewer processors (it is obvious that in this case some processors will become idle). Second, one processor receives the full information on the coarse grid problem, uses some iterative or direct method and solves the obtained system, and distributes the solution to all processors.

In order to separate the issue of MG convergence reduction due to large number of blocks from the influence of parallel smoothers, we present numerical results when two-dimensional Poisson problem is solved. It is approximated by the standard central-differences. The obtained system of linear equations is solved by the MG algorithm using V(1,1)-cycles and the point red–black Gauss–Seidel iterative algorithm as smoother. Thus the smoother is exactly the same for any number of blocks. Maximum 20 iterations are done to solve the discrete problem on the

Table 4 Results for the red–black point Gauss–Seidel smoother for different numbers of blocks.

Blocks	Grid per block	Sub=10	Sub=9	Sub=8	Sub=7	Sub=6 (50 it.)
1 (1×1)	1024×1024	62.0	62.2	62.0	61.9	83.7
4 (2×2)	512×512	—	46.0	46.5	46.2	61.8
16 (4×4)	256×256	—	—	42.2	42.2	55.9
64 (8×8)	128×128	—	—	—	42.1	55.8

coarsest grid. The initial grid 1024×1024 is split into many equally sized blocks. Results in Table 4 are obtained on the standard node of PC cluster consisting of Pentium 4 processors (3.2GHz), level 1 cache 16KB, level 2 cache 1MB, interconnected via Gigabit Smart Switch. Here CPU time is given for different numbers of blocks, which are arranged in 2D topology. *Sub* denotes a number of subgrids used in MG coarsening step.

The given results show that the computational speed of the block MG algorithm even improves for larger number of blocks. This fact can be explained by a better usage of level 1 and level 2 cache, as block data is reused much more efficiently when the number of blocks is increased. Starting from $Sub = 6$, the coarse level of the problem is too large to be solved efficiently by the red–black Gauss–Seidel iterative algorithm with 20 iterations. In order to regain the convergence rate of the MG algorithm, we needed at least 50 iterations. Thus for this problem, we can use efficiently up to 256 processors.

Table 5 shows the performance of the parallel block MG algorithm on IBM p5-575 cluster of IBM Power5 processors with peak CPU productivity 7.8 Gflops/s, connected with IBM high-performance switch. Here for each number of blocks and processors, the CPU time T_p and the value of the algorithmic speed-up coefficient $S_p = T_1/T_p$ are presented.

A superlinear speedup of the parallel algorithm is obtained for larger numbers of processors due to the better cache memory utilization in IBM p5-575 processors for smaller arrays. This fact was also established for parallel CG iterative algorithms in [3] and for a parallel solver used to simulate flows in porous media filters [4].

Results of similar experiments on the PC cluster are presented in Table 6.

Table 5 Performance of the parallel block MG algorithm on IBM p5-575.

Blocks	$p=1$	$p=2$	$p=4$	$p=8$	$p=16$	$p=32$
1	81.6	—	—	—	—	—
4	64.6	31.6 (2.0)	15.3 (4.2)	—	—	—
16	57.4	28.1 (2.0)	13.1 (4.4)	5.75 (10.)	2.89 (20.0)	—
64	56.9	27.6 (2.1)	13.0 (4.4)	5.92 (9.6)	3.40 (16.9)	1.6 (34.5)

Table 6 Performance of the parallel block MG algorithm on PC cluster.

Blocks	$p=1$	$p=2$	$p=4$	$p=8$	$p=16$
1	62.1	—	—	—	—
4	46.0	24.3 (1.9)	13.4 (3.4)	—	—
16	42.2	21.6 (1.9)	11.7 (3.6)	6.26 (6.7)	3.48 (12.1)

As it follows from the computational results given above, the measured speedups of the parallel block MG algorithm agree well with the predictions given by the scalability analysis.

6 Conclusions

We have presented a simple block parallelization approach for the geometric MG algorithm, used to solve a poroelastic model. The system of differential equations is approximated by the stabilized finite difference scheme. A special smoother is proposed, which makes the full MG algorithm very robust. It is shown that the convergence rate of the MG algorithm is not sensitive to the number of blocks used to decompose the grid into subgrids.

Practical experience has shown that special care must be taken for solving accurately the discrete problems on the coarsest grid, but this issue starts to be critical only for the number of blocks greater than 64. In fact, very satisfactory parallel efficiencies have been obtained for test problems. In addition, the block version of the MG algorithm helps to achieve better cache memory utilization during computations.

Acknowledgments R. Čiegis was supported by the Agency for International Science and Technology Development Programmes in Lithuania within the EUREKA Project E!3691 OPTCABLES and by the Lithuanian State Science and Studies Foundation within the project on B-03/2007 "Global optimization of complex systems using high performance computing and GRID technologies". F. Gaspar and C. Rodrigo were supported by Diputación General de Aragón and the project MEC/FEDER MTM2007-63204.

References

1. Brandt, A.: Multigrid Techniques: 1984 Guide with Applications to Fluid Dynamics. GMD-Studie Nr. 85, Sankt Augustin, Germany (1984)
2. Brezina, M., Tong, Ch., Becker,R.: Parallel algebraic multigrid for structural mechanics. SIAM J. Sci. Comput. **27**, 1534–1554 (2006)
3. Čiegis, R.: Parallel numerical algorithms for 3D parabolic problem with nonlocal boundary condition. Informatica **17**(3), 309–324 (2006)

4. Čiegis, R., Iliev, O., Lakdawala, Z.: On parallel numerical algorithms for simulating industrial filtration problems. Computational Methods in Applied Mathematics **7**(2), 118–134 (2007)
5. Gaspar, F.J., Lisbona, F.J., Oosterlee, C.W.: A stabilized difference scheme for deformable porous media and its numerical resolution by multigrid methods. Computing and Visualization in Science **11**, 67–76 (2008)
6. Gaspar, F.J., Lisbona, F.J., Oosterlee, C.W., Vabishchevich, P.N.: An efficient multigrid solver for a reformulated version of the poroelasticity system. Comput. Methods Appl. Mech. Eng. **196**, 1447–1457 (2007)
7. Gaspar, F.J., Lisbona, F.J., Oosterlee, C.W., Wienands, R.: A systematic comparison of coupled and distributive smoothing in multigrid for the poroelsaticity system. Numer. Linear Algebra Appl. **11**, 93–113 (2004)
8. Hackbusch, W.: Multigrid Methods and Applications. Springer, Berlin (1985)
9. Haase, G., Kuhn, M., Laanger, U.: Parallel multigrid 3D Maxwell solvers. Parallel Comp. **27**, 761–775 (2001)
10. Haase, G., Kuhn, M., Reitzinger, S.: Parallel algebraic multigrid methods on distributed memory computers. SIAM J. Sci. Comput. **24**, 410–427 (2002)
11. Jung, M.: On the parallelization of multi-grid methods using a non-overlapping domain decomposition data structure. Appl. Numer. Math. **23**, 119–137 (1997)
12. Krechel, A., Stüben, K.: Parallel algebraic multigrid based on subdomain blocking. Parallel Comp. **27**, 1009–1031 (2001).
13. Linden, J., Steckel, B., Stüben, K.: Parallel multigrid solution of the Navier-Stokes equations on general 2D-domains. Parallel Comp. **7**, 461–475 (1988)
14. Lonsdale, G., Schüller, A.: Multigrid efficiency for complex flow simulations on distributed memory machines. Parallel Comp. **19**, 23–32 (1993)
15. Oosterlee, C.W.: The convergence of parallel multiblock multigrid methods. Appl. Numer. Math. **19**, 115–128 (1995)
16. Samarskii, A.A.: The Theory of Difference Schemes. Marcel Dekker, Inc., New York–Basel (2001)
17. Trottenberg, U., Oosterlee, C.W., Schuller, A.: Multigrid. Academic Press, New York (2001)

A Parallel Solver for the 3D Simulation of Flows Through Oil Filters

Vadimas Starikovičius, Raimondas Čiegis, Oleg Iliev, and Zhara Lakdawala

Abstract The performance of oil filters used in automotive engines and other areas can be significantly improved using computer simulation as an essential component of the design process. In this chapter, a parallel solver for the 3D simulation of flows through oil filters is presented. The Navier–Stokes–Brinkmann system of equations is used to describe the coupled laminar flow of incompressible isothermal oil through open cavities and cavities with filtering porous media. The space discretization in the complicated filter geometry is based on the finite-volume method.

Two parallel algorithms are developed on the basis of the sequential numerical algorithm. First, a data (domain) decomposition method is used to develop a parallel algorithm, where the data exchange is implemented with MPI library. The domain is partitioned between the processes using METIS library for the partitioning of unstructured graphs. A theoretical model is proposed for the estimation of the complexity of the proposed parallel algorithm. A second parallel algorithm is obtained by the OpenMP parallelization of the linear solver, which takes 90% of the total CPU time. The performance of implementations of both algorithms is studied on multicore computers.

1 Introduction

The power of modern personal computers is increasing constantly, but not enough to fulfill all scientific and engineering computational demands. In such cases,

Vadimas Starikovičius · Raimondas Čiegis
Vilnius Gediminas Technical University, Saulėtekio al. 11, LT-10223 Vilnius, Lithuania
e-mail: vs@sc.vgtu.lt · rc@fm.vgtu.lt

Oleg Iliev · Zhara Lakdawala
Fraunhofer ITWM, Fraunhofer-Platz 1, D-67663 Kaiserslautern, Germany
e-mail: {iliev · lakdawala}@itwm.fhg.de

parallel computing may be the answer. Parallel computing not only gives access to increasing computational resources but it also becomes economically feasible.

Filtering out solid particles from liquid oil is very essential for automotive engines (as well as for many other applications). An oil filter can be described shortly as a filter box (which could be of complicated shape) with inlet/s for dirty oil and outlet/s for filtrated oil. The inlet/s and outlet/s are separated by a filtering medium, which is a single or a multilayer porous media. Optimal shape design for the filter housing, achieving optimal pressure drop–flow rate ratio, etc., requires detailed knowledge about the flow field through the filter.

This chapter aims to discuss parallelization of the existing industrial sequential software SuFiS (Suction Filter Simulation), which is extensively used for the simulation of fluid flows in industrial filters. Two parallel algorithms are developed. First, the domain or data parallelization paradigm [10] is used to build a parallel algorithm and the MPI library [11] to implement the data communication between processes. A second parallel algorithm is obtained by the OpenMP [12] parallelization of the linear solver only, as it takes 90% of the total CPU time.

The rest of the chapter is organized as follows. In Sect. 2, first the differential model is formulated. Next, we describe the fractional time step and the finite volume (FV) discretization used for the numerical solution. And finally, the subgrid method is briefly presented as the latest development of numerical algorithms used in SuFiS. Developed parallel algorithms are described in Sect. 3. A theoretical model, which estimates the complexity of the domain decomposition parallel algorithm, is presented. The performance of two algorithms (approaches) is studied and compared on multicore computers. Some final conclusions are given in Sect. 4.

2 Mathematical Model and Discretization

The Brinkmann model (see, e.g., [2]), describing the flow in porous media, Ω_p, and the Navier–Stokes equations (see, e.g., [7]), describing the flow in the pure fluid region, Ω_f, are reformulated into a single system of PDEs. This single system governs the flow in the pure liquid and in the porous media and satisfies the interface conditions for the continuity of the velocity and the continuity of the normal component of the stress tensor. The Navier–Stokes–Brinkmann system of equations describing laminar, incompressible, and isothermal flow in the whole domain reads (a detailed description of all coefficients is given in [4]):

$$\begin{cases} \underbrace{\frac{\partial \rho \mathbf{u}}{\partial t} + (\rho \mathbf{u}\cdot\nabla)\mathbf{u} - \nabla\cdot(\tilde{\mu}\nabla\mathbf{u})}_{\text{Navier–Stokes}} + \overbrace{\tilde{\mu}\tilde{\mathbf{K}}^{-1}\mathbf{u} + \underbrace{\nabla p = \tilde{\mathbf{f}}}}^{\text{Darcy law}}, \\ \nabla\cdot\mathbf{u} = 0. \end{cases} \tag{1}$$

The tilde-quantities are defined using fictitious region method:

$$\tilde{\mu} = \begin{cases} \mu & \text{in } \Omega_f, \\ \mu_{eff} & \text{in } \Omega_p, \end{cases} \quad \tilde{\mathbf{f}} = \begin{cases} \mathbf{f}_{NS} & \text{in } \Omega_f, \\ \mathbf{f}_B & \text{in } \Omega_p, \end{cases} \quad \tilde{\mathbf{K}}^{-1} = \begin{cases} 0 & \text{in } \Omega_f, \\ \mathbf{K}^{-1} & \text{in } \Omega_p. \end{cases}$$

Here $\mathbf{u}$ and p stand for the velocity vector and the pressure, respectively, and $\rho, \mu, \mathbf{K}$ denote the density, the viscosity, and the permeability tensor of the porous medium, respectively.

No slip boundary conditions are the boundary conditions on the solid wall, given flow rate (i.e., velocity) is the boundary condition at the inflow, and soft boundary conditions are prescribed at the outflow.

2.1 Time Discretization

The choice of the time discretization influences the accuracy of the numerical solution and the stability of the algorithm (e.g., the restrictions on the time step). Let $\mathbf{U}$ and P be the discrete functions of velocity and pressure. We denote the operators corresponding with discretized convective and diffusive terms in the momentum equations by $\mathsf{C}(\mathbf{U})\mathbf{U}$ and $\mathsf{D}\mathbf{U}$, respectively. The particular form of these operators will be discussed below. Further, we denote by G the discretization of the gradient, and by G^T the discretization of the divergence operator. Finally, we denote by $\mathsf{B}\mathbf{U}$ the operator corresponding with Darcy term, namely $\tilde{\mu}\tilde{\mathbf{K}}^{-1}\mathbf{u}$, in the momentum equations. Below, we use the superscript n for the values at the old time level, and $(n+1)$ or no superscript for the values at the new time level. Notation $\mathbf{U}^*$ is used for the prediction of the velocity, τ stands for the time step, $\tau = t^{n+1} - t^n$. The following fractional time step discretization scheme is defined:

$$\begin{aligned} &(\rho\mathbf{U}^* - \rho\mathbf{U}^n) + \tau(\mathsf{C}(\mathbf{U}^n) - \mathsf{D} + \mathsf{B})\ \mathbf{U}^* = \tau\mathsf{G}P^n, \\ &(\rho\mathbf{U}^{n+1} - \rho\mathbf{U}^*) + \tau\left(\mathsf{B}\mathbf{U}^{n+1} - \mathsf{B}\mathbf{U}^*\right) = \tau\left(\mathsf{G}P^{n+1} - \mathsf{G}P^n\right), \\ &\mathsf{G}^T\rho\mathbf{U}^{n+1} = 0. \end{aligned} \tag{2}$$

The pressure correction equation should be defined in a special way in order to take into account the specifics of the flow in the porous media (see a detailed discussion of this topic in [4, 5]):

$$\mathsf{G}^T\left(\mathsf{I} + \frac{\tau}{\rho}\mathsf{B}\right)^{-1}\tau\mathsf{G}P_c = -\mathsf{G}^T\rho\mathbf{U}^*, \tag{3}$$

here $P_c = P^{n+1} - P^n$ is the pressure correction, and I is the 3×3 identity matrix.

After the pressure correction equation is solved, the pressure is updated, $P^{n+1} = P^n + P_c$, and the new velocity is calculated:

$$\rho \mathbf{U}^{n+1} = \rho \mathbf{U}^* + (\mathsf{I} + \frac{\tau}{\rho}\mathsf{B})^{-1} \tau \mathsf{G} P_c. \quad (4)$$

2.2 Finite Volume Discretization in Space

The geometrical information about the computational domain is usually provided in a CAD format. The governing equations are discretized by the finite volume method (see [7]) on the generated Cartesian grid. Cell-centered grid with collocated arrangement of the velocity and pressure is used. The Rhie–Chow interpolation is used to avoid the spurious oscillations, which could appear due to the collocated arrangement of the unknowns.

The discretization of the convective and the diffusive (viscous) terms in the pure fluid region is done by well-known schemes (details can be found in [6, 7, 8]). Special attention is paid to discretization near the interfaces between the fluid and the porous medium. Conservativity of the discretization is achieved here by choosing finite volume method as a discretization technique.

A description of the sequential SuFiS numerical algorithm is given in Algorithm 1. During each iteration at steps (6) and (7), four systems of linear equations are solved. For many problems, this part of the algorithm requires 90% of total CPU time. We use the BiCGSTAB algorithm, which solves the non-symmetric linear system $\mathsf{A}\mathbf{x} = \mathbf{f}$ by using the preconditioned BiConjugate Gradient Stabilized method [1].

Algorithm 1 Sequential SuFiS numerical algorithm.

Sufis ()
begin
(1) Initiate computational domain
(2) ok = true; k = 0;
(3) **while** (ok) **do**
(4) k = k+1; $U^k = U^{k-1}$; $P^k = P^{k-1}$;
(5) **for** (i=0; i < MaxNonLinearIter; i++) **do**
(6) Compute velocities from momentum equations
$Q_j U_j^* = F_j^i - \partial_{x_j} P^k, \quad j = 1,2,3;$
(7) Solve equation for the pressure correction
$L_h P_c = R^i;$
(8) Correct the velocities
$U_j^k = U_j^* + \alpha_u (D^{-1} \nabla_h P_c)_j, \quad j = 1,2,3;$
(9) Correct the pressure
$P^k = P^k + \alpha_p P_c;$
end do
(10) **if** (final time step) ok = false;
end do
end

2.3 Subgrid Approach

The latest developments of the numerical solution algorithms in SuFiS include the subgrid approach. The subgrid approach is widely used in modeling turbulent flows, aiming to provide the current grid discretization with information about the unresolved by the grid finer scale vortices. The idea here is to apply similar approach in solving laminar incompressible Navier–Stokes–Brinkman equations in complex domains. If solving on a very fine grid is not possible due to memory and CPU time restrictions, the subgrid approach still allows one to capture some fine grid details when solving on a coarser grid. Numerical upscaling approach is used for this purpose. The domain is resolved by a fine grid, which captures the geometrical details, but is too big for solving the problem there, and by a coarse grid, where the problem can be solved. A special procedure selects those coarse cells, which contain complicated fine scale geometry. For example, coarse cells that are completely occupied by fine fluid cells are not marked, whereas coarse cells that contain mixture of fluid, solid, and porous fine cells are marked. For all the marked coarse cells, an auxiliary problem on fine grid is solved locally. The formulation of the problem comes from the homogenization theory, where auxiliary cell problems for Stokes problem are solved in order to calculate Darcy permeability at a macro scale. In this way, we calculate permeabilities for all the marked coarse cells, and solve Navier–Stokes–Brinkman system of equations on the coarse scale. The use of such effective permeability tensors allows us to get much more precise results compared with solving the original system on the coarse grid. The first tests on the model problems approved those expectations.

Similar approach is used in [3]. However, the authors solve only for the pressure on the coarse grid, while for the velocity they solve on the fine grid. In our case, this is not possible due to the size of the industrial problems we are solving. Moreover, luckily in our case this is not needed, because our main aim is to calculate accurately the pressure drop through the filter, and for some of the applications the details of the velocity are not important. It should be noted that for problems where the details of the solution are important, we plan to modify our approach so that it works as multilevel algorithm with adaptive local refinement. This means that the pressure field on the coarse scale will be used to locally recover the velocities, and this will be done iteratively with the upscaling procedure.

3 Parallel Algorithms

Next we consider two types of parallel algorithms for parallelization of the sequential SuFiS algorithm.

3.1 DD Parallel Algorithm

The first one is based on the domain (data) decomposition (DD) method. The Navier–Stokes–Brinkmann system of equations (1) is solved in a complicated 3D region. A discrete grid is described as a general non-structured set of finite-volumes. The goal of the DD method is to define a suitable mapping of all finite-volumes V to the set of p processes

$$V = V_1 \cup V_2 \cup \ldots \cup V_p,$$

where V_j defines the elements mapped to jth process. The load balancing problem should be solved during the implementation of this step. First, it is aimed that each process has about the same number of elements, as this number would define the computational complexity for all parts of the SuFiS algorithm.

Due to the stencil of discretization, the computational domains of processes overlap. The information belonging to the overlapped regions should be exchanged between processes. This is done by the additional halo layers of so-called ghost-cells. The time costs of such data exchanges are contributing to the additional costs of the parallel algorithm. Thus, a second goal of defining the optimal data mapping is to minimize the overlapping regions. In our parallel solver, we are using the multilevel partitioning method from METIS software library [9]. This is one of the most efficient partitioning heuristics having a linear time complexity. The communication layer of the DD parallel algorithm is implemented using MPI library.

Next, we estimate the complexity of the parallel algorithm. First, we note that initialization costs do not depend on the number of iterations and the number of time steps. Therefore, they can be neglected for problems where a long transition time is simulated. The matrices and right-hand-side vectors are assembled element by element. This is done locally by each process. The time required to calculate all coefficients of the discrete problem is given by $W_{p,coeff} = c_1 n/p$, here n is the number of elements in the grid.

All ghost values of the vectors belonging to overlapping regions are exchanged between processes. The data communication is implemented by an *odd–even* type algorithm and can be done in parallel between different pairs of processes. Thus, we can estimate the costs of data exchange operation as $W_{exch} = \alpha + \beta m$, where m is the number of items sent between two processes, α is the message startup time, and β is the time required to send one element of data.

The sequential BiCGSTAB algorithm is modified to the parallel version in a way such that its convergence properties are not changed during the parallelization process. The only exceptions are due to inevitable round-off errors and implementation of the preconditioner $\mathbf{B}$. In the parallel algorithm, we use a block version of the Gauss–Seidel preconditioner, when each process computes $\mathbf{B}^{-1}$ by using only a local part of matrix $\mathbf{A}$.

Four different operations of the BiCGSTAB algorithm require different data communications between processes. Computation of vector *saxpy* ($\mathbf{y} = \alpha \mathbf{x} + \mathbf{y}$) operation is estimated by $W_{p,saxpy} = c_2 n/p$ (i.e., no communications), computation of matrix-vector multiplication by

$$W_{p,mv} = \frac{c_3 n}{p} + 2(\alpha + \beta m(p)),$$

and computation of inner product or norm by

$$W_{p,dot} = \frac{c_4 n}{p} + R(p)(\alpha + \beta).$$

In the latter, MPI Allreduce function is used. The computation of the preconditioner **B** and application of this preconditioner is done locally by each process without any communication operation; the cost of this step is given by $W_{p,D} = c_5 n/p$. Summing up all the estimates, we obtain the theoretical model of the complexity of the parallel SuFiS algorithm:

$$\begin{aligned} W_p =& K\big(c_6 \frac{n}{p} + c_7(\alpha + \beta m(p))\big) \\ &+ N\big(c_8 \frac{n}{p} + c_9 R(p)(\alpha + \beta) + c_{10}(\alpha + \beta m(p))\big), \end{aligned} \tag{5}$$

where K is the number of steps in the outer loop of SuFiS algorithm, and N is a total number of BiCGSTAB iterations. The theoretical complexity model presented above was tested experimentally by solving one real industrial application of the oil filters and showed good accuracy [4].

In this work, we are presenting the results of restructured parallel solver implementing DD algorithm. The new data structures have allowed us to significantly reduce the memory requirements of the solver, mainly due to removal of auxiliary 3D structured grid, which was used before as a reference grid [4]. Our test problem with a real industrial geometry has 483874 finite volumes. The maximum number of BiCGSTAB iterations is taken equal to 600. It is not sufficient for the full convergence of the pressure correction equation, but as we are interested to find only a stationary solution, such strategy is efficient solving industrial applications. This also means that the parallel linear solver for pressure correction equation in all experiments is doing the same number of iterations despite the possible differences in the convergence. Only three time steps of SuFiS algorithm are computed in all experiments to get the performance of parallel algorithms.

First, the performance of DD parallel algorithm was tested on distributed memory parallel machine (Vilkas cluster at VGTU). It consists of Pentium 4 processors (3.2GHz, level 1 cache 16KB, level 2 cache 1MB, 800MHz FSB) interconnected via Gigabit Smart Switch (http://vilkas.vgtu.lt). Obtained performance results are presented in Table 1. Here for each number of processes p, the wall clock time T_p, the values of the algorithmic speed-up coefficient $S_p = T_1/T_p$, and efficiency $E_p = S_p/p$ are presented.

Table 1 Performance results of DD parallel algorithm on VGTU Vilkas cluster.

	$p=1$	$p=2$	$p=4$	$p=8$	$p=12$	$p=16$
T_p	456	234	130	76.9	53.8	41.8
S_p	1.0	1.95	3.51	5.93	8.48	10.9
E_p	1.0	0.97	0.88	0.74	0.71	0.68

Table 2 Performance results of DD parallel algorithm on ITWM Hercules cluster (one process per node).

	$p=1$	$p=2$	$p=4$	$p=8$	$p=12$	$p=16$
T_p	335.3	164.0	79.1	37.4	23.9	17.6
S_p	1.0	2.04	4.24	8.96	14.3	19.05
E_p	1.0	1.02	1.06	1.12	1.17	1.19

As we can see, the scalability of the DD parallel algorithm is robust. According to the theoretical model of its complexity (5), it scales better for the systems with better interconnect network: with smaller α, β, $R(p)$. This can be seen from the results in Table 2, which were obtained on ITWM Hercules cluster: dual nodes (PowerEdge 1950) with dual core Intel Xeon (Woodcrest) processors (2.3GHz, L1 32+32 KB, L2 4MB, 1333MHz FSB) interconnected with Infiniband DDR. The superlinear speedup is explained also by the growing number of cache hits with increasing p.

Next, we have performed performance tests of the DD parallel algorithm on the shared memory parallel machine - multicore computer. The system has Intel(R) Core(TM)2 Quad processor Q6600. Four processing cores are running at 2.4 GHz each and sharing 8 MB of L2 cache and a 1066 MHz Front Side Bus. Each of the four cores can complete up to four full instructions simultaneously. The performance results are presented in Table 3.

The results of the test-runs show that the speedup of 1.51 is obtained for two processes and it saturates for a larger number of processes; algorithm is not efficient even for 3 or 4 processes. In order to get more information on the run time behavior of the parallel MPI code, we have used a parallel profiling tool: Intel(R) Trace Analyzer and Collector. It showed that for $p=2$, all MPI functions took 3.05 seconds (1.00 and 2.05 seconds for first and second processes accordingly), and for $p=4$ processes 8.93 seconds (1.90, 2.50, 1.17, 3.36 s). Thus the communication part of the algorithm is implemented very efficiently. It scales – grows (as in our theoretical model), but not linearly. Because the load balance of the data is also equal to one, the bottleneck of the algorithm can arise only due to the conflicts in memory access between different processes.

The same tests on the better shared memory architecture – single node of ITWM Hercules cluster (2x2 cores) – gave us slightly better numbers, but qualitatively the

Table 3 Performance results of DD parallel algorithm on the multicore computer: Intel Core 2 Quad processor Q6600 @ 2.4GHz.

	$p=1$	$p=2$	$p=3$	$p=4$
T_p	281.9	186.3	183.6	176.3
S_p	1.0	1.51	1.54	1.60
E_p	1.0	0.76	0.51	0.40

Table 4 Performance results of DD parallel algorithm on the multicore computer: ITWM Hercules cluster's single node (PowerEdge 1950).

	$p = 1$	$p = 2$	$p = 4$
T_p	335.3	185.9	153.2
S_p	1.0	1.80	2.19
E_p	1.0	0.90	0.55

same picture. From Table 4, we can see that the use of all 4 cores on the node is not efficient. The run-time is almost the same as using two separate nodes with 2 processes (Table 2).

3.2 OpenMP Parallel Algorithm

The second approach is to use OpenMP application program interface to see if our analysis is correct or we can get something better with special programming tools for shared memory computers. As stated above, the solver for the linear system of equations typically requires up to 90% of total computation time. Thus there is a possibility to parallelize only the linear system solver: BiCGStab routine. Parallel algorithm is obtained quite easily by putting special directives in saxpy, dot product, matrix–vector multiplication, and preconditioner operations. Then the discretization is done sequentially only on the master thread while the linear systems of equations are solved in parallel.

The first important result following from computational experiments is that the direct application of the OpenMP version of the parallel algorithm is impossible, as the asynchronous block version of the Gauss–Seidel preconditioner is not giving a converging iterations. Thus we have used a diagonal preconditioner in all computational experiments presented in this subsection. In Table 5, we compare the results obtained using the DD parallel algorithm with MPI and OpenMP parallel algorithm. As one can see, we got the same picture: we obtain a reasonable speed-up only for 2 processes. The use of 3 and 4 processes is inefficient on the multicore computer used in our tests.

Table 5 Performance results of DD and OpenMP parallel algorithms with the diagonal preconditioner on the multicore computer: Intel Core 2 Quad processor Q6600 @ 2.4GHz.

		$p = 1$	$p = 2$	$p = 3$	$p = 4$
	T_p	198.0	139.9	138.4	139.1
DD MPI	S_p	1.0	1.42	1.43	1.42
	E_p	1.0	0.71	0.72	0.71
	T_p	202.3	155.0	155.3	151.9
OpenMP	S_p	1.0	1.31	1.31	1.33
	E_p	1.0	0.65	0.65	0.66

Table 6 Profiling results of the OpenMP parallel algorithm on the multicore computer: Intel Core 2 Quad processor Q6600 @ 2.4GHz.

Section	$p = 1$	$p = 2$	$p = 3$	$p = 4$
saxpy	38.6	31.3	31.8	32.1
dot	17.8	13.6	13.3	12.9
mult	80.0	48.1	46.4	43.8
precond	47.1	41.8	43.2	42.9

Next, we present the information collected by the profiling of the OpenMP parallel algorithm. Our goal was to see and to compare the performance of parallelization of different sections of the linear solver. In Table 6, the summary of execution times of four constituent parts of the algorithm are given. Here *saxpy* denotes the CPU time of all saxpy type operations, *dot* is the overall time of all dot products and norm computations, *mult* denotes the CPU time of all matrix–vector multiplications, and *precond* is the time spent in the implementation of the diagonal preconditioner.

It follows from the results in Table 6 that the considerable speedup is achieved only for the matrix–vector multiplication part, but even here the speedup is significant only for two processes. This again can be explained by the memory access bottleneck. The ratio of computations to data movement is better in matrix–vector multiplication, therefore we get better parallelization for that operation.

4 Conclusions

The results of the performance tests show that fine-tuned MPI implementation of domain decomposition parallel algorithm for SuFiS (Suction Filter Simulation) software can perform very well not only on distributed memory systems but is also difficult to outperform on shared memory systems. MPI communication inside the shared memory is well optimized in current implementations of MPI libraries.

The problems with scalability on some of the shared memory systems (e.g., multicore computers) are caused not by MPI but by hardware architecture and cannot be overcome with the use of simple shared memory programming tools.

Acknowledgments R. Čiegis and V. Starikovičius were supported by the Lithuanian State Science and Studies Foundation within the project B-03/2007 on "Global optimization of complex systems using high performance computing and GRID technologies".

References

1. Barrett, R., Berry, M., Chan, T.F., Demmel, J., Donato, J., Dongarra, J., Eijkhout, V., Pozo, R., Romine, C., der Vorst, H.V.: Templates for the Solution of Linear Systems: Building Blocks for Iterative Methods, 2nd Edition. Kluwer Academic Publishers, Boston (1990)

2. Bear, J., Bachmat, Y.: Introduction to Modeling of Transport Phenomena in Porous Media. SIAM, Philadelphia, PA (1994)
3. Bonfigli, G., Lunati, I., Jenny, P.: Multi-scale finite-volume method for incompressible flows. URL http://www.ifd.mavt.ethz.ch/research/group_pj/projects/project_msfv_ns/
4. Čiegis, R., Iliev, O., Lakdawala, Z.: On parallel numerical algorithms for simulating industrial filtration problems. Computational Methods in Applied Mathematics **7**(2), 118–134 (2007)
5. Čiegis, R., Iliev, O., Starikovičius, V., Steiner, K.: Numerical algorithms for solving problems of multiphase flows in porous media. Mathematical Modelling and Analysis **11**(2), 133–148 (2006)
6. Ferziger, J.H., Peric, M.: Computational Methods for Fluid Dynamics. Springer-Verlag, Berlin (1999)
7. Fletcher, C.A.J.: Computational Techniques for Fluid Dynamics. Springer-Verlag, Berlin (1991)
8. Iliev, O., Laptev, V.: On numerical simulation of flow through oil filters. Comput. Vis. Sci. **6**, 139–146 (2004)
9. Karypis, G., Kumar, V.: A fast and high quality multilevel scheme for partitioning irregular graphs. SIAM Journal on Scientific Computing **20**(1), 359–392 (1999)
10. Kumar, V., Grama, A., Gupta, A., Karypis, G.: Introduction to Parallel Computing: Design and Analysis of Algorithms. Benjamin/Cummings, Redwood City, CA (1994)
11. MPI: A message-passing interface standard. URL http://www.mpi-forum.org/
12. OpenMP: Application program interface. URL http://www.openmp.org

High-Performance Computing in Jet Aerodynamics

Simon Eastwood, Paul Tucker, and Hao Xia

Abstract Reducing the noise generated by the propulsive jet of an aircraft engine is of great environmental importance. The 'jet noise' is generated by complex turbulent interactions that are demanding to capture numerically, requiring fine spatial and temporal resolution. The use of high-performance computing facilities is essential, allowing detailed flow studies to be carried out that help to disentangle the effects of numerics from flow physics. The scalability and efficiency of algorithms and different codes are also important and are considered in the context of the physical problem being investigated.

1 Introduction

A dramatic increase in air transport is predicted in the next 20 years. Due to this, urgent steps must be taken to reduce the noise from aircraft, and in particular the jet engine. The propulsive jet of a modern turbofan engine produces a substantial part of the engine noise at take-off. The ability to accurately predict this noise and develop methods of controlling it are highly desirable for the aircraft industry. The 'jet noise' is generated by complex turbulent interactions that are demanding to capture numerically. Secundov et al. [16] outline the difficult task of using Reynolds Averaged Navier Stokes (RANS) equations to model the substantially different flow physics of isothermal, non-isothermal, subsonic and supersonic jets for a range of complex nozzle geometries. The surprising lack of understanding of the flow physics of even basic jets is outlined.

In contrast with the problems of RANS, Large Eddy Simulation (LES), even on quite coarse grids [17], can predict the correct trends. As discussed by Pope [14], a compelling case for the use of LES can be made for momentum, heat, and mass

Simon Eastwood · Paul Tucker · Hao Xia
Whittle Laboratory, University of Cambridge, UK
e-mail: se282@cam.ac.uk

transfer in free shear flows. In this case, the transport processes of interest are effected by the resolved large-scale motions with a cascade of energy from the resolved large scales to the statistically isotropic and universal small scales. As noted by Suzuki and Colonius [5], in jet shear layers the small scales have little influence on the large, suggesting that the modeling of back scatter, as facilitated by more advanced LES models, is relatively unimportant. In a turbulent jet, most downstream noise is generated by large structures at the tip of the potential core. There are therefore strong reasons to expect LES to be successful in capturing the major noise sources.

This work uses LES in conjunction with high-performance computing to investigate the jet noise problem. For LES, different forms of the Navier–Stokes equations can be discretized in many ways giving numerical traits and contamination. Once an appropriate form has been found, there is a vast array of temporal and spatial discretizations. Other possibilities such as staggered grids, cell centered or cell vertex codes, codes with explicit smoothers, mesh singularity treatments, and boundary conditions all have significant influence on the solution. With jets, the far field boundary conditions have an especially strong influence. The discretizations must also be represented on a grid where flow non-alignment with the grid and cell skewing can produce further dissipation. Spatial numerical influences can even be non-vanishing with grid refinement [4, 7]. The use of a subgrid scale model for LES is also problematic, with a significant body of literature suggesting that disentangling numerical influences from those of the subgrid scale model is difficult. For most industrially relevant LES simulations, the numerics will be playing a significant role that will cloud the subgrid scale model's role. Clearly, to successfully gain a high-quality LES simulation in conjunction with productive use of high-performance computing facilities is not a trivial task. To explore the use of LES, simulations are made with a range of codes having very different discretizations. For two of the three codes tested, no subgrid scale model is used. The diffusion inherent in the codes is used to drain energy from the larger scales. Following Pope [14], the term Numerical LES (NLES) is used with the implication being that the numerical elements of the scheme are significantly active with no claim being made that the numerical contributions represent an especially good subgrid scale model. The third code is neutrally dissipative hence the use of an LES model is necessary. Here, a Smagorinsky model is used.

Farfield sound and complex geometries are then explored. When modeling complex geometries, a further challenge for LES is near wall modeling. The near wall region contains LES challenging streak like structures. Typically, with LES, grid restrictions mean that the boundary layer momentum thickness is an order of magnitude higher than for experimental setups. Modeling the correct momentum thickness correctly is important as this is what governs the development of the jet. Hence in this work a near wall $k-l$ RANS model is used near to the wall. This approach is successfully used for modeling axisymmetric jet nozzles [6]. With its algebraically prescribed length scale, it is well suited to capturing the law of the wall and importantly the viscous sublayer. As noted by Bodony and Lele [1], capturing the correct initial shear layer thickness is of greater importance than the subgrid scale model used.

2 Numerical Background

2.1 HYDRA and FLUXp

HYDRA uses a cell vertex storage system with an explicit time stepping scheme. The flux at the interface is based on the flux difference ideas of Roe [15] combining central differencing of the non-linear inviscid fluxes with a smoothing flux based on one-dimensional characteristic variables. Normally with Roe's scheme, the flux at a control volume face Φ_f (inviscid flux) is expressed as

$$\Phi_f = \frac{1}{2}((\Phi_L + \Phi_R)) - \frac{1}{2}|A|[\phi_L - \phi_R], \tag{1}$$

where $A = \frac{\partial \Phi}{\partial \phi}$ (ϕ represents primitive variables), Φ_L and Φ_R represent values interpolated to the control volume face based on information to the left and right hand side of the face. Because the reconstruction of Φ_L and Φ_R is an expensive process, following Morgan et al. [11] Φ_L and Φ_R are simply taken as the adjacent nodal values, i.e., $(\Phi_L + \Phi_R)/2$ represents a standard second order central difference. The evaluation of this central difference term is computationally cheap. The smoothing term is also approximated in a second order fashion as

$$\frac{1}{2}|A|[\phi_L - \phi_R] \approx \frac{1}{2}\varepsilon|A|[\tilde{\nabla}_L^2\phi - \tilde{\nabla}_R^2\phi], \tag{2}$$

where $\tilde{\nabla}_L^2$ and $\tilde{\nabla}_R^2$ are undivided Laplacian's evaluated at the node locations L and R. ε is a tunable constant, the standard *HYDRA* value being 0.5. A static, spatially varying ε field is used such that the smoothing is higher toward the farfield boundaries. The term $|A|$ involves differences between the local convection velocity and speed of sound. At high Mach numbers, $|A|$ becomes relatively small hence the smoothing is small, such as in the turbulent region at the jet exit. Smoothing is primarily applied to lend stability to the scheme but also provides the necessary numerical dissipation for NLES.

FLUXp is another second order unstructured solver which is cell-centered and finite volume based. The spatial discretization features linear reconstruction and flux difference splitting. The dual-time integration consists of the physical time part and the pseudo part. The implicit second order backward Euler scheme and the explicit 4-stage Runge-Kutta scheme are applied to them respectively. The flux difference splitting employed is the modified Roe scheme to gain a local solution to the Riemann problem, as in (1).

Here again the parameter ε is tunable and typically ranges from 0 to 1 standing for the pure central difference and the original Roe scheme, respectively. In practice, for the subsonic jet flow ε can be set as low as 0.05. The higher order accuracy is achieved by the linear reconstruction

$$\phi_L = \phi_C + \delta\phi\, dx, \tag{3}$$

where ϕ_C is at the cell center, ϕ_L is at a left face, and $\delta\phi$ is the gradient. The advantage of linear reconstruction is that it takes into account the mesh cell shape. *FLUXp* has been successfully applied to simulate synthetic jet flows [21].

2.2 Boundary and Initial Conditions

A pressure inlet based on a subsonic inflow is set. The flow transitions to turbulence naturally. No perturbations are applied to the inlet flow. Away from the nozzle on the left-hand side of the domain, and on cylindrical boundaries, a free stream boundary condition is used. At the far right-hand boundary a subsonic outflow condition is used. No slip and impermeability conditions are applied on the solid wall boundaries. The far field boundaries have 1-D characteristic based non-reflecting boundary conditions [8]. The Riemann invariants are given by

$$R^{\pm} = u \pm \frac{2}{\gamma - 1} c \tag{4}$$

based on the eigenvalues $\lambda = u + c$ and $\lambda = u - c$. The tangent velocities and entropy are related to three eigenvalues, $\lambda = u$, so that there are five variables in total, corresponding with the number of degrees of freedom in the governing equations. For the subsonic inflow, only R^- is defined from inside the domain with all other variables defined from ambient conditions. For the subsonic outflow, R^- is defined from ambient condition while the other four are obtained from within the domain.

2.3 Ffowcs Williams Hawkings Surface

In order to derive the far field sound, the governing equations could be solved to the far field. However, this would require a large domain and impractical solution times. An alternative is to solve within a domain that includes the acoustic source and the region immediately surrounding that source. At the edge of this domain, pressure information from the flow field is stored. This information is then propagated to the far field in order to calculate the noise. The surface upon which data is stored is called a Ffowcs Williams Hawking (FWH) surface. The approximate position of the FWH surface relative to the jet flow is illustrated by Fig. 1a. The FWH surface should completely surround the turbulent source region and be in irrotational flow. However in a jet, eddies are convected downstream. Hence it becomes impossible

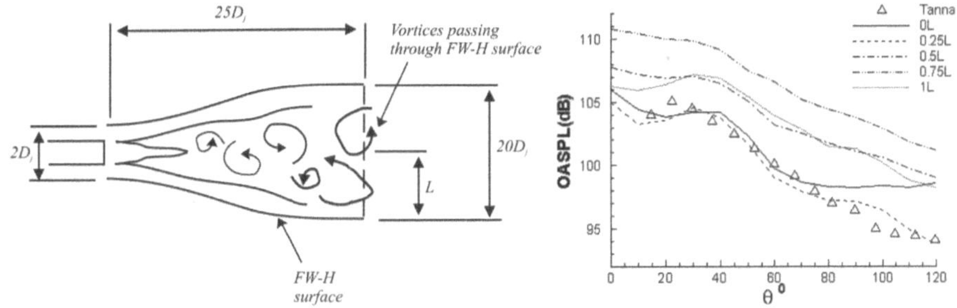

(a) Ffowcs Williams Hawkings surface.

(b) Variation of sound pressure level with Ffowcs Williams Hawkings closing disk.

Fig. 1 Ffowcs Williams Hawkings information.

to have a complete surface that does not cross irrotational flow on the jet centerline. This is referred to as the closing disk problem. By varying the degree to which the closing disk covers the irrotational region, significant variation in far field sound can be gained (up to 20%). This is shown in Fig. 1b, which plots the sound pressure level at $100D$ (D is the jet diameter) for various lengths of closing disk. $L = 1.0$ corresponds to a full closing disk, and $L = 0.0$ corresponds to no closing disk. The results are calculated using *HYDRA*. In the remainder of this work, no closing disk is used so that no element of 'tuning' can be introduced.

3 Computing Facilities

The codes have been run using three machines:

- The University of Wales Swansea, IBM BlueC machine. This comprises 22 IBM p575 servers with 16 Power5 64-bit 1.9GHz processors for each server.
- The University of Cambridge, Darwin machine. The Cambridge High Performance Computing Cluster Darwin was implemented in November 2006 as the largest academic supercomputer in the UK, providing 50% more performance than any other academic machine in the UK. The system has in total 2340 3.0 GHz Intel Woodcrest cores that are provided by 585 dual socket Dell 1950 1U rack mount server nodes giving four cores in total, forming a single SMP unit with 8GB of RAM (2GB per core) and 80GB of local disk.
- The UK's flagship high-performance computing service, HPCx. The HPCx machines comprises 96 IBM POWER5 eserver nodes, corresponding with 1536 processors. Up to 1024 processors are available for one job. Each eserver has 32GB of memory. Access to this machine is provided under the UK Applied Aerodynamics Consortium (UKAAC).

4 Code Parallelization and Scalability

The Rolls-Royce *HYDRA* solver has been parallelized using a method based on the OPlus library originally developed at Oxford University [3]. The aim of the library is to separate the necessary parallel programming from the application programming. A standard sequential FORTRAN DO-loop can be converted to a parallel loop by adding calls to OPLUS routines at the start of the loop. All message passing is handled by the library. Thus a single source FORTRAN code can be developed, debugged, and maintained on a sequential machine and then executed in parallel. This is of great benefit since parallelizing unstructured grid codes is otherwise a time consuming and laborious process. Furthermore, considerable effort can be devoted to optimizing the OPlus library for a variety of machines, using machine-specific low-level communications libraries as necessary. Hills [10] describes further work done to achieve parallel performance. High parallel scalability, (see Fig. 2) is achieved for fully unsteady unstructured mesh cases. Here, the speedup is relative to the performance on two 32 processor nodes and the ideal linear speedup is also shown. *HYDRA* parallelization tests are run on HPCx. Codes that demonstrate excellent scaling properties on HPCx can apply for 'Seals of Approval' that qualify them for discounted use of CPU time. As a result of the scaling performance demonstrated, *HYDRA* was awarded a gold seal of approval [10]. In this work, both *HYDRA* and *FLUXp* have been used for complex geometries. *FLUXp* is also parallelized using MPI. Its performance is illustrated in Fig. 2 and has been tested up to 25 processors on a local cluster [21]. Here the speedup is relative to the serial performance. Hence, *HYDRA* has more promise for massively parallel computing due to the high scalability and development. However, for meshes up to 1×10^7 cells, run on ~ 64 processors, *FLUXp* is useful.

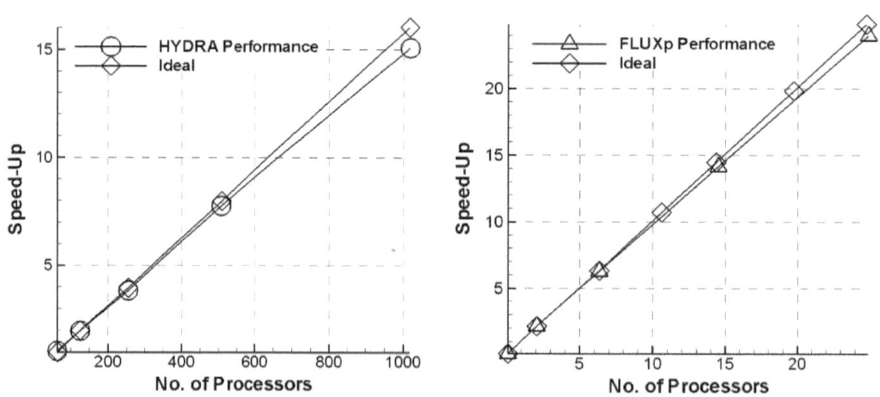

(a) HYDRA parallel performance [10]. (b) FLUXp parallel performance [21].

Fig. 2 Scaling for unsteady unstructured mesh calculation.

5 Axisymmetric Jet Results

5.1 Problem Set Up and Mesh

To test the eddy resolving capabilities of the code, a direct comparison is made with a $Ma = 0.9, Re = 1 \times 10^4$ jet (based on the jet diameter, D, and outlet velocity, U_j). A steady laminar solution was used as an initial guess but prior to this a $k - \varepsilon$ solution was tried. However, the sudden removal of the strong eddy viscosity in the shear layers, when switching to the eddy resolving mode, created a large force imbalance. This resulted in regions of high supersonic flow and numerous shock structures. Hence, the more unstable laminar velocity profiles, which provide a natural route to turbulence transition and ultimate jet development, were considered a better choice. For evolving the flow, the simulation is initially run for $t^* > 300(t^* = tUo/D)$. The flow is then allowed to settle for a dimensionless time period of 150 units before time averaging is started. Hence, here the flow is evolved of about 100 time units. The statistics are then gathered over a period for about 200 time units. Comparison is made with the mean velocity and turbulence statistics measurements of Hussein, Capp, and George [9], Panchapeksen and Lumley [13], Bridges and Wernet [2], and the centerline velocity decay measurements of Morse [12].

Figure 3 shows three views of the axisymmetric mesh which has 5×10^6 cells. Figure 3a shows an isometric view of the mesh while frame (b) shows an x–y view of the mesh. Here, the embedded H-Block mesh is used with *HYDRA* and *FLUXp*. The *VU40* mesh does not have the embedded H-Block, hence there is an axis singularity. The mesh uses hexahedral cells. Previous code studies [20], show that tetrahedral cells do not perform well for wave propagation problems. For *VU40*, the centerline H-Block treatment is not used. For these cases, a velocity profile is imposed at the

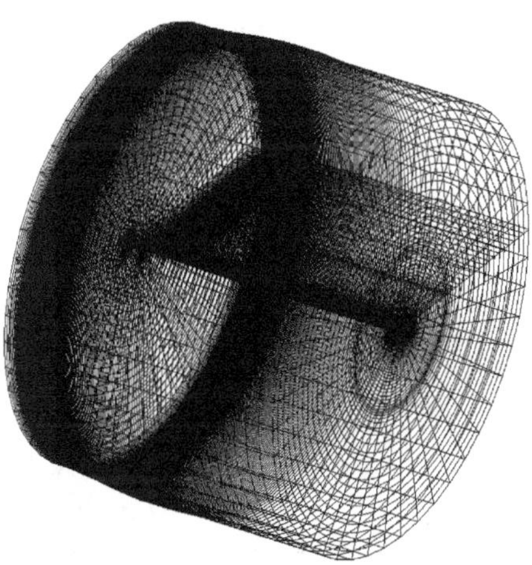

(a) Three-dimensional view of cojenmeshmesh.

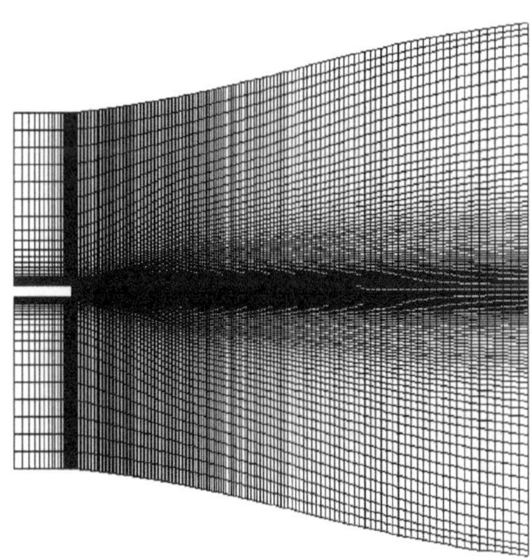

(b) xy plane view of mesh.

Fig. 3 Three dimensional view of mesh.

inlet. By not modeling the nozzle, the variability of near wall performance of the different codes is avoided. This method is used successfully by other workers [17].

5.2 Results

Figure 4 shows the near field results for the three codes. Along with *HYDRA* and *FLUXp*, results for the staggered grid *VU40* [19] code are shown for comparison. Frame (*a*) plots the centerline velocity decay normalized by the jet exit velocity. The results are compared with the measurements of Morse [12] and Bridges and Wernet [2]. All three results lie within the scatter of the measurements. Frames (*b*) and (*c*) plot, respectively, the centerline normal stress and peak shear layer shear stress. The normal stress on the jet centerline at the potential core end ($x/D = 10.0$) is responsible for the low-frequency downstream noise. The shear layer stress is responsible for the higher frequency sideline noise. Hence, for aeroacoustics it is important that these quantities are captured correctly. Frame (*d*) plots the radial profile of the Normal Stress at $x/D = 15.0$. The peak Normal Stress value for *VU40* results seems to be in better agreement with the measurements than *HYDRA* or *FLUXp*. However, results for *HYDRA* and *FLUXp* are still within 20% of the measurements. Frame (*e*) plots the farfield sound at $100D$ for *HYDRA* and *FLUXp* codes.

The measurements of Tanna [18] are also presented. (No closing disk is used, i.e., $L = 0.0$.)

6 Complex Geometries

Figure 5 shows the two complex geometries that are considered. Frame (*a*) shows a co-flowing short cowl nozzle while frames (*b*) and (*c*) show photographs of the single jet flow chevron nozzle. For frame (*b*), the chevrons penetrate the flow at 5^0, whereas for frame (*c*) the chevrons penetrate the flow at 18^0. These are referred to as SMC001 and SMC006, respectively. The co-flowing geometry is run using *HYDRA*, and the chevron mesh results for *FLUXp* are included in the paper.

6.1 Mesh and Initial Conditions

Figure 6, frame (*a*) shows an $x{-}y$ slice through the centerline of the short cowl nozzle domain. Frame (*b*) shows a zoomed view around the nozzle. As with the axisymmetric jet, it is desirable to keep the numerical accuracy high and minimize the degradation of acoustic waves as they pass through the domain. On the centerline, the nozzle is made blunt, as shown in frame (*c*), so that an embedded H-Block can be used. The mesh needs to be fine in the most active turbulent region up to

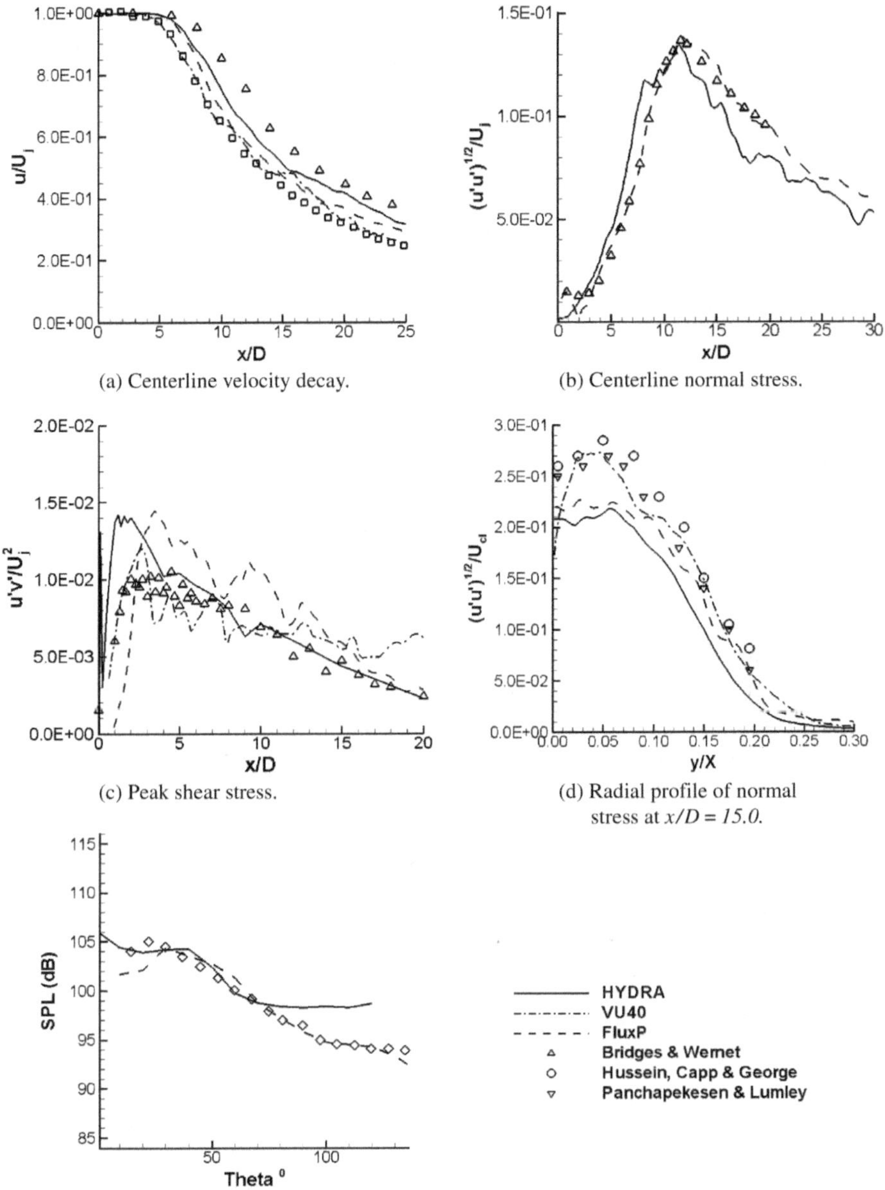

(a) Centerline velocity decay.

(b) Centerline normal stress.

(c) Peak shear stress.

(d) Radial profile of normal stress at $x/D = 15.0$.

(e) Sound pressure level at $100D_j$.

Fig. 4 Axisymmetric jet results.

(a) Short cowl co-flow nozzle. (b) 5° chevron nozzle. (c) 18° chevron nozzle.

Fig. 5 Complex geometries.

the FWH surface. There are also high demands for resolution near the nozzle walls and in the shear layers. Furthermore, a reasonable size far-field mesh (beyond the FWH surface) is required to prevent spurious reflections into the acoustic region. The first off wall grid node is placed at $y^+ = 1$. Pressure is set at the inlet such that the core flow is at $240m/s$ and the co-flow at $220m/s$. The flow is isothermal. At first, NLES is used throughout the domain. Following this, the $k-l$ model is used near the nozzle wall for $y^+ \approx 60$. To improve the solution speed, firstly a coarser mesh (5×10^6 cells) is generated to gain a flow solution. This solution was interpolated onto a 12×10^6 cell mesh using an octree-based search procedure. Typically, for the fine grid, one non-dimensional timestep can be completed in 48 hours, using 64 processors on the University of Wales, Swansea, IBM BlueC machine.

Figure 7 shows the chevron nozzle mesh which also contains 12×10^6 cells. Again, the first off wall grid node is placed at $y^+ = 1$ with the RANS-NLES interface being at $y^+ \approx 60$. Frame (a) shows the full domain while frame (b) shows the near nozzle region. Frame (c) plots an xz profile of the mesh on the centerline. As with the other meshes, an embedded H-Block is used. The velocity inlet is $284m/s(Ma = 0.9)$ and the flow is isothermal.

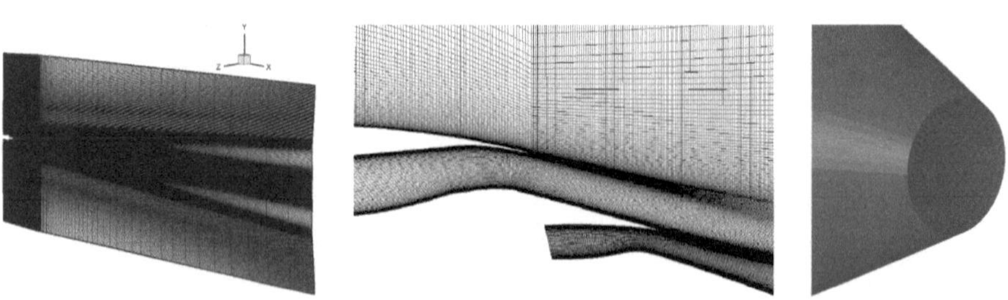

(a) xy plane of full domain. (b) xy plane of nozzle. (c) Blunt edge on jet centerline.

Fig. 6 Short cowl co-flowing jet nozzle.

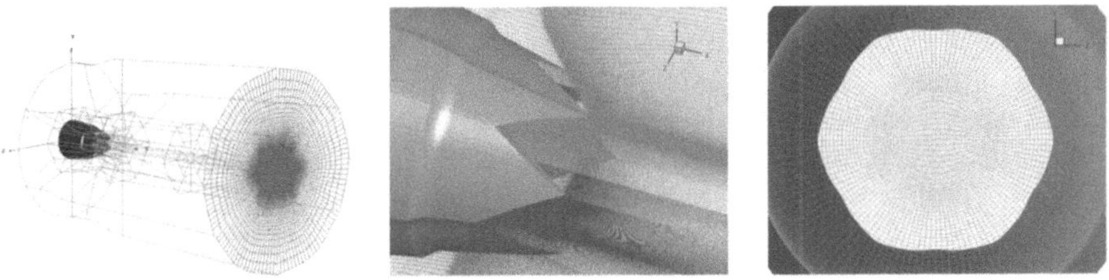

(a) Full domain. (b) Near nozzle region of mesh. (c) xy view of mesh centerline.

Fig. 7 Chevron mesh.

6.2 Results

Figure 8 presents results for the co-flowing nozzle calculated using *HYDRA*. Plotted are radial profiles of streamwise velocity at $x/D = 1.0, 2.0, 4.0$ and 10.0. Solid lines show the NLES results, while the symbols are PIV results from the University of Warwick [6]. The rig used here is small ($D = 0.018M$) such that the Reynolds number (300,000) and hence momentum thickness of the nozzle can perhaps be matched between the NLES and experiment. The results are encouraging although it is evident that at $x/D = 10.0$, the results have not been averaged for long enough. Hence turbulence statistics, though available, are not particularly revealing yet. Frame (*e*) shows isosurfaces of density. Here, the outer shear layer between the co-flow and ambient flow generates toroidal vortex structures, which are visualized. Frame (*f*)

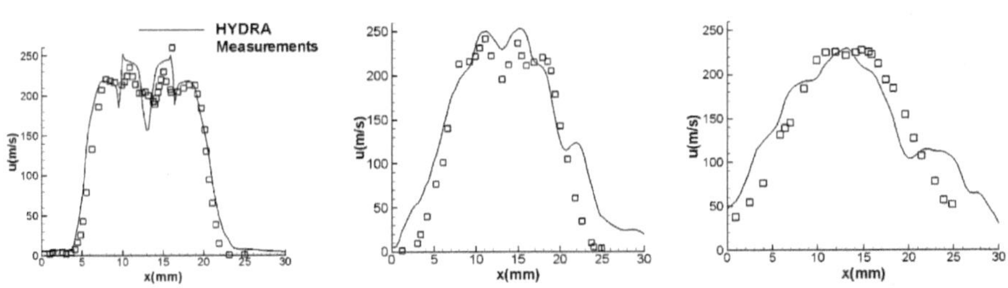

(a) $x/D = 1.0$. (b) $x/D = 2.0$. (c) $x/D = 4.0$.

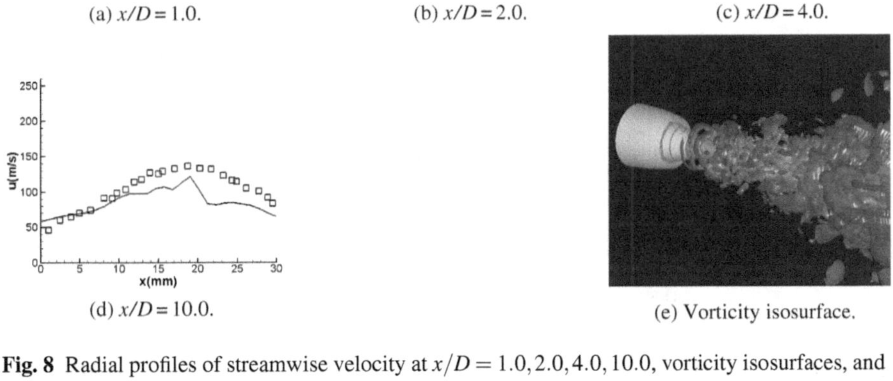

(d) $x/D = 10.0$. (e) Vorticity isosurface.

Fig. 8 Radial profiles of streamwise velocity at $x/D = 1.0, 2.0, 4.0, 10.0$, vorticity isosurfaces, and instantaneous velocity contours.

shows instantaneous streamwise velocity contours, where a faster potential core region can be seen. Due to the similar velocities of the core and co-flows, the jet development is similar to that of a single flow jet.

Instantaneous velocity contours are presented in Fig. 9.

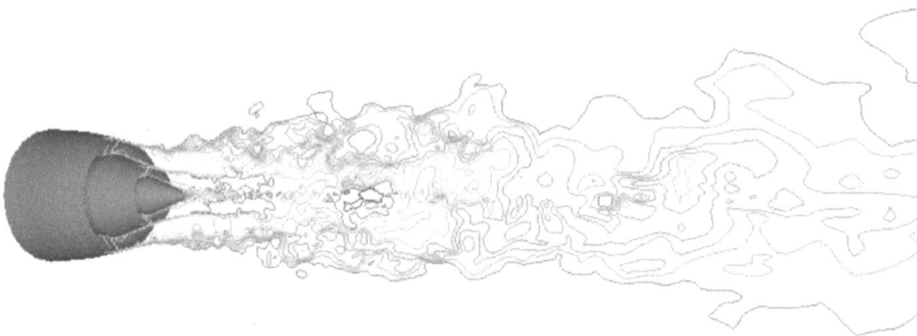

Fig. 9 Instantaneous velocity contours.

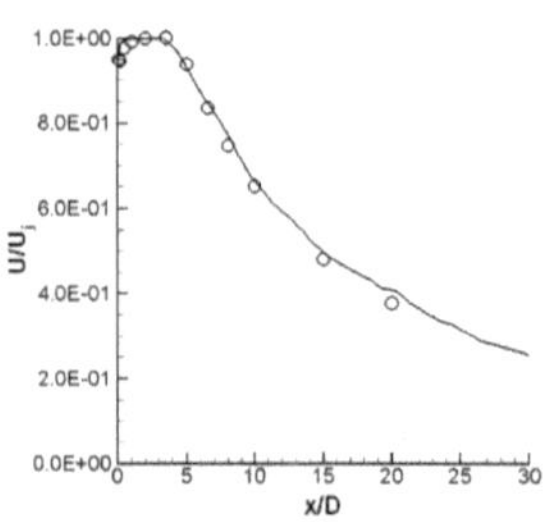

(a) Centerline velocity decay.

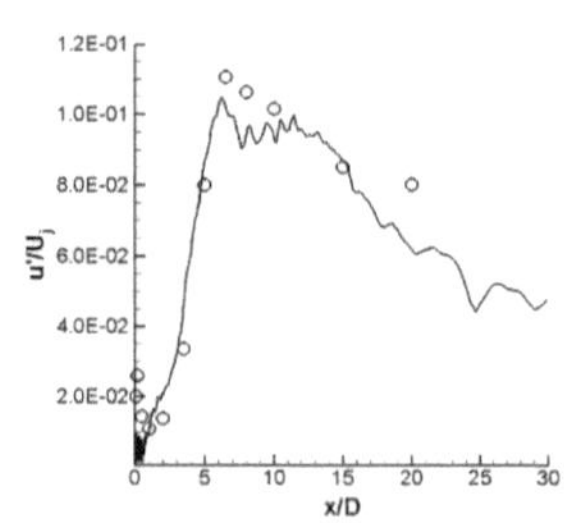

(b) Centerline normal stress.

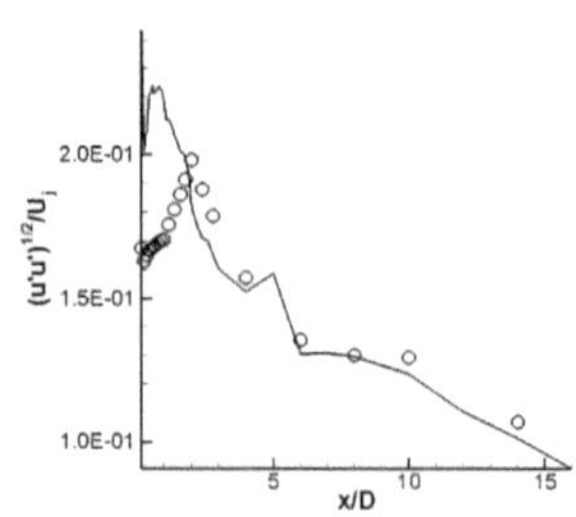

(c) Peak normal stress in shear layer.

Fig. 10 Chevron mesh results.

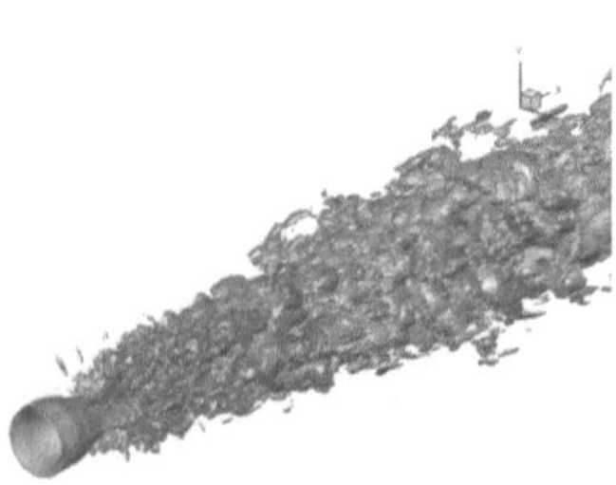

(a) Vorticity isosurfaces.

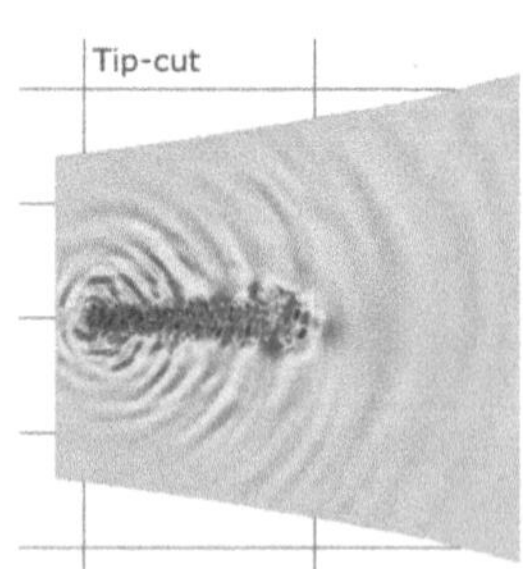

(b) Tip cut of pressure–time contours.

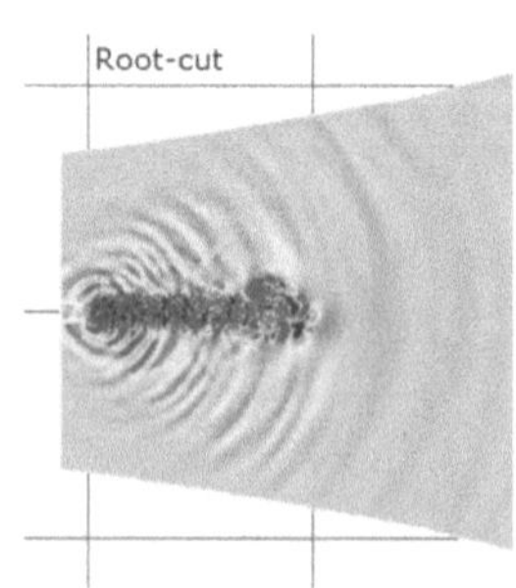

(c) Root cut of pressure–time contours.

Fig. 11 Chevron mesh results. (a) Vorticity isosurfaces, (b) *xy* plane at chevron tip of pressure–time contours, (c) *xy* plane at chevron root of pressure–time contours.

Figure 10 presents results for the SMC006 chevron nozzle (18^0) using the *FLUXp* code. The NLES results are represented by the solid lines and the PIV measurements by the symbols. Frame (*a*) plots the centerline velocity decay normalized by the jet exit velocity. Frame (*b*) plots the centerline normal stress, and frame (*c*) plots the shear layer normal stress.

Figure 11 shows isosurfaces of vorticity to help visualize the flow. Frames (*b*) and (*c*) plot tip and root cut *xy* contours of pressure time derivative. The propagation of acoustic waves to the boundary without reflection can be seen.

7 Conclusions

Jet noise is generated by complex turbulent interactions that are demanding to capture numerically. Large Eddy Simulation (LES) related techniques are attractive for solving such turbulent flows but require high-performance computing. For successful results, careful attention must be paid to boundary and initial conditions, mesh quality, numerical influences, subgrid scale modeling, and discretization types. With such attention, results show that industrial codes that tend to produce an excess of dissipation can be used successfully to predict turbulent flows and associated acoustic fields for single flow axisymmetric jets, chevron nozzles, and co-flowing jets. Using the OPlus environment, almost linear scalability on up to 1000 processors will allow massively parallel computation on meshes up to 50×10^6 cells.

Acknowledgments The authors would like to acknowledge RollsRoyce for the use of *HYDRA* and associated support, Dr. Nick Hills of Surrey University for the information regarding the parallel performance of *HYDRA*, the computing support from Cambridge University, University of Wales, Swansea, and computing time on HPCx under the Applied Aerodynamics Consortium.

References

1. Bodony, D., Lele, S.: Review of the current status of jet noise predictions using Large Eddy Simulation. In: 44th AIAA Aerospace Science Meeting & Exhibit, Reno, Nevada, Paper No. 200648 (2006)
2. Bridges, J., Wernet, M.: Measurements of aeroacoustic sound sources in hot jets. AIAA 2003-3131 9th AIAA Aeroacoustics Conference, Hilton Head, South Carolina, 12-14 May (2003)
3. Burgess, D., Crumpton, P., Giles, M.: A parallel framework for unstructured grid solvers. In: K.M. Decker, R.M. Rehmann (eds.) Programming Environments for Massively Parallel Distribued Systems, pp. 97–106. Birkhauser (1994)
4. Chow, F. Moin, P.: A further study of numerical errors in large eddy simulations. Journal of Computational Physics **184**, 366–380 (2003)
5. Colonius, T., Lele, S.K., Moin, P.: The scattering of sound waves by a vortex: numerical simulations and analytical solutions. J. Fluid Mechanics **260**, 271–298 (1994)
6. Eastwood, S., Tucker, P., Xia, H., Carpenter, P., Dunkley P.: Comparison of LES to PIV and LDA measurement of a small scale high speed coflowing jet. In: AIAA-2008-0010, 46th AIAA Aerospace Sciences Meeting and Exhibit (2008)

7. Ghosal, S.: An analysis of numerical errors in large eddy simulations of turbulence. Journal of Computational Physics **125**, 187–206 (2002)
8. Giles, M.: Nonreflecting boundary conditions for Euler equation calculations. AIAA Journal **28**(12), 2050–2058 (1990)
9. Hussein, H.J., Capp, S.P., George, S.K.: Velocity measurements in a high-Reynolds-number, momentum-conserving, axisymmetric turbulent jet. J. Fluid Mechanics **258**, 31–75 (1994)
10. Hills, N.: Achieving high parallel performance for an unstructured unsteady turbomachinery CFD code. The Aeronautical Journal, UK Applied Aerodynamics Consortium Special Edition **111**, 185–193 (2007)
11. Morgan, K., Peraire, J., Hassan, O.: The computation of three dimensional flows using unstructured grids. Computer Methods in Applied Mechanics and Engineering **87**, 335–352 (1991)
12. Morse, A.P.: Axisymmetric turbulent shear flows with and without swirl. Ph.D. thesis, University of London (1980)
13. Panchapakesan, N.R., Lumley, J.L.: Turbulence measurements in axisymmetric jets of air and helium, Part I, Air Jet. J. Fluid Mechanics **246**, 197–223 (1993)
14. Pope, S.: Ten questions concerning the Large Eddy Simulation of turbulent jets. New Journal of Physics (2004). DOI 10.1088/1367-2630/6/1/035
15. Roe, P.: Approximate Riemann solvers, parameter vectors and difference schemes. Journal of Computational Physics **43**, 357–372 (1981)
16. Secundov, A., Birch, S., Tucker, P.: Propulsive jets and their acoustics. Philosophical Transactions of the Royal Society **365**, 2443–2467 (2007)
17. Shur, M.L., Spalart, P.R., Strelets, M., Travin, A.K.: Towards the prediction of noise from jet engines. Int. J. of Heat and Fluid Flow **24**, 551–561 (2003)
18. Tanna, H.: An experimental study of jet noise, Part 1: turbulent mixing noise. Journal of Sound and Vibration **50**(3), 405–428 (1977)
19. Tucker, P.: Computation of Unsteady Internal Flows. Kluwer Academic Publishers (2001)
20. Tucker, P., Coupland, J., Eastwood, S., Xia, H., Liu, Y., Loveday, R., Hassan, O.: Contrasting code performance for computational aeroacoustics of jets. In: 12th AIAA Aeroacoustics Conference, 8-12th May 2006, Cambridge, MA, USA (2006)
21. Xia, H.: Dynamic grid Detached-Eddy Simulation for synthetic jet flows. Ph.D. thesis, Univ. of Sheffield, UK (2005)

Parallel Numerical Solver for the Simulation of the Heat Conduction in Electrical Cables

Gerda Jankevičiūtė and Raimondas Čiegis

Abstract The modeling of the heat conduction in electrical cables is a complex mathematical problem. To get a quantitative description of the thermo-electrical characteristics in the electrical cables, one requires a mathematical model for it. In this chapter, we develop parallel numerical algorithms for the heat transfer simulation in cable bundles. They are implemented using MPI and targeted for distributed memory computers, including clusters of PCs. The results of simulations of two-dimensional heat transfer models are presented.

1 Introduction

In modern cars, electrical and electronic equipment is of great importance. For engineers the main aim is to determine optimal conductor cross-sections in electro cable bundles in order to minimize the total weight of cables. To get a quantitative description of the thermo-electrical characteristics in the electrical cables, one requires a mathematical model for it. It must involve the different physical phenomena occurring in the electrical cables, i.e., heat conduction, convection and radiation effects, description of heat sources due to current transitions, etc.

The aim of this chapter is develop robust and efficient parallel numerical methods for solution of the problem of heat conduction in electrical cables. Such parallel solver of a direct problem is an essential tool for solving two important optimization problems:

- fitting of the mathematical model to the experimental data;
- determination of the optimal cross-sections of wires in the electrical cables,

when the direct problem is solved many times in the minimization procedure.

Gerda Jankevičiūtė · Raimondas Čiegis
Vilnius Gediminas Technical University, Saulėtekio al. 11, LT–10223, Vilnius, Lithuania
e-mail: {Gerda.Jankeviciute · rc}@fm.vgtu.lt

The rest of the chapter is organized as follows. In Sect. 2, we first formulate the problem. Then the finite volume (FV) discretization method used for solving the flow problem is described. A parallel algorithm is described in Sect. 3, which is based on the parallel domain (data) decomposition method. MPI library is used to implement the data communication part of the algorithm. A theoretical model, which estimates the complexity of the parallel algorithm, is proposed. The results of computational experiments corresponding with a cluster of workstations are presented and the efficiency of the parallel algorithm is investigated. The computational results of experiments are presented, where some industrial cables are simulated. Some final conclusions are given in Sect. 4.

2 The Model of Heat Conduction in Electrical Cables and Discretization

In domain $D \times (0, t_F]$, where

$$D = \{X = (x_1, x_2): \quad x_1^2 + x_2^2 \leq R^2\},$$

we solve the nonlinear non-stationary problem, which describes a distribution of the temperature $T(X,t)$ in electrical cable. The mathematical model consists of the parabolic differential equation [1]:

$$\rho(X)c(X,T)\frac{\partial T}{\partial t} = \sum_{i=1}^{2} \frac{\partial}{\partial x_i}\left(k(X)\frac{\partial T}{\partial x_i}\right) + f(X,T), \ (X,t) \in D \times (0, t_F], \quad (1)$$

subject to the initial condition

$$T(X,0) = T_a, \quad X \in \bar{D} = D \cup \partial D, \quad (2)$$

and the nonlinear boundary conditions of the third kind

$$k(X,T)\frac{\partial T}{\partial \eta} + \alpha_k(T)\big(T(X,t) - T_a\big) + \varepsilon\sigma\big(T^4 - T_a^4\big) = 0, \quad X \in \partial D. \quad (3)$$

The following continuity conditions are specified at the points of discontinuity of coefficients

$$[T(x,t)] = 0, \quad \left[k\frac{\partial T}{\partial x_i}\right] = 0.$$

Here $c(X,T)$ is the specific heat capacity, $\rho(X)$ is the density, and $k(X)$ is the heat conductivity coefficient. The density of the energy source $f(X,T)$ is defined as

$$f = \left(\frac{I}{A}\right)^2 \rho_0\big(1 + \alpha_\rho(T - 20)\big),$$

where I is the density of the current, A is a area of the cross-section of the cable, ρ_0 is the specific resistivity of the conductor, and T_a is the temperature of the environment.

Robustness of numerical algorithms for approximation of the heat conduction equation with discontinuous diffusion coefficients is very important for development of methods to be used in simulation of various properties of electrical cables. The differential problem is approximated by the discrete problem by using the finite volume method, which is applied on the vertex centered grids.

In our chapter, we use grids $\bar{D}_h$, which are not aligned with the interfaces where the diffusion coefficient is discontinuous and do not coincide with the boundary of the computational domain (but approximate it). First, we define an auxiliary grid $\widetilde{D}_h = \Omega_h \cap \bar{D}$, which is defined as intersection of the equidistant rectangular grid Ω_h with the computational domain $\bar{D}$:

$$\Omega_h = \{X_{ij} = (x_{1i}, x_{2j}): \quad x_{1i} = L_1 + ih_1, \quad i = 0, \ldots, I, \quad x_{1I} = R_1,$$
$$x_{2j} = L_2 + jh_2, \quad j = 0, \ldots, J, \quad x_{2J} = R_2\}.$$

For each node $X_{ij} \in \widetilde{D}_h$, we define a set of neighbors

$$N(X_{ij}) = \{X_{kl}: \quad X_{i\pm1,j} \in \widetilde{D}_h, \ X_{i,j\pm1} \in \widetilde{D}_h\}.$$

The computational grid $\bar{D}_h = D_h \cup \partial D_h$ is obtained after deletion from $\widetilde{D}_h$ those nodes X_{ij} for which both neighbors in some direction do not belong to $\widetilde{D}_h$, i.e., $X_{i\pm1,j} \notin \widetilde{D}_h$ or $X_{i,j\pm1} \notin \widetilde{D}_h$ (see Fig. 1). The set of neighbors $N(X_{ij})$ is also modified in a similar way.

For each $X_{ij} \in \bar{D}_h$, a *control volume* is defined (see Fig. 1):

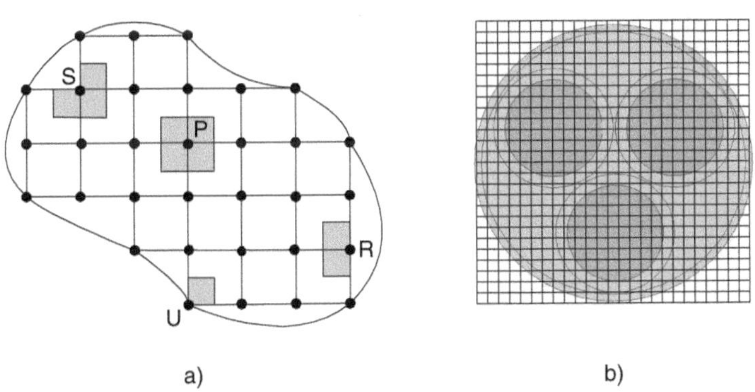

Fig. 1 Discretization: (a) discrete grid D_h and examples of control volumes, (b) basic grid Ω_h and the obtained discretization of the computational domain.

$$e_{ij} = \sum_{k=0}^{3} e_k(X_{ij})\delta_{ijk}.$$

For example, condition $e_1 \in \bar{D}$ is satisfied if all three vertexes $X_{i+1,j}, X_{i,j+1}, X_{i+1,j+1}$ belong to $\bar{D}_h$. In $\bar{D}_h$, we define discrete functions

$$U_{ij}^n = U(x_{1i}, x_{2j}, t^n), \quad X_{ij} \in \bar{D}_h,$$

where $t^n = n\tau$ and τ is the discrete time step. Integrating the differential equation over the control volume e_{ij} and approximating the obtained integrals with an individual quadrature for each term, the differential problem is discretized by the conservative scheme [3, 4, 7]

$$S_{ij}\rho_{ij}c_{ij}(U_{ij}^n)\frac{U_{ij}^n - U_{ij}^{n-1}}{\tau} = \sum_{k=0}^{3} \delta_{ijk}J_{ijk}(U_{ij}^n)U_{ij}^n + S_{ij}f_{ij}(U_{ij}^n), \quad X_{ij} \in \bar{D}_h. \tag{4}$$

Here $J_{ijk}^n(U_{ij}^n)U_{ij}^n$ are the heat fluxes through a surface of the control volume $e_{ijk} = e_k(X_{ij})$, for example:

$$\begin{aligned} J_{ij0}(V_{ij})U_{ij} =& \frac{h_2}{2}\Big(-k_{i-1/2,j}\frac{U_{ij} - U_{i-1,j}}{h_1} + (1-\delta_{ij1})\alpha_G(V_{ij})(U_{ij} - T_a)\Big) \\ &+ \frac{h_1}{2}\Big(k_{i,j+1/2}\frac{U_{i,j+1} - U_{ij}}{h_2} + (1-\delta_{ij3})\alpha_G(V_{ij})(U_{ij} - T_a)\Big). \end{aligned} \tag{5}$$

S_{ij} denotes the measure (or area) of e_{ij}. The diffusion coefficient is approximated by using the harmonic averaging formula

$$k_{i\pm1/2,j} = \frac{2k(X_{i\pm1,j})k(X_{ij})}{k(X_{i\pm1,j}) + k(X_{ij})}, \quad k_{i,j\pm1/2} = \frac{2k(X_{i,j\pm1})k(X_{ij})}{k(X_{i,j\pm1}) + k(X_{ij})}.$$

The derived finite difference scheme (4) defines a system of nonlinear equations. By using the predictor–corrector method, we approximate it by the linear finite-difference scheme of the same order of accuracy.

- **Predictor** ($\forall X_{ij} \in \bar{D}_h$):

$$S_{ij}\rho_{ij}c_{ij}(U_{ij}^{n-1})\frac{\widetilde{U}_{ij}^n - U_{ij}^{n-1}}{\tau} = \sum_{k=0}^{3} \delta_{ijk}J_{ijk}(U_{ij}^{n-1})\widetilde{U}_{ij}^n + S_{ij}f_{ij}(U_{ij}^{n-1}). \tag{6}$$

- **Corrector** ($\forall X_{ij} \in \bar{D}_h$):

$$S_{ij}\rho_{ij}c_{ij}(\widetilde{U}_{ij}^n)\frac{U_{ij}^n - U_{ij}^{n-1}}{\tau} = \sum_{k=0}^{3} \delta_{ijk}J_{ijk}(\widetilde{U}_{ij}^n)U_{ij}^n + S_{ij}f_{ij}(\widetilde{U}_{ij}^n). \tag{7}$$

It is easy to check that the matrix A, arising after the linearization of the proposed finite volume scheme, satisfies the maximum principle [7]. In numerical experiments, we use the BiCGStab iterative method with the Gauss–Seidel type preconditioner [8].

3 Parallel Algorithm

The parallel algorithm is based on the domain decomposition method. The discrete grid $\tilde{D}_h$ is distributed among p processors. The load balancing problem is solved at this step: each processor must obtain the same number of grid points, and the sizes of overlapping regions of subdomains (due to the stencil of the grid, which requires information from the neighboring processors) is minimized. Such a partition of the grid is done by using METIS software library [5].

In order to get a scalable parallel algorithm, we implement both, discretization and linear algebra, steps in parallel. But the main part of CPU time is spent in solving the obtained systems of linear equations by the BiCGStab iterative method. The convergence rate of iterative methods depends essentially on the quality of the preconditoner. The Gauss–Seidel algorithm is sequential in its nature, thus for the parallel algorithm we use a Jacobi version of the Gauss–Seidel preconditioner. Such an implementation depends only on the local part of the matrix, and data communication is required. Due to this modification, the convergence rate of the parallel linear solver can be worse than the convergence rate of the sequential iterative algorithm. Thus the complexity of the full parallel algorithm does not coincide with the complexity of the original algorithm.

A theoretical model is proposed in [2] for the estimation of the complexity of the parallel BiCGStab iterative algorithm, and a scalability analysis can be done on the basis of this model. It gives a linear isoefficiency function of the parallel algorithm [6] for solution of a heat conduction problem.

Results of computational experiments are presented in Table 1. The discrete problem was solved on 600×600 and 900×900 reference grids. Computations were performed on Vilkas cluster of computers at VGTU, consisting of Pentium 4 processors (3.2GHz), level 1 cache 16kB, level 2 cache 1MB, interconnected via Gigabit Smart Switch.

Table 1 Results of computational experiments on Vilkas cluster.

	$p=1$	$p=2$	$p=4$	$p=8$	$p=12$	$p=16$
$T_p(600)$	564.5	341.6	176.6	107.1	80.0	74.0
$S_p(600)$	1.0	1.65	3.19	5.27	7.06	7.63
$E_p(600)$	1.0	0.83	0.80	0.66	0.59	0.48
$T_p(900)$	2232	1282	676.3	382.4	263.7	218.5
$S_p(900)$	1.0	1.74	3.30	5.84	8.46	10.2
$E_p(900)$	1.0	0.87	0.82	0.73	0.71	0.64

It follows from the presented results that the efficiency of the parallel algorithm is improved when the size of the problem is increased.

4 Conclusions

The new parallel solver improves the modeling capabilities of the heat conduction in electrical cables, first, by giving a possibility to solve larger discrete problems, i.e., improving the approximation accuracy, and second, by reducing the CPU time required to find a stationary solution for the fixed values of the parameters. These capabilities are very important for solution of the inverse problems, i.e., fitting the model to the existing experimental data and optimization of the cross-sections of wires in the electrical cables.

The parallel algorithm implements the finite-volume scheme and uses the domain partitioning of the physical domain. The results of computational experiments show that the code is scalable with respect to the number of processors and the size of the problem.

Further studies must be carried out to improve the linear solver, including multi-grid algorithms. The second task is to study the efficiency of the algorithm on new multicore processors.

Acknowledgments The authors were supported by the Agency for International Science and Technology Development Programmes in Lithuania within the EUREKA Project E!3691 OPT-CABLES and by the Lithuanian State Science and Studies Foundation within the project on B-03/2007 "Global optimization of complex systems using high performance computing and GRID technologies".

References

1. Čiegis, R., Ilgevičius, A., Liess, H., Meilūnas, M., Suboč, O.: Numerical simulation of the heat conduction in electrical cables. Mathematical Modelling and Analysis **12**(4), 425–439 (2007)
2. Čiegis, R., Iliev, O., Lakdawala, Z.: On parallel numerical algorithms for simulating industrial filtration problems. Computational Methods in Applied Mathematics **7**(2), 118–134 (2007)
3. Emmrich, E., Grigorieff, R.: Supraconvergence of a finite difference scheme for elliptic boundary value problems of the third kind in fractional order sobolev spaces. Comp. Meth. Appl. Math. **6**(2), 154–852 (2006)
4. Ewing, R., Iliev, O., Lazarov, R.: A modified finite volume approximation of second order elliptic with discontinuous coefficients. SIAM J. Sci. Comp. **23**(4), 1334–13350 (2001)
5. Karypis, G., Kumar, V.: Parallel multilevel *k*-way partitioning scheme for irregular graphs. SIAM Review **41**(2), 278–300 (1999)
6. Kumar, V., Grama, A., Gupta, A., Karypis, G.: Introduction to Parallel Computing: Design and Analysis of Algorithms. Benjamin/Cummings, Redwood City, CA (1994)
7. Samarskii, A.A.: The Theory of Difference Schemes. Marcel Dekker, Inc., New York–Basel (2001)
8. Van der Vorst, H.: Bi-cgstab: A fast and smoothly converging variant of bi-cg for the solution of nonsymmetric linear systems. SIAM J. Sci. Statist. Comput. **13**(3), 631–644 (1992)

Orthogonalization Procedure for Antisymmetrization of *J*-shell States

Algirdas Deveikis

Abstract An efficient procedure for construction of the antisymmetric basis of j-shell states with isospin is presented. The basis is represented by one-particle coefficients of fractional parentage (CFPs) employing a simple enumeration scheme of many-particle states. The CFPs are those eigenvectors of the antisymmetrization operator matrix that correspond with unit eigenvalues. The approach is based on an efficient algorithm of construction of the idempotent matrix eigenvectors. The presented algorithm is faster than the diagonalization routine `rs()` from EISPACK for antisymmetrization procedure applications and is also amenable to parallel calculations.

1 Introduction

For the *ab initio* no-core nuclear shell-model calculations, the approach based on CFPs with isospin could produce matrices that can be orders of magnitude less in dimension than those in the m-scheme approach [5]. In general, the CFPs may be defined as the coefficients for the expansion of the antisymmetric A-particle wave function in terms of a complete set of vector-coupled parent-state wave functions with a lower degree of antisymmetry. For large-scale shell-model calculations, it is necessary to simplify the j-shell states classification and formation algorithms. A simple method of the CFPs calculation, based on complete rejection of higher-order group-theoretical classification of many-particle antisymmetrical states, was presented in [4]. In this approach, many-particle antisymmetrical states in a single j-shell are characterized only by a well-defined set of quantum numbers: the total angular momentum J, the total isospin T, and one additional integer quantum number $\Delta = 1, \ldots, r$, which is necessary for unambiguous enumeration of the states.

Algirdas Deveikis
Department of Applied Informatics, Vytautas Magnus University, Kaunas, Lithuania
e-mail: a.deveikis@if.vdu.lt

Here r is the rank of the corresponding antisymmetrization-operator matrix (the degeneracy of the JT state). This method of the CFPs calculation was implemented in the very quick, efficient, and numerically stable computer program `CFPjOrbit` [3] that produces results possessing only small numerical uncertainities.

In this chapter, we present the `CFPjOrbit` algorithm of construction of the idempotent matrix eigenvectors and investigation results of procedure for construction of antisymmetrization operator matrix Y. The new results of CFPs calculations for up to the 9 nucleons in the $j = 11/2$-shell are presented. The approach is based on a simple enumeration scheme for antisymmetric A-particle states and an efficient method for constructing the eigenvectors of an idempotent matrix Y. The CFPs are those eigenvectors of the idempotent matrix Y that correspond with unit eigenvalues. The algorithm presented in this paper is aimed to the finding of eigenvectors of a specific class of symmetric real projection (idempotent) matrix. The eigenvalues of the matrix are known in advance, and the eigenvectors corresponding with the eigenvalues equal to one form an orthogonal system. In fact, the algorithm built such an orthogonal system of vectors. A number of well-known orthogonalization algorithms is devoted to this problem, for example: Householder, Gram–Schmidt, and modified Gram–Schmidt [7]. The presented algorithm is the modification of the Cholesky factorization [9].

Direct diagonalization may be regarded as an alternative way of producing the CFPs. So, for evaluation of efficiency of the presented algorithm for spectral decomposition of an antisymmetrization operator matrix Y, test calculations were also performed using the diagonalization routine `rs()` from EISPACK. The developing of effective numerical algorithms for solution of eigenvalue problems is permanently the active field of research. We could reference, for example, the well-known LAPACK library routines DSYEV, DSYEVD, and DSYEVR, as well as its parallel version—ScaLAPACK library routines PDSYEV, PDSYEVD, and PDSYEVR.

2 Antisymmetrization of Identical Fermions States

The antisymmetric wave function of an arbitrary system of identical fermions may be obtained by straightforward calculation of the A-particle antisymmetrization operator matrix A on the basis of appropriate functions with a lower degree of antisymmetry. Actually, only antisymmetrizers of states antisymmetrical in the base of functions antisymmetrical with regard to the permutations of the first $(A-p)$ particles as well as the last p particles are used. So, the operator A may be replaced by the simpler operator Y [1].

$$A_{1,\ldots,A} = A_{A-p+1,\ldots,A} A_{1,\ldots,A-p} Y_{A,p} A_{1,\ldots,A-p} A_{A-p+1,\ldots,A}, \tag{1}$$

The operator $A_{1,\ldots,A-p}$ antisymmetrizes functions only with respect to the $1,\ldots,A-p$ variables, and the operator $A_{A-p+1,\ldots,A}$ antisymmetrizes functions only with respect to the $A-p+1,\ldots,A$ variables. The general expression for the operator Y is

$$Y_{A,p} = \binom{A}{p}^{-1} \times \left\{ 1 + \sum_{k=1}^{\min(A-p,p)} (-1)^k \binom{A-p}{k} \binom{p}{k} P_{A-p+1-k,A+1-k} \cdots P_{A-p,A} \right\}. \quad (2)$$

Here the antisymmetrizer is expressed in terms of transposition operators P of the symmetrical group S_A.

The functions for the calculation of an antisymmetrization operator $Y_{A,1}$ matrix may be expanded in terms of a complete set of the angular-momentum-coupled grandparent-state wave functions with a lower degree of antisymmetry

$$\Phi_{j^A \Delta JTM_JM_T}(x_1,\ldots,x_{A-1};x_A) = \sum_{\overline{\overline{\Delta JT}}} \left\langle j^{A-2}\overline{\overline{\Delta JT}}; j \| j^{A-1}\overline{\Delta JT} \right\rangle \times \left\{ \Phi_{j^{A-1}\overline{\Delta JT}}(x_1,\ldots,x_{A-2};x_{A-1}) \otimes \varphi_{elj}(x_A) \right\}_{JTM_JM_T}.$$

The quantum numbers with the single line over them are those for the parent state and those with two lines stand for the grandparent state. The J and T without the line are the total angular momentum and isospin quantum numbers, respectively, of the A-particle nuclear state; M_J and M_T are the magnetic projection quantum numbers of J and T, respectively; and $\langle \ldots ; \ldots \| \ldots \rangle$ is the coefficient of fractional parentage. The j^{A-2} indicates that there are $(A-2)$ nucleons in the grandparent state. The $\overline{\overline{\Delta}}$ stands for the classification number of the grandparent state, which has the same total angular momentum $\overline{\overline{J}}$ and isospin $\overline{\overline{T}}$ as other grandparent states. Similarly on the right-hand side, the j^{A-1} and $\overline{\Delta}$ characterize the parent state. A semicolon means that the corresponding wave function is antisymmetric only with respect to permutations of the variables preceding the semicolon. In the case of the one-particle CFP, only j stands after semicolon. The single-particle variables are $x_i \equiv \mathbf{r}_i\sigma_i\tau_i$ (a set of the corresponding radius-vector, spin, and isospin variables). $\{\ldots \otimes \ldots\}$ is a vector-coupled grandparent state function with the non-antisymmetrized two last particles. Here $\varphi_{elj}(x_i)$ are eigenfunctions of the single-particle harmonic-oscillator Hamiltonian in the $j-j$ coupled representation with isospin.

The expression for the matrix elements of $Y_{A,1}$ in the basis of functions antisymmetric with respect to permutations of the variables $x_1,\ldots,x_{A-1}$ is

$$\left\langle (j^{A-1}\overline{\Delta JT}; j) j^A \Delta JT \left| Y_{A,1} \right| (j^{A-1}\overline{\Delta}' \overline{J}' \overline{T}'; j) j^A \Delta JT \right\rangle = \frac{1}{A} \delta_{\overline{\Delta JT}, \overline{\Delta}'\overline{J}'\overline{T}'} + \frac{A-1}{A} (-1)^{2j+J+T+\overline{J}'+\overline{T}'} \sqrt{[\overline{J},\overline{T},\overline{J}',\overline{T}']} \sum_{\overline{\overline{JT}}} \begin{Bmatrix} j & \overline{J} & \overline{\overline{J}} \\ j & \overline{J}' & J \end{Bmatrix} \begin{Bmatrix} 1/2 & \overline{T} & \overline{\overline{T}} \\ 1/2 & \overline{T}' & T \end{Bmatrix} \times \sum_{\overline{\overline{\Delta}}} \left\langle j^{A-2}\overline{\overline{\Delta JT}}; j \| j^{A-1}\overline{\Delta JT} \right\rangle \left\langle j^{A-2}\overline{\overline{\Delta JT}}; j \| j^{A-1}\overline{\Delta}'\overline{J}'\overline{T}' \right\rangle. \quad (3)$$

Here $[k] \equiv 2k+1$, and the double bar indicates the grandparent state. The usual notation for the standard $6j$ vector coefficient [8] is employed in (3). The separation of sum over the $\overline{\overline{\Delta}}$ is very important for computational reasons – it allows one to fill the matrix Y in blocks, with dimensions corresponding with the degeneracy of the parent states. At the same time, the time-consuming calculation of $6j$ vector coefficient may be performed only once for every block. The antisymmetrizer Y is a projection operator $Y^2 = Y$. Consequently, the matrix Y has the following properties: the matrix elements are less than one in absolute value, the trace of the matrix is equal to its rank r

$$SpY = r, \tag{4}$$

the sum of the squares of the matrix elements of any row or column is equal to the corresponding diagonal element

$$Y_{ii} = \sum_{j=1}^{N} Y_{ij}^2, \tag{5}$$

and eigenvalues are only ones and zeros. The matrix Y is a symmetric, real projection matrix: $Y^+ = Y = Y^*$, $YY = Y$, and the spectral decomposition [4] is given by

$$Y_{N\times N} = F_{N\times r} F^+_{r\times N}. \tag{6}$$

The subscripts indicate the dimension of the corresponding matrix, N equals the dimension of the basis, and r is the rank of the matrix Y. Any column of the matrix F is an eigenvector of matrix Y corresponding with a unit eigenvalue

$$Y_{N\times N} F_{N\times r} = F_{N\times r}. \tag{7}$$

The orthonormality condition for eigenvectors is

$$F^+_{r\times N} F_{N\times r} = I_{r\times r}, \tag{8}$$

i.e., the columns with the same $\Delta = 1,\ldots,r$ are normalized to unity and are orthogonal. The matrix elements of F are the coefficients of fractional parentage. For calculation of the matrix F, it is sufficient to calculate only r linearly independent columns of the matrix Y. We may construct the matrix Y' (with the first r rows and columns linearly independent) by rotation of the original Y matrix with the matrix T:

$$Y' = T^+ Y T. \tag{9}$$

Here T is the matrix multiplication of the row and column permutation-operator matrices T_i

$$T = \prod_{i=1}^{P} T_i. \tag{10}$$

Finally, the required eigenvectors F of the matrix Y may be found from the eigenvectors F' of the matrix Y'

$$F = TF'. \tag{11}$$

The spectral decomposition of the antisymmetrization-operator matrices is not uniquely defined. Thus, it is possible to make a free choice for an orthogonal matrix $G_{r\times r}$, defined by

$$Y' = F'GG^{+}F'^{+} = F''F''^{+}, \tag{12}$$

An orthogonal matrix G has $r(r-1)/2$ independent parameters, so we can choose it so that it allows us to fix the corresponding number of matrix elements of F'

$$F'_{ij} = 0, \quad \text{if} \quad 1 \leq i < j \leq r. \tag{13}$$

The matrix elements of F' that are not from the upper triangle (13) may be calculated by the formulae [4]

$$\begin{cases} F'^{2}_{ii} = Y'_{ii} - \sum\limits_{k=1}^{i-1} F'^{2}_{ik}, \\ F'_{ij} = \frac{1}{F'_{jj}} \left\{ Y'_{ij} - \sum\limits_{k=1}^{i-1} F'_{ik} F'_{jk} \right\}, \end{cases} \tag{14}$$

for every value $i = 1, \ldots, r$ and the corresponding set of $j = i+1, \ldots, N$. It is convenient to choose positive values for the F'_{ii}, as the overall sign of the CFP vector is arbitrary. We may number the columns of the matrix F' (and similarly the columns of F) by positive integers $\Delta' = 1, \ldots, r$.

The first equation of (14) may serve for eigenvectors linear independency test

$$Y'_{ii} - \sum_{k=1}^{i-1} F'^{2}_{ik} = 0. \tag{15}$$

The pseudocode describing the method of construction of the idempotent matrix eigenvectors is outlined in Algorithm 1.

Here P is the number of permutations of Y and F matrices, ε_1 is non-negative parameter specifying the error tolerance used in the Y matrix diagonal elements zero value test, ε_2 is non-negative parameter specifying the error tolerance used in the eigenvectors linear independency test, and the pseudocode conventions of [2] are adopted.

3 Calculations and Results

The practical use of the presented antisymmetrization procedure and Algorithm 1 of construction of the idempotent matrix eigenvectors is shown for the problem

Algorithm 1 Construction of the idempotent matrix eigenvectors.

```
INPUT: Y
for i ← 1 to N
    do L : if i ≤ r
            then while |Y_ii| < ε_1
                    do P ← P+1
                       if P ≥ N
                          then stop with error
                       T ← T T_P
                       Y ← T^+ Y T
        for j ← 1 to r
            do if j < i
                  then F_ij ← (1/F_jj) { Y_ij − \sum_{k=1}^{i-1} F_ik F_jk }
               if j = i
                  then if | Y_ii − \sum_{k=1}^{i-1} F_ik^2 | < ε_2
                          P ← P+1
                          if P ≥ N
                             then stop with error
                          T ← T T_P
                          Y ← T^+ Y T
                          break L
                  else F_ii ← \sqrt{ Y_ii − \sum_{k=1}^{i-1} F_ik^2 }
               if j > i
                  then F_ij ← 0
 if P > 0
   then F ← T F
OUTPUT: F
```

of construction of the antisymmetric basis of $j = 11/2$-shell states with isospin. The calculations were performed for up to 9 nucleons. Extensive numerical computations illustrate the performance of Algorithm 1 and provide a comparison with the corresponding EISPACK routine `rs()` for diagonalization of real symmetric matrices. The test calculations were performed on Pentium 1.8GHz PC with 512MB RAM. The FORTRAN90 programs for construction of idempotent matrices, computation of eigenvectors, and diagonalization were run on Fortran PowerStation 4.0.

The simple enumeration scheme of many-particle states and efficient procedure of filling the Y matrix in blocks, with dimensions corresponding with the degeneracy of the parent states, were used for construction of antisymmetrization operator matrix Y. The accuracy of procedure may be illustrated by the results of Table 1. In Table 1 the discrepancy M_r is defined as the difference between the rank r of the matrix Y and the sum of its diagonal elements. The discrepancy M_0 is the largest difference between the diagonal elements of the matrix Y and the sums of its elements from the rows where stand the diagonal elements under consideration. Because r may be obtained from combinatorial analysis, the condition (4), as well as (5), may

Table 1 The characteristics of Y matrices for up to 9 nucleons in the shell with $j = 11/2$. Columns: N is the matrix dimension; r is the matrix rank; M_r is the discrepancy of (4); M_0 is the discrepancy of (5).

N	r	M_r	M_0
1013	85	2.27E-13	3.36E-15
2012	157	9.95E-13	1.98E-15
3147	314	4.55E-13	4.08E-15
4196	326	3.41E-13	3.37E-15
5020	404	1.42E-12	3.91E-15
6051	563	1.25E-12	3.79E-15

be useful as a check of the calculations. It should be pointed out that the accuracy of the available $6j$ subroutine does not influence severely the numerical accuracy of calculations, as the largest values of $J = 67/2$ and $T = 9/2$ used for construction of matrices Y are no so large [6].

We present the running time for the EISPACK routine `rs()`, compared with the running time of Algorithm 1 in Table 2. The eigenvectors computation stage is the most time-consuming part, and thus efficient implementation is crucial to the large-scale calculations. One can see that Algorithm 1 exhibits running time of $O(N^2P)$. The routine `rs()` has running time of $O(N^3)$. So, the speedup of Algorithm 1 strongly depends on the number of permutations of Y and F matrices. The smallest speedup is 1.15 for the matrix Y with $N = 170$ and $r = 34$, when the number of permutations is 131. The largest speedup is 68.23 for the matrix Y with $N = 412$ and $r = 33$, when the number of permutations is 4. For construction of the antisymmetric basis for up to 9 nucleons in the $j = 11/2$-shell, the total running time of the routine `rs()` is 161974.14 seconds. At the same time, the total running time of Algorithm 1 is 36103.84 seconds. So, for the antisymmetrization problem under consideration, Algorithm 1 is faster by about 4.5 times than the EISPACK routine `rs()`.

Because the number of permutations of Y and F matrices is so crucial to the effectiveness of Algorithm 1, we present some permutations characteristics in

Table 2 The time for computations of eigenvectors, in seconds. Columns: N is the matrix dimension; r is the matrix rank; P is the number of permutations of Y and F matrices; rs() is the computing time of the diagonalization routine `rs()` from EISPACK; F is the computing time of Algorithm 1; Spd. denotes the obtained speedup.

N	r	P	rs()	F	Spd.
1013	85	30	43.9	2.08	21.1
2012	157	239	334	66.3	5.03
3147	314	220	1240	171	7.25
4196	326	206	2990	378	7.91
5020	404	305	4980	1050	4.74
6051	563	329	8710	2170	4.01

Table 3 The number of permutations of Y and F matrices for up to 9 nucleons in the shell with $j = 11/2$. Columns: $D_{\min} - D_{\max}$ is the range of dimensions of Y matrices $\geq D_{\min}$ and $< D_{\max}$ ($\leq$ for 6051); $\#Y$ is the number of Y matrices; $P_{\min} - P_{\max}$ is the range of permutation times of Y and F matrices; P_{avr} is the average number of permutations of Y and F matrices.

$D_{\min} - D_{\max}$	$\#Y$	$P_{\min} - P_{\max}$	P_{avr}
4000 – 6051	21	206 – 513	362
3000 – 4000	12	188 – 270	226
2000 – 3000	21	106 – 292	165
1000 – 2000	70	30 – 360	106

Table 4 The accuracy of eigenvectors computations. Columns: N is the matrix dimension; r is the matrix rank; accuracy of orthonormality condition (8) obtained by diagonalization routine `rs()` from EISPACK is denoted by M_{orth} and using Algorithm 1 by F_{orth}; discrepancy of eigenvalue equation (7) obtained by diagonalization routine `rs()` from EISPACK is denoted by M_{eig} and using Algorithm 1 by F_{eig}.

N	r	M_{orth}	F_{orth}	M_{eig}	F_{eig}
1013	85	5.22E-15	9.84E-13	3.52E-15	1.19E-13
2012	157	8.44E-15	4.70E-13	4.91E-15	3.21E-13
3147	314	1.62E-14	1.71E-12	6.07E-15	4.85E-13
4196	326	1.53E-14	1.51E-12	5.97E-15	3.18E-13
5020	404	1.47E-14	9.60E-13	9.41E-15	3.00E-13
6051	563	1.95E-14	1.51E-12	8.94E-15	4.94E-13

Table 3. One can see that the numbers of permutations P are on average about an order of magnitude smaller than the dimensions of corresponding matrices Y. This may explain the obtained significant speedup of Algorithm 1 over the routine `rs()`.

The accuracy of eigenvectors computations is illustrated in Table 4. Here the accuracy of orthonormality condition (8) is defined as the difference between the zero or one and the corresponding sum of products of eigenvectors. The discrepancy of eigenvalue equation (7) is the largest difference between the corresponding matrices elements values on the right and left side of the equation. In fact, direct diagonalization is up to two orders of magnitude more accurate than is Algorithm 1, however, Algorithm 1 accuracy is very high and is more than sufficient for large-scale shell-model calculations [5].

4 Conclusions

An efficient antisymmetrization procedure for j-shell states with isospin is presented. The application of the procedure is illustrated by $j = 11/2$-shell calculations for up to 9 nucleons. At the present time, we are not aware of another approach that can generate the antisymmetric basis for 7, 8, and 9 nucleons in the $j = 11/2$-shell.

The construction of antisymmetrization operator matrices Y is performed by numerically stable and quick computational procedure. The presented Algorithm 1 of construction of the idempotent matrix eigenvectors is approximately 4.5 times faster than the EISPACK routine `rs()` for applications to the antisymmetrization of considered j-shell states. On top of that, Algorithm 1 is simple and easy to implement.

The important advantage of our procedure is the fact that one does not have to compute the whole Y matrix. It is quit sufficient to calculate only r linearly independent rows (or columns) of Y matrix. Because the rank r of the Y matrix is about an order of magnitude less than its dimension N, our procedure may save considerably on the computation resources.

Another distinct feature of the presented procedure for construction of the idempotent matrix eigenvectors is that a precise arithmetic can be applied. Instead of calculations with real numbers, which are connected with serious numerical instabilities for large values of quantum numbers [8], the calculations could be performed with numbers represented in the root rational fraction form $a\sqrt{b}\Big/\left(c\sqrt{d}\right)$, where a, b, c, and d are integers.

Finally, the construction of the eigenvectors of the matrix Y with Algorithm 1 is also amenable to parallel calculations. Only the first r rows of the eigenvectors matrix F of the matrix Y couldn't be calculated independently according to the Algorithm 1. When the first r rows of the eigenvectors matrix F are obtained, the remaining part of the columns of this matrix may be processed in parallel. Because for j-shell states the rank r may be obtained from combinatorial analysis and r is about an order of magnitude less than the dimension N of the matrix Y, the largest part: $N-r$ of the rows of the eigenvectors matrix F may be processed in parallel.

References

1. Coleman, A.J.: Structure of Fermion density matrices. Rev. Mod. Phys. **35**, 668–689 (1963)
2. Cormen, T.H., Leiserson, C.E., Rivest, R.L., Stein, C.: Introduction to Algorithms. MIT Press, Cambridge (2001)
3. Deveikis, A.: A program for generating one-particle and two-particle coefficients of fractional parentage for the single j-orbit with isospin. Comp. Phys. Comm. **173**, 186–192 (2005)
4. Deveikis, A., Bončkus, A., Kalinauskas, R.K.: Calculation of coefficients of fractional parentage for nuclear oscillator shell model. Lithuanian Phys. J. **41**, 3–12 (2001)
5. Deveikis, A., Kalinauskas, R.K., Barrett, B.R.: Calculation of coefficients of fractional parentage for large-basis harmonic-oscillator shell model. Ann. Phys. **296**, 287–298 (2002)
6. Deveikis, A., Kuznecovas, A.: Analytical scheme calculations of angular momentum coupling and recoupling coefficients. JETP Lett. **4**, 267–272 (2007)
7. Golub, A.M., Van Loan, C.F.: Matrix Computations. The Johns Hopkins University Press, Baltimore London (1996)
8. Jutsys, A.P., Savukynas, A.: Mathematical Foundations of the Theory of Atoms. Mintis, Vilnius (1973)
9. Watkins, D.S.: Fundamentals of Matrix Computations. Wiley, New York (2002)

Parallel Direct Numerical Simulation of an Annular Gas–Liquid Two-Phase Jet with Swirl

George A. Siamas, Xi Jiang, and Luiz C. Wrobel

Abstract The flow characteristics of an annular swirling liquid jet in a gas medium have been examined by direct solution of the compressible Navier–Stokes equations. A mathematical formulation is developed that is capable of representing the two-phase flow system while the volume of fluid method has been adapted to account for the gas compressibility. The effect of surface tension is captured by a continuum surface force model. Analytical swirling inflow conditions have been derived that enable exact definition of the boundary conditions at the domain inlet. The mathematical formulation is then applied to the computational analysis to achieve a better understanding on the flow physics by providing detailed information on the flow development. Fully 3D parallel direct numerical simulation (DNS) has been performed utilizing 512 processors, and parallelization of the code was based on domain decomposition. The numerical results show the existence of a recirculation zone further down the nozzle exit. Enhanced and sudden liquid dispersion is observed in the cross-streamwise direction with vortical structures developing at downstream locations due to Kelvin–Helmholtz instability. Downstream the flow becomes more energetic, and analysis of the energy spectra shows that the annular gas–liquid two-phase jet has a tendency of transition to turbulence.

1 Introduction

Gas–liquid two-phase flows are encountered in a variety of engineering applications such as propulsion and fuel injection in combustion engines. A liquid sheet spray process is a two-phase flow system with a gas, usually air, as the continuous phase and a liquid as the dispersed phase in the form of droplets or ligaments. In many

George A. Siamas · Xi Jiang · Luiz C. Wrobel
Brunel University, Mechanical Engineering, School of Engineering and Design
Uxbridge, UB8 3PH, UK
e-mail: {George.Siamas · Xi.Jiang · Luiz.Wrobel}@brunel.ac.uk

applications, the gas phase is compressible whereas the liquid phase exhibits incompressibility by nature. The two phases are coupled through exchange of mass, momentum, and energy, and the interactions between the phases can occur in different ways, at different times, involving various fluid dynamic factors. Understanding of the fluid dynamic behavior of liquid sheets in gas environments is essential to effectively control the desired transfer rates.

The process of atomization is very complex and the mechanisms governing the liquid breakup still remain unclear. It is very difficult to understand the physics behind liquid breakup using theoretical and/or experimental approaches, due to the complex mixing and coupling between the liquid and gas phases and the broad range of time and length scales involved. Researchers have tried to tackle this complex two-phase flow problem in the past but their studies were focused on experimental visualizations and simplified mathematical models, which are insufficient to reveal and describe the complex details of liquid breakup and atomization [2, 10, 14]. In terms of obtaining fundamental understanding, DNS is advantageous over the Reynolds-averaged Navier–Stokes modeling approach and large-eddy simulations. It can be a very powerful tool that not only leads to a better understanding of the fluid mechanics involved, but also provides a useful database for the potential development of physical models able to overcome the problems of the current breakup models.

In this study, an annular liquid jet in a compressible gas medium is investigated using a Eulerian approach with mixed-fluid treatment [3] for the governing equations describing the gas–liquid two-phase flow system where the gas phase is treated as fully compressible and the liquid phase as incompressible. The flow characteristics are examined by direct solution of the compressible, time-dependent, non-dimensional Navier–Stokes equations using highly accurate numerical schemes. The interface dynamics are captured using an adjusted volume of fluid (VOF) and continuum surface force (CSF) models [1, 8]. Fully 3D parallel simulation is performed, under the MPI environment, and the code is parallelized using domain decomposition.

2 Governing Equations

The flow field is described in a Cartesian coordinate system, where the z-axis is aligned with the streamwise direction of the jet whereas the x–y plane is in the cross-streamwise direction. In the Eulerian approach with mixed-fluid treatment adopted [3], the two phases are assumed to be in local kinetic and thermal equilibrium, i.e., the relative velocities and temperatures are not significant, while the density and viscosity are considered as gas–liquid mixture properties and they are functions of the individual densities and viscosities of the two phases [7], given as

$$\rho = \Phi \rho_l + (1 - \Phi)\rho_g, \qquad (1)$$

$$\mu = \Phi \mu_l + (1 - \Phi)\mu_g. \tag{2}$$

In this study, the original VOF method has been adapted to solve an equation for the liquid mass fraction Y rather than the volume fraction Φ in order to suit the compressible gas phase formulation [12, 21, 20]. From their definitions, a relation between liquid volume fraction and mass fraction can be derived as

$$\Phi = \frac{\rho_g Y}{\rho_l - (\rho_l - \rho_g)Y}. \tag{3}$$

The gas–liquid interface dynamics are resolved using a continuum surface force (CSF) model [1], which represents the surface tension effect as a continuous volumetric force acting within the region where the two phases coexist. The CSF model overcomes the problem of directly computing the surface tension integral that appears in the Navier–Stokes momentum equations, which requires the exact shape and location of the interface. In the CSF model, the surface tension force in its non-dimensional form can be approximated as $\sigma\kappa/We\nabla\Phi$, with σ representing surface tension and We the Weber number. The curvature of the interface is given by

$$\kappa = -\nabla \cdot \left(\frac{\nabla \Phi}{|\nabla \Phi|} \right). \tag{4}$$

The flow system is prescribed by the compressible, non-dimensional, time-dependent Navier–Stokes equations, which include the transport equation for the liquid concentration per unit volume. The conservation laws are given as

$$\frac{\partial \rho_g}{\partial t} + \frac{\partial (\rho_g u)}{\partial x} + \frac{\partial (\rho_g v)}{\partial y} + \frac{\partial (\rho_g w)}{\partial z} = 0, \tag{5}$$

$$\frac{\partial (\rho u)}{\partial t} + \frac{\partial (\rho u^2 + p - \tau_{xx})}{\partial x} + \frac{\partial (\rho uv - \tau_{xy})}{\partial y} + \frac{\partial (\rho uw - \tau_{xz})}{\partial z} - \frac{\sigma\kappa}{We}\frac{\partial Y}{\partial x} = 0, \tag{6}$$

$$\frac{\partial (\rho v)}{\partial t} + \frac{\partial (\rho uv - \tau_{xy})}{\partial x} + \frac{\partial (\rho v^2 + p - \tau_{yy})}{\partial y} + \frac{\partial (\rho vw - \tau_{yz})}{\partial z} - \frac{\sigma\kappa}{We}\frac{\partial Y}{\partial y} = 0, \tag{7}$$

$$\frac{\partial (\rho w)}{\partial t} + \frac{\partial (\rho uw - \tau_{xz})}{\partial x} + \frac{\partial (\rho uw - \tau_{yz})}{\partial y} + \frac{\partial (\rho w^2 + p - \tau_{zz})}{\partial z} - \frac{\sigma\kappa}{We}\frac{\partial Y}{\partial z} = 0, \tag{8}$$

$$\begin{aligned} \frac{\partial E_T}{\partial t} &+ \frac{\partial \left[(E_T + p)u + q_x - u\tau_{xx,g} - v\tau_{xy,g} - w\tau_{xz,g}\right]}{\partial x} + \\ &+ \frac{\partial \left[(E_T + p)v + q_y - u\tau_{xy,g} - v\tau_{yy,g} - w\tau_{yz,g}\right]}{\partial y} \\ &+ \frac{\partial \left[(E_T + p)w + q_z - u\tau_{xz,g} - v\tau_{yz,g} - w\tau_{zz,g}\right]}{\partial z} = 0, \end{aligned} \tag{9}$$

$$\frac{\partial(\rho Y)}{\partial t}+\frac{\partial}{\partial x}\left[\rho uY-\frac{1}{ReSc}\left(\mu\frac{\partial Y}{\partial x}\right)\right]+\frac{\partial}{\partial y}\left[\rho vY-\frac{1}{ReSc}\left(\mu\frac{\partial Y}{\partial y}\right)\right]$$
$$+\frac{\partial}{\partial z}\left[\rho wY-\frac{1}{ReSc}\left(\mu\frac{\partial Y}{\partial z}\right)\right]=0, \tag{10}$$

where the subscript g corresponds to the gas phase. The heat flux components are denoted by q and the viscous stress components by τ. The total energy of the gas with e representing the internal energy per unit mass can be given as $E_T=\rho_g\left[e+\left(u^2+v^2+w^2\right)/2\right]$. Re and Sc are the Reynolds and Schmidt numbers, respectively. The governing equations are accompanied by the ideal gas law defined as $p=\rho_g T/\left(\gamma Ma^2\right)$, where p is the gas pressure; T: temperature; γ: ratio of specific heats of the compressible gas; and Ma: Mach number.

3 Computational Methods

3.1 Time Advancement, Discretization, and Parallelization

The governing equations are integrated forward in time using a third-order compact-storage fully explicit Runge–Kutta scheme [23] while the time step was limited by the Courant–Friedrichs–Lewy (CFL) condition for stability. During the time advancement, the density and viscosity of the gas–liquid two-phase flow system are calculated according to (1)–(2), using the volume fraction Φ calculated from (3). However, the liquid mass fraction Y in (3) needs to be calculated from the solution variable ρY first. Using q to represent ρY at each time step, the liquid mass fraction Y can be calculated as

$$Y=\frac{\rho_l q}{\rho_l\rho_g-(\rho_l-\rho_g)q}. \tag{11}$$

Equation (11) can be derived from (1)-(3). At each time step, (11) is used first to calculate the liquid mass fraction, (3) is then used to calculate the liquid volume fraction, and (1)–(2) are finally used to update the mixture density and viscosity.

The spatial differentiation is performed using a sixth-order accurate compact (*Padé*) finite difference scheme with spectral-like resolution [15], which is of sixth-order at the inner points, fourth order at the next to the boundary points, and third order at the boundary. For a general variable ϕ_j at grid point j in the y-direction, the scheme can be written in the following form for the first and second derivatives

$$\phi'_{j-1}+3\phi'_j+\phi'_{j+1}=\frac{7}{3}\frac{\phi_{j+1}-\phi_{j-1}}{\Delta\eta}+\frac{1}{12}\frac{\phi_{j+2}-\phi_{j-2}}{\Delta\eta}, \tag{12}$$

$$\phi''_{j-1}+\frac{11}{2}\phi''_j+\phi''_{j+1}=6\frac{\phi_{j+1}-2\phi_j+\phi_{j-1}}{\Delta\eta^2}+\frac{3}{8}\frac{\phi_{j+2}-2\phi_j+\phi_{j-2}}{\Delta\eta^2}, \tag{13}$$

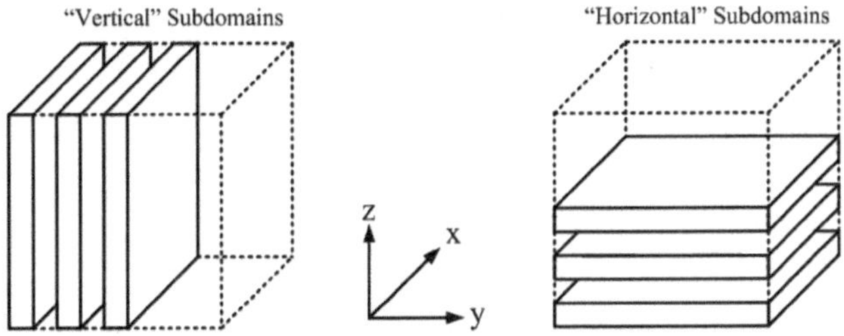

Fig. 1 Vertical and horizontal subdomains.

where $\Delta\eta$ is the mapped grid distance in the y-direction, which is uniform in space (grid mapping occurs when a non-uniform grid is used). The left-hand side of (12) and (13) leads to a tridiagonal system of equations whose solutions are obtained by solving the system using a tridiagonal matrix algorithm.

To perform parallel computations, the whole physical domain is divided into several subdomains (vertical and horizontal) using a domain decomposition method, as shown in Fig. 1. The use of both horizontal and vertical slices enables correct calculation of the derivatives in all three directions. The horizontal slices are used to calculate the x- and y-derivatives and the vertical slices are used to calculate the z-derivatives. To enable calculations in three dimensions, the flow data are interlinked by swapping the flow variables from the x–z planes to the x–y planes and vice versa. An intermediate array was used to facilitate the variables for swapping the data between the vertical and horizontal subdomains. Data exchange among the utilized processors is achieved using standard Message Passing Interface (MPI).

3.2 Boundary and Initial Conditions

The three-dimensional computational domain is bounded by the inflow and the outflow boundaries in the streamwise direction and open boundaries with the ambient field in the jet radial (cross-streamwise) direction. The non-reflecting characteristic boundary conditions [22] are applied at the open boundaries, which prevent wave reflections from outside the computational domain. The non-reflecting boundary conditions are also used at the outflow boundary in the streamwise direction. The spurious wave reflections from outside the boundary have been controlled by using a sponge layer next to the outflow boundary [11]. The strategy of using a sponge layer is similar to that of "sponge region" or "exit zone" [16], which has been proved to be very effective in controlling wave reflections through the outflow boundary. The results in the sponge layer are unphysical and therefore are not used in the data analysis.

Based on the concept of Pierce and Moin [17] for numerical generation of equilibrium swirling inflow conditions, analytical forms of the axial and azimuthal velocity components are derived that enable simple and precise specification of the desired swirl level [13]. The analytical profiles of axial and azimuthal velocities are given as

$$w = -\frac{1}{4}\frac{f_x}{\mu}\left(r^2 - \frac{R_i^2 - R_o^2}{\ln R_i - \ln R_o}\ln r + \frac{R_i^2 \ln R_o - R_o^2 \ln R_i}{\ln R_i - \ln R_o}\right), \tag{14}$$

$$u_\theta = -\frac{1}{3}\frac{f_\theta}{\mu}\left(r^2 - \frac{R_i^2 + R_i R_o + R_o^2}{R_i + R_o}r + \frac{R_i^2 R_o^2}{R_i + R_o}\frac{1}{r}\right), \tag{15}$$

where $r = \sqrt{x^2 + y^2}$ is the radial distance, and R_i and R_o are the inner and outer radii of the annular jet, respectively. In (14)–(15), f_x and f_θ can be defined by the maximum velocities at the inflow boundary. For a unit maximum velocity, the constant f_x is defined as

$$f_x = -\frac{8\mu\,(\ln R_o - \ln R_i)}{R_o^2 - R_i^2 + \ln\left[\frac{R_i^2 - R_o^2}{2(\ln R_i - \ln R_o)}\right](R_i^2 - R_o^2) - 2R_i^2 \ln R_o + 2R_o^2 \ln R_i}. \tag{16}$$

Adjustment of the constant f_θ will define the desired degree of swirl. For known u_x and u_θ, the swirl number can be conveniently calculated from the following relation

$$S = \frac{\int_{R_i}^{R_o} u_x u_\theta r^2\, dr}{R_o \int_{R_i}^{R_o} u_x^2 r\, dr}. \tag{17}$$

From the azimuthal velocity u_θ, the cross-streamwise velocity components at the inflow can be specified as

$$u = -\frac{u_\theta y}{r}, \qquad v = \frac{u_\theta x}{r}.$$

Helical/flapping perturbations combined with axial disturbances were used to break the symmetry in space and induce the roll-up and pairing of the vortical structures [4]. The velocity components at the jet nozzle exit $z = 0$ are given as

$$u = \overline{u} + A\sin(m\varphi - 2\pi f_0 t), \quad v = \overline{v} + A\sin(m\varphi - 2\pi f_0 t),$$

$$w = \overline{w} + A\sin(m\varphi - 2\pi f_0 t),$$

where A is the amplitude of the disturbance, m is the mode number, φ is the azimuthal angle, and f_0 the excitation frequency. The amplitude of the disturbance is 1% of the maximum value of the streamwise velocity. The Strouhal number (St) of the unsteady disturbance is 0.3, which has been chosen to be the unstable mode leading to the jet preferred mode of instability [9]. Two helical disturbances with mode numbers $m = \pm 1$ are superimposed on the temporal disturbance [4].

4 Results and Discussion

The simulation parameters correspond with the injection of diesel fuel into compressed air at around $15MPa$ and $300K$, where the diesel surface tension was taken to be approximately $0.025N/m$. The Reynolds number is taken to be 2000, the Weber number 240, the Mach number 0.4, the Schmidt number 0.76, and the swirl number 0.4. The non-dimensional lengths of the computational box are $L_x = L_y = L_z = 10$. The grid system is of $512 \times 512 \times 512$ nodes with a uniform distribution in each direction. Fully 3D parallel DNS computation has been performed, under the MPI environment, on an IBM pSeries 690 Turbo Supercomputer utilizing 512 processors. An increase of the number of processors from 256 to 512 almost decreases by half the computing time needed. The results presented next are considered to be grid and time-step independent and are discussed in terms of the instantaneous and time-averaged flow properties.

4.1 Instantaneous Flow Data

Figure 2 shows the instantaneous isosurfaces of enstrophy $\Omega = \left(\omega_x^2 + \omega_y^2 + \omega_z^2\right)/2$, liquid volume fraction Φ, and x-vorticity ω_x at the non-dimensional time of $t = 30.0$. The individual vorticity components are defined as

$$\omega_x = \frac{\partial w}{\partial y} - \frac{\partial v}{\partial z}, \qquad \omega_y = \frac{\partial u}{\partial z} - \frac{\partial w}{\partial x}, \qquad \omega_z = \frac{\partial v}{\partial x} - \frac{\partial u}{\partial y}.$$

From Figs. 2a and 2b, it is evident that the dispersion of the liquid is dominated by large-scale vortical structures formed at the jet primary stream due to Kelvin–Helmholtz instability. In view of the Eulerian approach with mixed-fluid

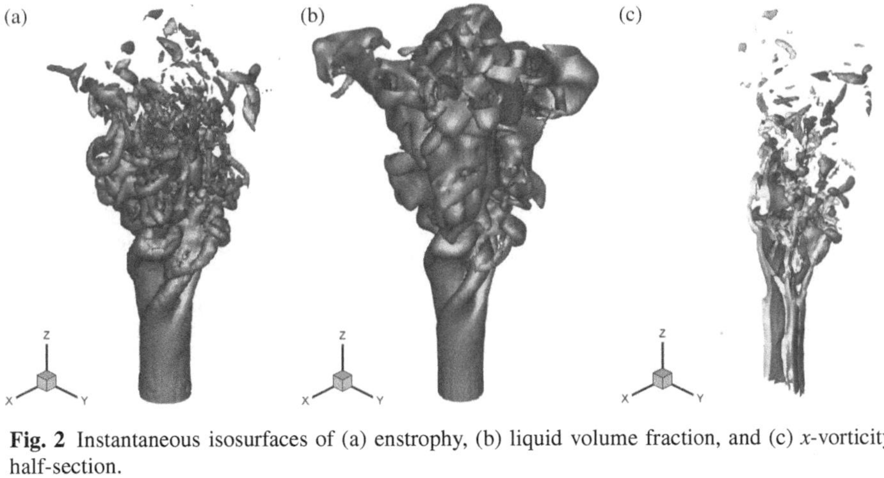

Fig. 2 Instantaneous isosurfaces of (a) enstrophy, (b) liquid volume fraction, and (c) x-vorticity half-section.

treatment [3] adopted in this study, the Rayleigh–Taylor instability is not expected to play a significant role in the flow development. As the flow develops, a more disorganized flow field appears further downstream, characterized by small scales indicating a possible transition to weak turbulence. The vortical structures have elongated finger-type shapes before collapsing to smaller structures. The unsteady behavior of the jet is characterized by the formation of streamwise vorticity, which is absent in idealized axisymmetric and planar simulations [11, 12, 21, 20]. The streamwise vorticity is mainly generated by 3D vortex stretching, which spatially distributes the vorticity and thus the dispersion of the liquid. Fig. 2c shows the x-vorticity component for a half-section of the annular jet. It is interesting to notice the presence of both negative and positive vorticity. This results in the formation of counter-rotating vortices in the cross-streamwise direction, which are enhanced by the presence of swirl. The rather large liquid dispersion in the cross-streamwise direction, especially at downstream locations, is primarily owed to the swirling mechanism.

The instantaneous velocity vector maps at $t = 30.0$, at various streamwise planes, are shown in Fig. 3. For clarity reasons, the vector plots are only shown for a limited

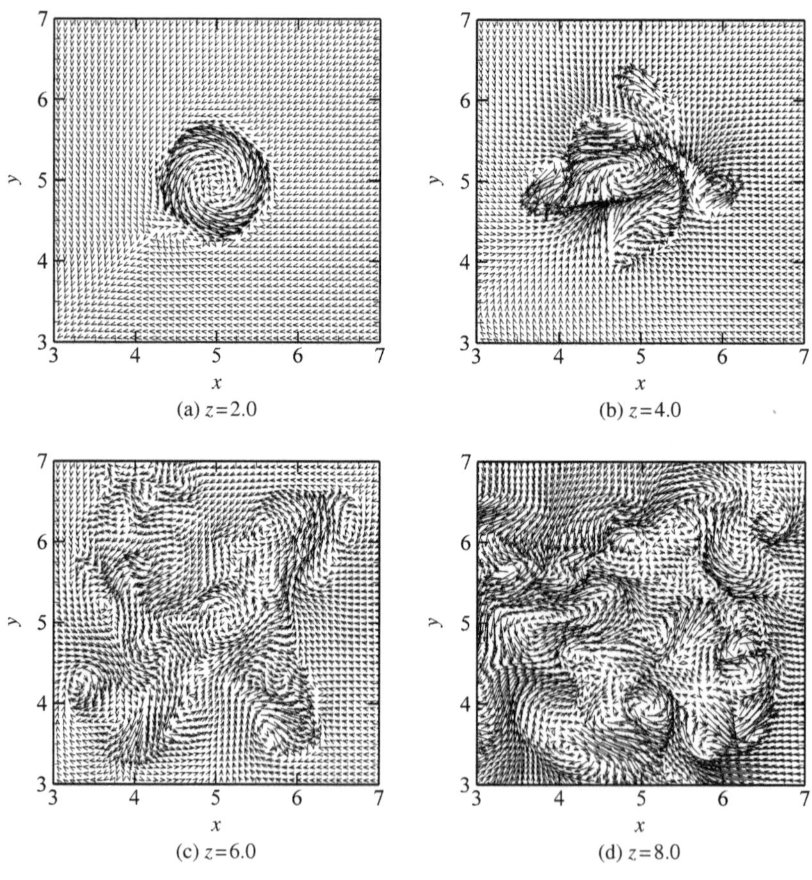

Fig. 3 Instantaneous velocity vector maps at various streamwise slices.

number of grid points, which is significantly less than the total number of grid points. For the time instant considered here, the flow is fully developed as the jet has passed through the computational domain several times. In Fig. 3a, the velocity map shows a rotating pattern due to the swirl applied at the inflow. At further downstream locations, the velocity distributions are very complex and at $z = 4.0$ and $z = 6.0$ form "star-type" shapes. At $z = 8.0$, the velocity field becomes very irregular, with minimal compactness compared with the other z-slices, due to the collapsing of the large-scale vortical structures to small-scale ones, as also noticed in Fig. 2.

To show how the velocity profiles affect the liquid dispersion, the liquid volume fraction distributions are shown in Fig. 4.

As expected, the liquid distribution is increased in the cross-streamwise direction as the flow progresses from the inlet to further downstream locations. The annular configuration of the liquid, as shown in Fig. 4a, is quickly broken in disorganized patterns that expand in both x- and y-directions as shown in Figs. 4b–4d. This behavior of the liquid is boosted by the swirl, which tends to suddenly increase the

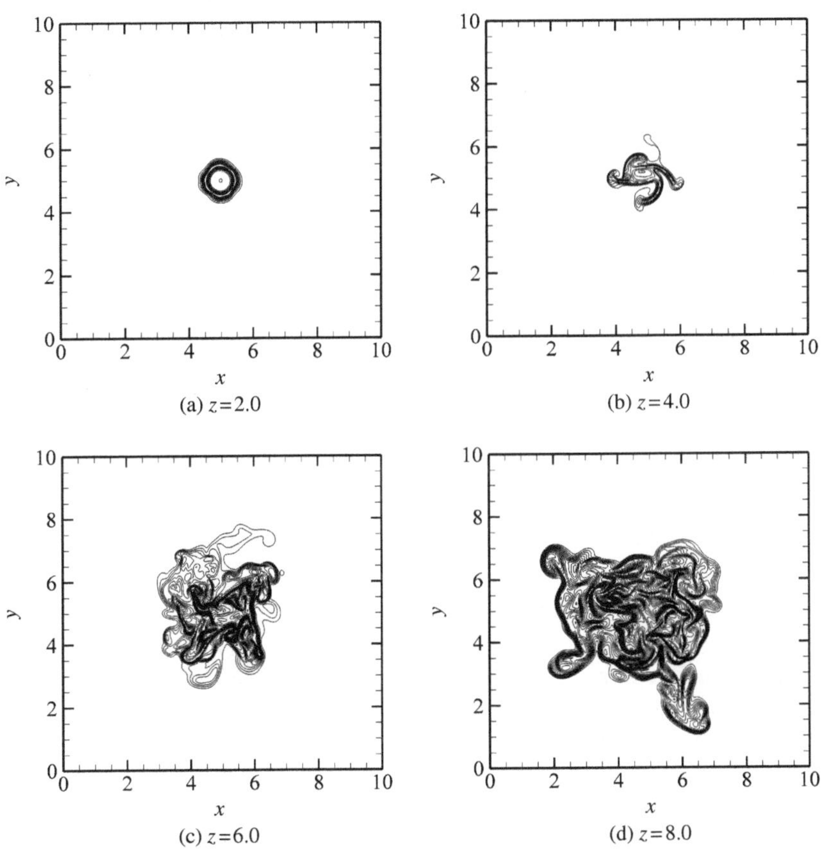

Fig. 4 Instantaneous liquid volume fraction contours at various streamwise slices.

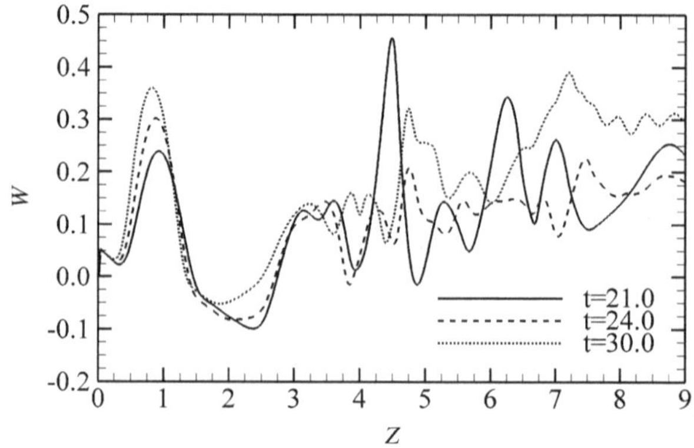

Fig. 5 Instantaneous centerline velocity profiles at different time instants.

liquid dispersion as also observed by Ibrahim and McKinney [10]. The presence of swirl gives significant rise to centrifugal and Coriolis forces, which act against the contracting effects of surface tension, causing the liquid sheet to move outwards in the radial direction. The experimental results of Ramamurthi and Tharakan [18] are in good agreement with the tendencies observed in Fig. 4.

Figure 5 shows the instantaneous centerline velocity profiles at different time instants. Significant negative velocity regions are present especially between $z = 1.0$ and $z = 3.0$. This indicates the presence of a recirculation zone [12, 21] having a geometric center at $z = 2.0$. The formation of recirculation zones in annular jets was also experimentally observed by Sheen et al. [19] and Del Taglia et al. [5]. Further downstream, significant velocity fluctuations are evident indicating the formation of large-scale vortical structures in the flow field. After $z = 7.0$, the large velocity peaks and troughs show a relative decrease in magnitude, which is primarily owed to the collapsing of large-scale structures to smaller ones.

4.2 Time-Averaged Data, Velocity Histories, and Energy Spectra

Additional analysis is presented in this subsection in an effort to better understand the flow physics and changes that occur in the flow field. The annular gas–liquid two-phase flow exhibits intrinsic instability that leads to the formation of vortical structures. To further examine the fluid dynamic behavior of the jet, time-averaged properties, velocity histories, and energy spectra are shown next. Figure 6 shows the time-averaged streamwise velocity contour at the $x = 5.0$ slice. The most important feature in Fig. 6 is the capturing of the recirculation zone, which is evident from the negative value of streamwise velocity in that region. It is interesting to note that the large-scale vortical structures at the downstream locations are an instantaneous flow

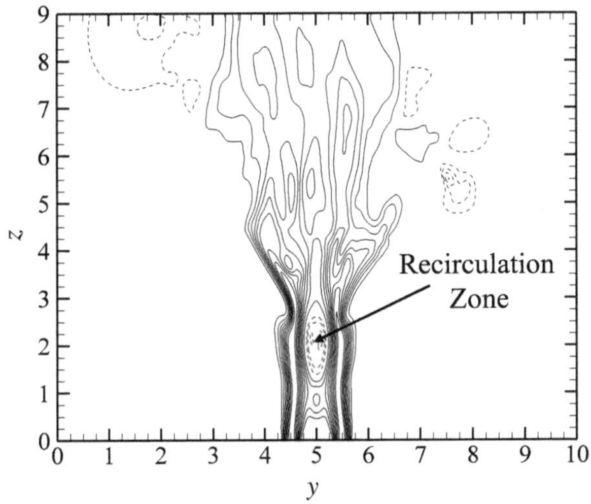

Fig. 6 Time-averaged streamwise velocity contour at $x = 5.0$ slice (solid line: positive; dashed line: negative).

characteristic and would not be present in time-averaged results, due to the fact that the vortical structures are continuously convected downstream by the mean flow.

Figure 7 shows, by means of time traces, the streamwise velocity histories at the centerline of the annular jet for two locations at $z = 2.0$ and $z = 6.0$.

In Fig. 7a, it is worth noticing that the velocity at $z = 2.0$ has negative values without showing any transition to positive values. This is because $z = 2.0$ is at the heart of the recirculation zone, as shown in Fig. 5, where only negative velocity is present. At location $z = 6.0$, downstream of the recirculation zone, larger velocity magnitudes are observed, while the negative velocity at $t = 10.0$ and $t = 15.0$

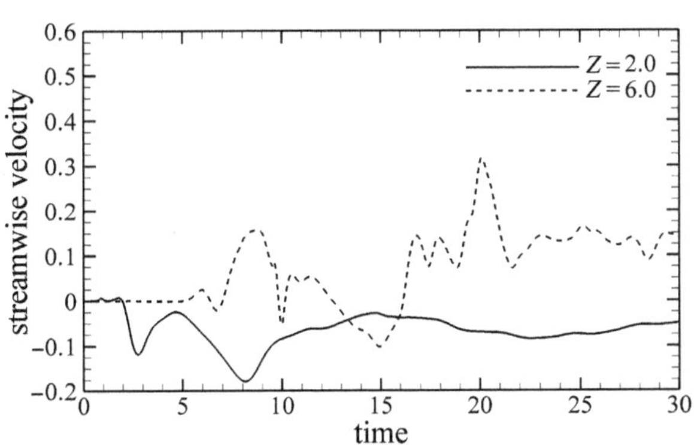

Fig. 7 Streamwise velocity histories at the jet centerline

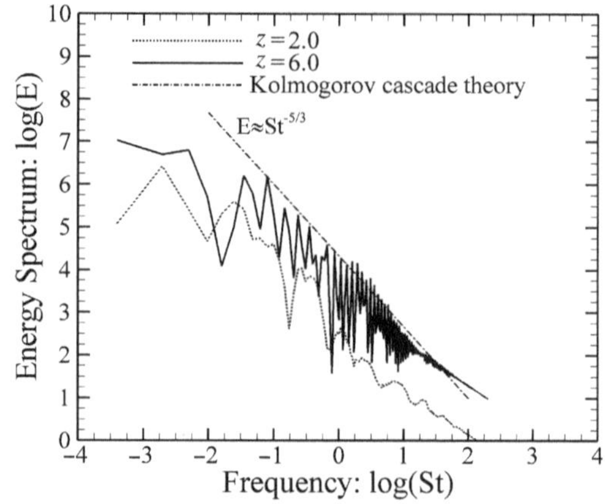

Fig. 8 Energy spectra of the instantaneous centerline velocity at different vertical locations.

is associated with the velocity reversals present in the flow field as identified by Siamas et al. [21]. The velocity fluctuations are relatively high, indicating strong vortex interaction in the flow field. In a vortical flow field, vortex interaction can lead to vortex merging/pairing, which subsequently leads to alterations of velocity periods. In Fig. 7, it is clear that the velocity amplitudes increase, as time progresses, especially at the downstream location $z = 6.0$. The increased velocity fluctuations and magnitudes indicate that strong vortex merging/pairing occurs further downstream while the flow field becomes more energetic. As a result, the development of Kelvin–Helmholtz instability tends to grow as the flow develops.

Figure 8 shows the energy spectra determined from the history of the instantaneous centerline streamwise velocity at different vertical locations of the flow field by using Fourier analysis of the Strouhal number and kinetic energy. The most important feature here is the development of high-frequency harmonics associated with the development of small scales, indicating emergence of small-scale turbulence. The transition to turbulence can be measured by the Kolmogorov cascade theory [6], which states a power law correction between the energy and the frequency of the form $E \approx St^{-5/3}$. In Fig. 8, the Kolmogorov power law is plotted together with the energy spectrum at the locations $z = 2.0$ and $z = 6.0$. The annular jet behavior approximately follows the Kolmogorov cascade theory indicating a possible transition to turbulence downstream.

5 Conclusions

Parallel direct numerical simulation of an annular liquid jet involving flow swirling has been performed. Code parallelization is based on domain decomposition and performed under the MPI environment. The mathematical formulation describing the gas–liquid two-phase flow system is based on a Eulerian approach with mixed-fluid treatment. The numerical algorithms include an adjusted VOF method for the computation of the compressible gas phase. The surface tension is resolved using a CSF model. Analytical equilibrium swirling inflow conditions have been derived that enable the exact definition of the boundary conditions at the inflow. High-order numerical schemes were used for the time-advancement and discretization.

The numerical results show that the dispersion of the liquid sheet is characterized by a recirculation zone downstream of the jet nozzle exit. Large-scale vortical structures are formed in the flow field due to the Kelvin–Helmholtz instability. The vortical structures interact with each other and lead to a more energetic flow field at further downstream locations, while analysis of the spectra shows that the jet exhibits a tendency of transition to turbulence.

Although advances in the modeling and simulation of two-phase flows have been made in recent years, the process of atomization and the exact mechanisms behind the liquid breakup still remain unclear. With the aid of computational tools like DNS, further in-depth understanding of these complex flows can be achieved, but the extremely high computational cost is always a drawback. DNS can serve as a basis for the development of databases and atomization models that will be able to overcome the problems associated with the current ones. These models are unable to correctly describe and predict the breakup of the liquid.

Acknowledgments This work made use of the facilities of HPCx, the UK's national high-performance computing service, which is provided by EPCC at the University of Edinburgh and by CCLRC Daresbury Laboratory, and funded by the Office of Science and Technology through EPSRC's High End Computing Programme. Computing time was provided by the UK Turbulence Consortium UKTC (EPSRC Grant No. EP/D044073/1).

References

1. Brackbill, J.U., Kothe, D.B., Zemach, C.: A continuum method for modelling surface tension. J. Comput. Phys. **100**, 335–354 (1992)
2. Choi, C.J., Lee, S.Y.: Droplet formation from thin hollow liquid jet with a core air flow. Atom. Sprays **15**, 469–487 (2005)
3. Crowe, C.T.: Multiphase Flow Handbook. Taylor & Francis (2006)
4. Danaila, I., Boersma, B.J.: Direct numerical simulation of bifurcating jets. Phys. Fluids **26**, 2932–2938 (2000)
5. Del Taglia, C., Blum, L., Gass, J., Ventikos, Y., Poulikakos, D.: Numerical and experimental investigation of an annular jet flow with large blockage. J. Fluids Eng. **126**, 375–384 (2004)
6. Grinstein, F.F., DeVore, C.R.: Dynamics of coherent structures and transition to turbulence in free square jets. Phys. Fluids **8**, 1237–1251 (1996)
7. Gueyffier, D., Li, J., Nadim, A., Scardovelli, R., Zaleski, S.: Volume-of-fluid interface tracking with smoothed surface stress methods for three-dimensional flows. J. Comput. Phys. **152**, 423–456 (1999)
8. Hirt, C.W., Nichols, B.D.: Volume of fluid (VOF) method for the dynamics of free boundaries. J. Comput. Phys. **39**, 201–225 (1981)
9. Hussain, A.K.M.F., Zaman, K.B.M.Q.: The preferred mode of the axisymmetric jet. J. Fluid Mech. **110**, 39–71 (1981)

10. Ibrahim, E.A., McKinney, T.R.: Injection characteristics of non-swirling and swirling annular liquid sheets. Proc. of the IMechE Part C J. Mech. Eng. Sc. **220**, 203–214 (2006)
11. Jiang, X., Luo, K.H.: Direct numerical simulation of the puffing phenomenon of an axisymmetric thermal plume. Theor. and Comput. Fluid Dyn. **14**, 55–74 (2000)
12. Jiang, X., Siamas, G.A.: Direct computation of an annular liquid jet. J. Algorithms Comput. Technology **1**, 103–125 (2007)
13. Jiang, X., Siamas, G.A., Wrobel, L.C.: Analytical equilibrium swirling inflow conditions for Computational Fluid Dynamics. AIAA J. **46**, 1015–1019 (2008)
14. Lasheras, J.C., Villermaux, E., Hopfinger, E.J.: Break-up and atomization of a round water jet by a high-speed annular air jet. J. Fluid Mech. **357**, 351–379 (1998)
15. Lele, S.K.: Compact finite-difference schemes with spectral like resolution. J. Comput. Phys. **103**, 16–42 (1992)
16. Mitchell, B.E., Lele, S.K., Moin, P.: Direct computation of the sound generated by vortex pairing in an axisymmetric jet. J. Fluid Mech. **383**, 113–142 (1999)
17. Pierce, C.D., Moin, P.: Method for generating equilibrium swirling inflow conditions. AIAA J. **36**, 1325–1327 (1999)
18. Ramamurthi, K., Tharakan, T.J.: Flow transition in swirled liquid sheets. AIAA J. **36**, 420–427 (1998)
19. Sheen, H.J., Chen, W.J., Jeng, S.Y.: Recirculation zones of unconfined and confined annular swirling jets. AIAA J. **34**, 572–579 (1996)
20. Siamas, G.A., Jiang, X.: Direct numerical simulation of a liquid sheet in a compressible gas stream in axisymmetric and planar configurations. Theor. Comput. Fluid Dyn. **21**, 447–471 (2007)
21. Siamas, G.A., Jiang, X., Wrobel, L.C.: A numerical study of an annular liquid jet in a compressible gas medium. Int. J. Multiph. Flow. **34**, 393–407 (2008)
22. Thompson, K.W.: Time dependent boundary conditions for hyperbolic systems. J. Comput. Phys. **68**, 1–24 (1987)
23. Williamson, J.H.: Low-storage Runge-Kutta schemes. J. Comput. Phys. **35**, 1-24 (1980)

Parallel Numerical Algorithm for the Traveling Wave Model

Inga Laukaitytė, Raimondas Čiegis, Mark Lichtner, and Mindaugas Radziunas

Abstract A parallel algorithm for the simulation of the dynamics of high-power semiconductor lasers is presented. The model equations describing the multisection broad-area semiconductors lasers are solved by the finite difference scheme, which is constructed on staggered grids. This nonlinear scheme is linearized applying the predictor–corrector method. The algorithm is implemented by using the ParSol tool of parallel linear algebra objects. For parallelization, we adopt the domain partitioning method; the domain is split along the longitudinal axis. Results of computational experiments are presented. The obtained speed up and efficiency of the parallel algorithm agree well with the theoretical scalability analysis.

1 Introduction

High-power, high-brightness, edge-emitting semiconductor lasers are compact devices and can serve a key role in different laser technologies such as free space communication [5], optical frequency conversion [17], printing, marking materials processing [23], or pumping fiber amplifiers [20].

A high-quality beam can be relatively easily obtained in the semiconductor laser with narrow width waveguide, where the lateral mode is confined to the stripe center. The dynamics of such lasers can be appropriately described by the Traveling Wave (TW) $(1+1)$-D model [3], which is a system of first-order PDEs with temporal and single (longitudinal) spatial dimension taken into account. Besides rather

Inga Laukaitytė · Raimondas Čiegis
Vilnius Gediminas Technical University, Saulėtekio al. 11, LT–10223, Vilnius, Lithuania
e-mail: {Inga.Laukaityte · rc}@fm.vgtu.lt

Mark Lichtner · Mindaugas Radziunas
Weierstrass Institute for Applied Analysis and Stochastics
Mohrenstarsse 39, 10117 Berlin, Germany
e-mail: {lichtner · radziunas}@wias-berlin.de

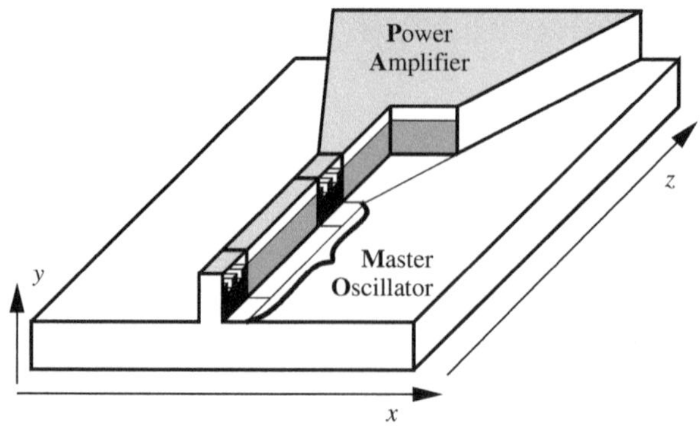

Fig. 1 Scheme of the device consisting of narrow Master Oscillator and tapered Power Amplifier parts.

fast numerical simulations, this model admits also more advanced optical mode [21] and numerical bifurcation analysis [22], which has proved to be very helpful when tailoring multisection lasers for specific applications. However, the beam power generated by such lasers usually can't exceed few hundreds of milliwatts, which is not sufficient for the applications mentioned above.

The required high output power from a semiconductor laser can be easily obtained by increasing its pumped stripe width. Unfortunately, such broad-area lasers are known to exhibit lateral and longitudinal mode instabilities, resulting in filamentations [18] that degrade the beam quality. To achieve an appropriate beam quality while keeping the beam power high, one should optimize the broad stripe laser parameters [16] or consider some more complex structures such as, e.g., tapered laser [25], schematically represented in Fig. 1.

The first three narrow sections of this device compose a Master Oscillator (MO), which, as a single laser, has stable single mode operation with good beam quality. However, the quality of the beam propagating through the strongly pumped Power Amplifier (PA) part can be degraded again due to carrier induced self-focusing or filamentation. Moreover, the non-vanishing field reflectivity from the PA output facet disturbs stable operation of the MO and can imply additional instabilities leading to mode hops with a sequence of unstable (pulsating) transition regimes [9, 24].

There exist different models describing stationary and/or dynamical states in the above-mentioned laser devices. The most complicated of them is resolving temporal-spatial dynamics of full semiconductor equations accounting for microscopic effects and is given by $(3+1)$-D PDEs (here we denote the three space coordinates plus the time coordinate). [10]. Other less complex three-dimensional models are treating some important functionalities phenomenologically and only resolve stationary states. Further simplifications of the model for tapered or broad-area lasers are made by averaging over the vertical y direction. The dynamical $(2+1)$-D models can be resolved orders of magnitudes faster allowing for parameter studies in acceptable time.

In the current chapter, we deal with a $(2+1)$-D dynamical PDE model similar to that one derived in [2, 3, 19]. Our model for the optics can be derived starting from the wave equation by assuming a TE polarized electric field (field vector pointing parallel to the x-axis in 1), a stable vertical wave guiding using the effective index method, slowly varying envelopes, and a paraxial approximation [2]. In addition to the $(1+1)-D$ longitudinal TW model [3], we take into account the diffraction and diffusion of fields and carriers in the lateral direction (described by Schrödinger and diffusion operators) as well as nonhomogeneous x-dependent device parameters, which capture the geometrical laser design. We are solving the model equations by means of finite-difference (FD) time-domain method.

The main aim of our chapter is to make numerical solution of the model as fast as possible, so that two or higher dimensional parameter studies become possible in reasonable time. By discretizing the lateral coordinate, we substitute our initial $(2+1)$-D model by J coupled $(1+1)$-D TW models [3]. For typical tapered lasers, J should be of order $10^2 - 10^3$. Thus, the CPU time needed to resolve $(2+1)$-D model is by 2 or 3 orders larger than CPU time needed to resolve a simple $(1+1)$-D TW model.

A possibility to reduce the CPU time is to use a nonuniform mesh in lateral direction. We have implemented this approach in the full FD method and without significant loss of precision were able to reduce the number of grid points (and CPU time) by factor 3.

Another, more effective way to reduce the computation time is to apply parallel computation techniques. It enables us to solve the given problems faster and/or to solve in real time problems of much larger sizes. In many cases, the latter is most important, as it gives us the possibility to simulate very complex processes with accurate approximations that require solving systems of equations with number of unknowns of order $10^6 - 10^8$ or even more.

The Domain Decomposition (DD) is a general paradigm used to develop parallel algorithms for solution of various applied problems described by systems of PDEs [12, 14]. For numerical algorithms used to solve systems of PDEs, usually the general template of such algorithms is fixed. Even more, the separation of algorithms itself and data structures used to implement these algorithms can be done. Therefore, it is possible to build general purpose libraries and templates that simplify implementation of parallel solvers, e.g., PETSc [1], Diffpack [14], and DUNE [4].

For structured orthogonal grids, the data structures used to implement numerical algorithms become even more simple. If the information on the stencil of the grid used to discretize differential equations is known in advance (or determined a posteriori from the algorithm), then it is possible to implement the data exchange among processors automatically. This approach is used in the well-known HPF project.

The new tool ParSol is targeted for implementation of numerical algorithms in C++ and semi-automatic parallelization of these algorithms on distributed memory parallel computers including clusters of PCs [11]. The library is available on the Internet at http://techmat.vgtu.lt/~alexj/ParSol/. ParSol presents very efficient and robust implementations of linear algebra objects such as arrays, vectors, and matrices [8]. Examples of different problems solved using the ParSol library are given in [7, 6].

In the current work, we apply the ParSol library for parallelization of the numerical schemes for broad-area or tapered lasers. The numerical experiments were performed on two clusters. The second cluster consists of SMP quad nodes enabling us to investigate the efficiency of the proposed parallel algorithm on multicore processors. The algorithm was implemented using MPI library, and the same code was used for shared memory data exchange inside SMP node and across distributed memory of different nodes.

The development, analysis, and implementation of numerical algorithms for solution of the full $(2+1)$-D dynamical PDE model on parallel computers is the main result of this chapter. We demonstrate a speedup of computations by a factor nearly proportional to the number of applied processors.

Our paper is organized as follows. In Sect. 2, we give a brief description of the mathematical model. The finite difference schemes for our model and a short explanation of the numerical algorithm are given in Sect. 3. Section 4 explains parallelization of our algorithm and gives estimations of the effectiveness of our approach. Some final conclusions are given in Sect. 5.

2 Mathematical Model

The model equations will be considered in the region

$$Q = \{(z,x,t): \ (z,x,t) \in (0,L) \times (-X,X) \times (0,T]\},$$

where L is the length of the laser, the interval $(-X,X)$ exceeds the lateral size of the laser, and T is the length of time interval where we perform integration.

The dynamics of the considered laser device is defined by spatio-temporal evolution of the counter-propagating complex slowly varying amplitudes of optical fields $E^{\pm}(z,x,t)$, complex dielectric dispersive polarization functions $p^{\pm}(z,x,t)$, and the real carrier density function $N(z,x,t)$. The optical fields are scaled so that

$$P(z,x,t) = |E^{+}(z,x,t)|^2 + |E^{-}(z,x,t)|^2$$

represents local photon density at the time moment t. All these functions are governed by the following $(2+1)$-D traveling wave model:

$$\frac{1}{v_g}\frac{\partial E^{\pm}}{\partial t} \pm \frac{\partial E^{\pm}}{\partial z} = -iD_f \frac{\partial^2 E^{\pm}}{\partial x^2} - i\beta(N,P)E^{\pm} - i\kappa^{\mp}E^{\mp} \qquad (1)$$
$$- \frac{g_p}{2}\left(E^{\pm} - p^{\pm}\right) + F_{sp}^{\pm},$$

$$\frac{\partial p^{\pm}}{\partial t} = i\omega_p p^{\pm} + \gamma_p (E^{\pm} - p^{\pm}), \tag{2}$$

$$\frac{\partial N}{\partial t} = \frac{\partial}{\partial x}\left(D_N \frac{\partial N}{\partial x}\right) + \frac{I}{ed} - (AN + BN^2 + CN^3) - \frac{v_g G(N) P}{1+\varepsilon P}, \tag{3}$$

where $i = \sqrt{-1}$ and propagation factor β, gain function $G(N)$, and index change function $\tilde{n}(N)$ are given by

$$\beta(N,P) = \delta + \tilde{n}(N) + \frac{i}{2}\left(\frac{G(N)}{1+\varepsilon P} - \alpha\right), \quad G(N) = g' n_{tr} \log\left(\frac{\max(N,n_*)}{n_{tr}}\right),$$

$$\tilde{n}(N) = \sigma n_{tr}\left(\sqrt{\max(N,n_*)/n_{tr}} - 1\right), \quad 0 < \frac{n_*}{n_{tr}} \ll 1.$$

One can also assume any other functional dependence of gain and index change on local carrier density N.

Well-posedness of the evolution equation (1)–(3) can be done in a similar way as in [15] by using additional $L^\infty - L^1$ estimates for the Schrödinger semigroup.

The coefficients $\kappa^{\pm}$, δ, α, ε, n_{tr}, g', σ, I, d, A, B, and C stand for complex field coupling due to the Bragg grating, static detuning, internal losses of the field, nonlinear gain compression, carrier density at transparency, differential gain, differential index change, current injection density, depth of active zone, and three recombination factors, respectively. g_p, ω_p, and γ_p represent Lorenzian fit of the gain profile in the frequency domain and denote amplitude, central frequency, and half width at half maximum of this Lorenzian. We note that almost all of the coefficients strongly depend on the spatial positions (z,x) in a discontinuous manner depending on the geometry of the laser device. For simplicity of notation, we are not showing this dependence explicitly.

Parameters D_f and D_N denote field diffraction and carrier diffusion. In general, they can also weakly depend on coordinates x and z. Their dependence on lateral coordinate x makes no harm when applying full finite difference approach (see Sect. 3, where we allow the dependence of carrier diffusion D_N on x), but implies great troubles when the split-step Fourier method is used. The factors v_g, e, denote group velocity and electron charge. Finally, the random function $F_{sp}^{\pm}$ represents spontaneous emission.

The fields $E^{\pm}$ at the laser facets $z = 0$ and $z = L$ satisfy the reflecting boundary conditions

$$E^+(0,x,t) = r_0(x) E^-(0,x,t), \quad E^-(L,x,t) = r_L(x) E^+(L,x,t), \tag{4}$$

where $r_{0,L}$ are complex reflectivity factors, $|r_{0,L}| \le 1$.

At the initial time moment, initial values of the fields, polarizations, and carrier densities are defined on $\bar{Q}_{z,x} = [0,L] \times [-X,X]$ as

$$E^{\pm}(z,x,0) = E^{\pm}_{in}(z,x), \; p^{\pm}(z,x,0) = p^{\pm}_{in}(z,x), \; N(z,x,0) = N_{in}(z,x). \tag{5}$$

The lateral boundary conditions are defined on $\bar{Q}_{z,t} = [0,L] \times (0,T]$ as

$$E^{\pm}(z,-X,t) = E^{\pm}(z,X,t) = 0, \quad N(z,-X,t) = N(z,X,t) = N_{bnd}. \tag{6}$$

3 Finite Difference Scheme

The interval $[-X,X]$ is partitioned non-uniformly

$$\omega_x = \{x_j : \ j = 0,\dots,J, \ x_0 = -X, \ x_J = X, \quad h_{x,j-1/2} = x_j - x_{j-1}\}.$$

Let us define discrete steps $h_z = L/M$, $h_t = T/K$, which are used to define uniform grids with respect to z and t coordinates. Let denote $z_i = ih_z$, $t^n = nh_t$. First we define the main discrete grid in the space domain

$$\omega_{zx} = \{(z_i, x_j) : i = 0,\dots,M, \ x_j \in \omega_x\}.$$

The discretization of problem (1)–(6) is done on staggered grids

$$\omega_E = \{(z_i, x_j, t^n) : \ i = 0,\dots,M, \ x_j \in \omega_x, \ n = 0,\dots,K\},$$

$$\omega_P = \{(z_{i-0.5}, x_j, t^n) : \ i = 1,\dots,M, \ x_j \in \omega_x, \ n = 0,\dots,K\},$$

$$\omega_N = \{(z_{i-0.5}, x_j, t^{n-0.5}) : \ i = 1,\dots,M, \ x_j \in \omega_x, \ n = 1,\dots,K\}.$$

Here subindex i is always an integer number (it should not lead to any misunderstanding with respect to $i = \sqrt{-1}$ in the PDEs of the mathematical model).

Staggered grids are very popular in solving CFD and porous media problems, they are also used to solve nonlinear optics problems [3, 21]. Such a selection of grids allows us to linearize the finite-difference scheme, which approximates a system of non-linear differential equations.

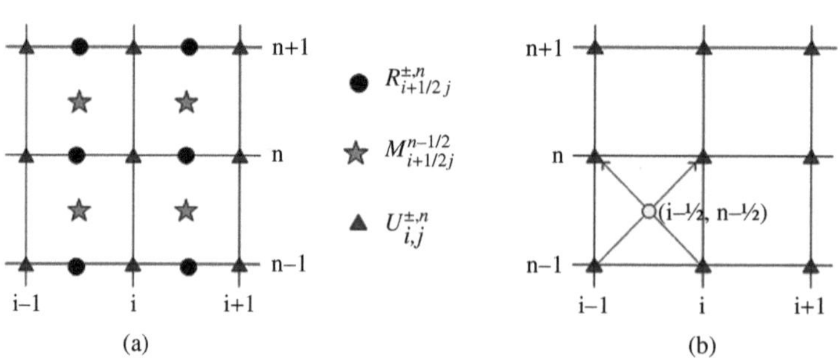

Fig. 2 Staggered grids at fixed lateral position x_j: (a) the domain of discrete functions, (b) the characteristics of transport equations (lateral x-axis is not represented in the figure).

The discrete functions $U_{i,j}^{\pm,n} = U^{\pm}(z_i, x_j, t^n)$, $R_{i-1/2,j}^{\pm,n} = R^{\pm}(z_{i-1/2}, x_j, t^n)$, and $M_{i-1/2,j}^{n-1/2} = M(z_{i-1/2}, x_j, t^{n-1/2})$ will be used to approximate $E^{\pm}$, $p^{\pm}$, and N on appropriate grids, respectively: see Fig. 2a. Approximation of differential equations is done by using the information about the characteristics of transport equations (see Fig. 2b) and applying the conservative finite volume averaging for mass conservation.

3.1 Discrete Transport Equations for Optical Fields

Transport equations (1) are approximated along characteristics, and time integration is implemented by using the Crank–Nicolson method

$$\partial_{ch} U_{i,j}^{\pm,n} = -iD_f \partial_x \partial_{\bar{x}} \bar{U}_{i-1/2,j}^{\pm,n-1/2} - i\beta\left(M_{i-1/2,j}^{n-1/2}, \bar{P}_{i-1/2,j}^{n-1/2}\right) \bar{U}_{i-1/2,j}^{\pm,n-1/2} \tag{7}$$

$$- i\kappa^{\mp} \bar{U}_{i-1/2,j}^{\mp,n-1/2} - \frac{g_p}{2}\left(\bar{U}_{i-1/2,j}^{\pm,n-1/2} - \bar{R}_{i-1/2,j}^{\pm,n-1/2}\right) + F_{sp,i-1/2,j}^{\pm,n-1/2},$$

where we use notation $h_{x,j} = 0.5(h_{x,j-1/2} + h_{x,j+1/2})$,

$$\partial_{ch} U_{i,j}^{+,n} = \frac{U_{i,j}^{+,n} - U_{i-1,j}^{+,n-1}}{h_z}, \quad \partial_{ch} U_{i,j}^{-,n} = \frac{U_{i-1,j}^{-,n} - U_{i,j}^{+,n-1}}{h_z},$$

$$\bar{U}_{i-1/2,j}^{+,n-1/2} = \frac{U_{i,j}^{+,n} + U_{i-1,j}^{+,n-1}}{2}, \quad \bar{U}_{i-1/2,j}^{-,n-1/2} = \frac{U_{i-1,j}^{-,n} + U_{i,j}^{+,n-1}}{2},$$

$$V_{i,j-1/2} := \partial_{\bar{x}} U_{i,j} = \frac{U_{i,j} - U_{i,j-1}}{h_{x,j-1/2}}, \quad \partial_x V_{i,j-1/2} = \frac{V_{i,j+1/2} - V_{i,j-1/2}}{h_{x,j}},$$

$$\bar{R}_{i-1/2,j}^{\pm,n-1/2} = \frac{R_{i-1/2,j}^{\pm,n} + R_{i-1/2,j}^{\pm,n-1}}{2}, \quad \bar{P}_{i-1/2,j}^{n-1/2} = |\bar{U}_{i-1/2,j}^{+,n-1/2}|^2 + |\bar{U}_{i-1/2,j}^{-,n-1/2}|^2.$$

Because the transport equations are approximated along characteristics, we take $h_z = v_g h_t$. The reflecting boundary conditions (4) are approximated by

$$U_{0,j}^{+,n} = r_0(x_j) U_{0,j}^{-,n}, \quad U_{M,j}^{-,n} = r_L(x_j) U_{M,j}^{+,n}, \quad n > 0,\ 0 \le j \le J. \tag{8}$$

The lateral boundary conditions are defined as

$$U_{i,0}^{\pm,n} = U_{i,J}^{\pm,n} = 0, \quad 0 \le i \le M. \tag{9}$$

3.2 Discrete Equations for Polarization Functions

Equations (2) are approximated by the exponentially fitted discrete equations

$$R^{\pm,n}_{i-1/2,j} = e^{-\widetilde{\gamma}_p h_t} R^{\pm,n-1}_{i-1/2,j} + \frac{1-e^{-\widetilde{\gamma}_p h_t}}{\widetilde{\gamma}_p} \gamma_p \bar{U}^{\pm,n-1/2}_{i-1/2,j}, \tag{10}$$

where $\widetilde{\gamma}_p = \gamma_p - i\omega_p$.

3.3 Discrete Equations for the Carrier Density Function

Equation (3) is approximated by the Crank–Nicolson type discrete scheme

$$\begin{aligned} \frac{M^{n+1/2}_{i-1/2,j} - M^{n-1/2}_{i-1/2,j}}{h_t} &= \partial_x \left(D^H_{N,j-1/2} \partial_{\bar{x}} \bar{M}^n_{i-1/2,j} \right) + \frac{I}{ed} \\ &\quad - \Gamma(\bar{M}^n_{i-1/2,j}) \bar{M}^n_{i-1/2,j} - \frac{v_g G(\bar{M}^n_{i-1/2,j}) P^n_{i-1/2,j}}{1+\varepsilon P^n_{i-1/2,j}}, \end{aligned} \tag{11}$$

where we denote

$$\begin{aligned} &\bar{M}^n_{i-1/2,j} = \frac{M^{n+1/2}_{i-1/2,j} + M^{n-1/2}_{i-1/2,j}}{2}, \quad \Gamma(M) = A + BM + CM^2, \\ &P^n_{i-1/2,j} = \frac{1}{4}\left(\left|U^{+,n}_{i,j} + U^{+,n}_{i-1,j}\right|^2 + \left|U^{-,n}_{i,j} + U^{-,n}_{i-1,j}\right|^2 \right), \\ &D^H_{N,i-1/2,j-1/2} = 2\left(\frac{1}{D_{N,i-1/2,j-1}} + \frac{1}{D_{N,i-1/2,j}} \right)^{-1}, \\ &D_{N,i-1/2,j} = D_N(x_j, z_{i-1/2}). \end{aligned}$$

The lateral boundary conditions are defined as

$$M^{n+1/2}_{i-1/2,0} = M^{n+1/2}_{i-1/2,J} = N_{bnd}, \quad 1 \le i \le M. \tag{12}$$

Approximation error of the discrete scheme is $\mathcal{O}(h_t^2 + h_z^2 + h_x^2)$.

3.4 Linearized Numerical Algorithm

Discrete scheme (7)–(12) is nonlinear. For its linearization, we use the predictor–corrector algorithm. Substitution of (10) into difference equation (7) yields the implicit discrete transport equations only for optical fields

$$
\begin{aligned}
\partial_{ch} U_{i,j}^{\pm,n} &= -iD_f \partial_x \partial_{\bar{x}} \bar{U}_{i-1/2,j}^{\pm,n-1/2} - i\kappa^{\mp} \bar{U}_{i-1/2,j}^{\mp,n-1/2} \\
&- \left(i\beta \left(M_{i-1/2,j}^{n-1/2}, \bar{P}_{i-1/2,j}^{n-1/2} \right) + \frac{g_p}{2} - \frac{g_p \gamma_p (1 - e^{-\tilde{\gamma}_p h_t})}{4\tilde{\gamma}_p} \right) \bar{U}_{i-1/2,j}^{\pm,n-1/2} \\
&+ g_p \left(1 + e^{-\tilde{\gamma}_p h_t} \right) R_{i-1/2,j}^{\pm,n-1}/4 + F_{sp,i-1/2,j}^{\pm,n-1/2}.
\end{aligned}
\tag{13}
$$

For each $i = 1,\ldots,M$, (13) are solved in two steps. In the first, predictor, step, we substitute the second argument of the propagation factor β in (13) by already known value

$$
P_{i-1/2,j}^{n-1} = \frac{1}{4}\left(|U_{i-1,j}^{+,n-1} + U_{i,j}^{+,n-1}|^2 + |U_{i,j}^{-,n-1} + U_{i-1,j}^{-,n-1}|^2 \right),
$$

and look for the grid function $\tilde{U}_{\cdot,\cdot}^{\pm,n}$, giving an intermediate approximation (*prediction*) for the unknown $U_{\cdot,\cdot}^{\pm,n}$ entering the nonlinear scheme (13). In the second, corrector, step, we use a corrected photon density approximation

$$
\tilde{P}_{i-1/2,j}^{n-1/2} = \frac{1}{4}\left(|U_{i-1,j}^{+,n-1} + \tilde{U}_{i,j}^{+,n}|^2 + |U_{i,j}^{-,n-1} + \tilde{U}_{i-1,j}^{-,n}|^2 \right).
$$

Being a more precise (*corrected*) approximation of the grid function $U_{\cdot,\cdot}^{\pm,n}$, the solution of the resulting linear scheme is used in the consequent computations.

Both prediction and correction steps are systems of linear equations with block-tridiagonal matrices. That is, these systems can be represented by

$$
\begin{cases}
V_0 = V_J = 0, \\
\mathscr{A}_j V_{j-1} + \mathscr{C}_j V_j + \mathscr{B}_j V_{j+1} = \mathscr{D}_j, \; 1 \le j < J,
\end{cases}
$$

where $V_j = (\tilde{U}_{i,j}^{+,n}, \tilde{U}_{i-1,j}^{-,n})^T$ (predictor step) or $V_j = (U_{i,j}^{+,n}, U_{i-1,j}^{-,n})^T$ (corrector step), and $\mathscr{A}_j$, $\mathscr{B}_j$, $\mathscr{C}_j$, are 2×2 matrices, and $\mathscr{D}_j$ is a two-component vector containing information about field values at the $(n-1)$-th time layer. These systems are effectively solved by means of the block version of the factorization algorithm.

The nonlinear scheme (11) is also solved in two steps. In the predictor step, we substitute the arguments of functions Γ and G by $M_{i-1/2,j}^{n-1/2}$ and look for the grid function $\tilde{M}_{\cdot,\cdot}^{n+1/2}$, giving an intermediate approximation of $M_{\cdot,\cdot}^{n+1/2}$ from nonlinear equations (11). In the corrector step, these arguments are substituted by $(\tilde{M}_{i-1/2,j}^{n+1/2} + M_{i-1/2,j}^{n-1/2})/2$. The solution of the resulting linear equations approximates $M_{\cdot,\cdot}^{n+1/2}$ and is used in the consequent computations. The obtained systems of linear equations with the boundary conditions (12) are solved by a standard factorization algorithm.

4 Parallelization of the Algorithm

The FDS is defined on the structured staggered grid, and the complexity of computations at each node of the grid is approximately the same. For such problems, the parallelization of the algorithm can be done by using domain decomposition (DD) paradigm [12].

4.1 Parallel Algorithm

The development of any DD-type parallel algorithm requires answers to two main questions. First, we need to select how the domain will be partitioned among processors. At this step, the main goal is to preserve the load balance of volumes of subdomains and to minimize the amount of edges connecting grid points of neigboring subdomains. The last requirement means the minimal costs of data communication among processors during computations. This property is especially important for clusters of PCs, where the ratio between computation and communication rates is not favorable. Let p be the number of processors. It is well-known that for 2D structured domains, the optimal DD is obtained if 2D topology of processor $p_1 \times p_2$ is used, where $p_j \sim \sqrt{p}$. But for the algorithm (13), we can't use such a decomposition straightforwardly as the matrix factorization algorithm for solution of the block three-diagonal system of linear equations is fully sequential in its nature. There are some modifications of the factorization algorithm with much better parallelization properties, but the complexity of such algorithms is at least 2 times larger than the original one (see, e.g., [26]). Thus in this chapter, we restrict to 1D block domain decomposition algorithms, decomposing the grid only in z direction (see Fig. 3).

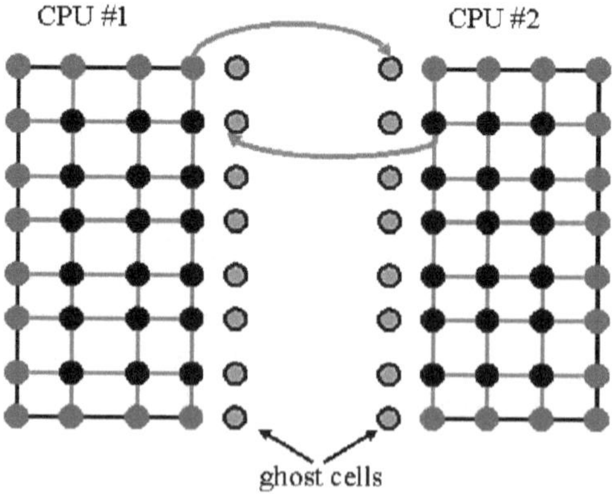

Fig. 3 Scheme of the 1D block domain decomposition (distribution with respect to the z coordinate.)

The second step in building a parallel algorithm is to define when and what data must be exchanged among processors. This information mainly depends on the stencil of the grid used to approximate differential equations by the discrete scheme. For algorithm (13), two different stencils are used to approximate waves moving in opposite directions.

Let us denote by $\omega_z(k)$ the subgrid belonging to the k-th processor

$$\omega_z(k) = \{z_i : \ i_{kL} \le i \le i_{kR}\}.$$

Here the local sub-domains are not overlapping, i.e., $i_{kL} = i_{k-1,R} + 1$.

In order to implement the computational algorithm, each processor extends its subgrid by ghost points

$$\widetilde{\omega}_z(k) = \{z_i : \ \widetilde{i}_{kL} \le i \le \widetilde{i}_{kR}\}, \ \ \tilde{i}_{kL} = \max(i_{kL} - 1, 0), \ \ \tilde{i}_R = \min(i_{kR} + 1, M).$$

Then after each predictor and corrector substeps, the k-th processor

- sends to $(k+1)$-th processor vector $U^{+,\cdot}_{i_{kR},\cdot}$ and receives from it vector $U^{-,\cdot}_{\tilde{i}_{kR},\cdot}$,
- sends to $(k-1)$-th processor vector $U^{-,\cdot}_{i_{kL},\cdot}$ and receives from it vector $U^{+,\cdot}_{\tilde{i}_{kL},\cdot}$.

Obviously, if $k = 0$ or $k = (p-1)$, then a part of the communications is not done. We note that vectors $R^{\cdot}_{i-1/2,\cdot}$, $M^{\cdot}_{i-1/2,\cdot}$ are computed locally by each processor and no communications of values at ghost points are required.

4.2 Scalability Analysis

In this section, we will estimate the complexity of the parallel algorithm. Neglecting the work done to update the boundary conditions on ω_{zx}, we get that the complexity of the serial algorithm for one time step is given by $W = \gamma M(J+1)$, where γ estimates the CPU time required to implement one basic operation of the algorithm.

The ParSol tool distributes among processors the grid ω_{zx} using 1D type distribution with respect to the z coordinate. The total size of this grid is $(M+1)(J+1)$ points. Then the computational complexity of parallel algorithm depends on the size of the largest local grid part, given to one processor. It is equal to

$$T_{p,comp} = \gamma\big(\lceil (M+1)/p \rceil + 1\big)(J+1),$$

where $\lceil x \rceil$ denotes a smallest integer number larger than or equal to x. This formula includes costs of extra computations involving ghost points.

Data communication time is given by $T_{p,comm} = 2\big(\alpha + \beta(J+1)\big)$, here α is the message startup time and β is the time required to send one element of data. We assume that communication between neighboring processors can be implemented in parallel. Thus the total complexity of the parallel algorithm is equal to

$$T_p = \gamma\big(\lceil (M+1)/p \rceil + 1\big)(J+1) + 2\big(\alpha + \beta(J+1)\big). \tag{14}$$

The scalability analysis of any parallel algorithm enables us to find the rate at which the size of problem W needs to grow with respect to the number of processors p in order to maintain a fixed efficiency E of the algorithm. Let $H(p,W) = pT_p - W$ be the total overhead of a parallel algorithm. Then the *isoefficiency* function $W = g(p,E)$ is defined by the implicit equation [12]:

$$W = \frac{E}{1-E} H(p,W). \tag{15}$$

The total overhead of the proposed parallel algorithm is given by

$$H(p,W) = \gamma(p+1)(J+1) + 2\alpha p + 2\beta p(J+1).$$

After simple computations, we get from (15) the following isoefficiency function, expressed with respect to the number of grid points in z coordinate:

$$M = \frac{E}{1-E}\left[\left(1 + 2\frac{\beta}{\gamma} + 2\frac{\alpha}{\gamma(J+1)}\right)p + 1\right].$$

Thus in order to maintain a fixed efficiency E of the parallel algorithm, it is sufficient to preserve the same number of grid $\omega_z(k)$ points per processor. The increase of J reduces the influence of the message startup time.

4.3 Computational Experiments

In this chapter, we restrict to computational experiments that are targeted for the efficiency and scalability analysis of the given parallel algorithm. Results of extensive computational experiments for simulation of the dynamics of multisection semiconductor lasers and analysis of their stability domain will be presented in a separate paper.

We have solved the problem (1)–(6) by using the discrete approximation (7)–(12). The dynamics of laser waves was simulated until 0.2 ns. The discretization was done on three discrete grids of $(M+1)\times(J+1)$ elements, with $(M=500, J=300)$, $(M=500, J=600)$, and $(M=1000, J=600)$ respectively. Note that an increase of M implies a proportional increase of the time steps K within the interval of computations.

The parallel algorithm was implemented by using the mathematical objects library ParSol [8, 11]. This tool not only implements some important linear algebra objects in C++, but also allows one to parallelize semiautomaticaly data parallel algorithms, similarly to HPF.

First, the parallel code was tested on the cluster of PCs at Vilnius Gediminas Technical University. It consists of Pentium 4 processors (3.2GHz, level 1 cache 16KB, level 2 cache 1MB) interconnected via Gigabit Smart Switch (http://vilkas.vgtu.lt). Obtained performance results are presented in Table 1. Here

Table 1 Results of computational experiments on Vilkas cluster.

	$p=1$	$p=2$	$p=4$	$p=8$	$p=16$
$S_p(500\times 300)$	1.0	1.93	3.81	7.42	14.2
$E_p(500\times 300)$	1.0	0.97	0.95	0.93	0.90
$S_p(500\times 600)$	1.0	1.93	3.80	7.43	14.4
$E_p(500\times 600)$	1.0	0.97	0.95	0.93	0.90
$S_p(1000\times 600)$	1.0	1.94	3.82	7.62	14.9
$E_p(1000\times 600)$	1.0	0.97	0.96	0.95	0.93

for each number of processors p, the coefficients of the algorithmic speedup $S_p = T_1/T_p$ and efficiency $E_p = S_p/p$ are presented. T_p denotes the CPU time required to solve the problem using p processors, and the following results were obtained for the sequential algorithm (in seconds):

$$T_1(500\times 300) = 407.3,\ \ T_1(500\times 600) = 814.2,\ \ T_1(1000\times 600) = 3308.4.$$

We see that experimental results scale according the theoretical complexity analysis prediction given by (14). For example, the efficiency of the parallel algorithm satisfies the estimate $E_{2p}(1000\times 600) \approx E_p(500\times 600)$.

Next we present results obtained on the Hercules cluster in ITWM, Germany. It consists of dual Intel Xeon 5148LV nodes (i.e., 4 CPU cores per node), each node has 8GB RAM, 80GB HDD, and the nodes are interconnected by 2x Gigabit Ethernet, Infiniband. In Table 2, the values of the speed-up and efficiency coefficients are presented for the discrete problem simulated on the discrete grid of size 640×400 and different configurations of nodes. In all cases, the nodes were used in the dedicated to one user mode. We denote by $n\times m$ the configuration, where n nodes are used with m processes on each node.

It follows from the presented results that the proposed parallel algorithm efficiently runs using both computational modes of the given cluster. In the case of $p\times 1$ configuration, the classic cluster of the distributed memory is obtained, and in the case of $n\times m$ configuration, the mixture model of the global memory inside one node and distributed memory across the different nodes is used. It seems that the usage of L1 and L2 cache memory also improved for larger numbers of processes, when smaller local problems are allocated to each node.

Table 2 Results of computational experiments on Hercules cluster.

	1×1	2×1	1×2	4×1	1×4	8×1	2×4	4×4	8×4
S_p	1.0	1.88	1.95	3.94	3.40	7.97	7.22	14.96	29.2
E_p	1.0	0.94	0.97	0.99	0.85	0.99	0.90	0.93	0.91

5 Conclusions

The new parallel algorithm for simulation of the dynamics of high-power semi-conductor lasers is presented. The code implements second-order accurate finite-difference schemes in space and time. It uses a domain decomposition paralleliza- tion paradigm for the effective partitioning of the computational domain. The parallel algorithm is implemented by using ParSol tool, which uses MPI for data communication. The computational experiments carried out have shown the scalability of the code in different clusters including SMP nodes with 4 cores.

Further studies must be carried out to test 2D data decomposition model in order to reduce the amount of data communicated during computations. The second problem is to consider a splitting type numerical disretization and to use FFT to solve the linear algebra sub-tasks arising after the discretization of the diffraction operator.

Acknowledgments R. Čiegis and I. Laukaitytė were supported by the Lithuanian State Science and Studies Foundation within the project on B-03/2007 "Global optimization of complex systems using high performance computing and GRID technologies". The work of M. Radziunas was supported by DFG Research Center MATHEON "Mathematics for key technologies: Modelling, simulation and optimization of the real world processes".

References

1. Balay, S., Buschelman, K., Eijkhout, V., Gropp, W.D., Kaushik, D., Knepley, M.G., Curfman McInnes, L., Smith, B.F., Zhang, H.: PETSc Users Manual. ANL-95/11 – Revision 2.3.0. Argonne National Laboratory (2005)
2. Balsamo, S., Sartori, F., Montrosset, I.: Dynamic beam propagation method for flared semiconductor power amplifiers. IEEE Journal of Selected Topics in Quantum Electronics **2**, 378–384 (1996)
3. Bandelow, U., Radziunas, M, Sieber, J., Wolfrum, M.: Impact of gain dispersion on the spatio-temporal dynamics of multisection lasers. IEEE J. Quantum Electron. **37**, 183–188 (2001)
4. Blatt, M., Bastian, P.: The iterative solver template library. In: B. Kågström, E. Elmroth, J. Dongarra, J. Wasniewski (eds.) Applied Parallel Computing: State of the Art in Scientific Computing, *Lecture Notes in Scientific Computing*, vol. 4699, pp. 666–675. Springer, Berlin Heidelberg New York (2007)
5. Chazan, P., Mayor, J.M., Morgott, S., Mikulla, M., Kiefer, R., Müller, S., Walther, M., Braunstein, J., Weimann, G.: High-power near difraction-limited tapered amplifiers at 1064 nm for optical intersatelite communications. IEEE Phot. Techn. Lett. **10**(11), 1542–1544 (1998)
6. Čiegis, Raim, Čiegis, Rem., Jakušev, A., Šaltenienė, G.: Parallel variational iterative algorithms for solution of linear systems. Mathematical Modelling and Analysis **12**(1) 1–16 (2007)
7. Čiegis, R., Jakušev, A., Krylovas, A., Suboč, O.: Parallel algorithms for solution of nonlinear diffusion problems in image smoothing. Mathematical Modelling and Analysis **10**(2), 155–172 (2005)
8. Čiegis, R., Jakušev, A., Starikovičius, V.: Parallel tool for solution of multiphase flow problems. In: R. Wyrzykowski, J. Dongarra, N. Meyer, J. Wasniewski (eds.) Sixth International conference on Parallel Processing and Applied Mathematics. Poznan, Poland, September 10-14, 2005, *Lecture Notes in Computer Science*, vol. 3911, pp. 312–319. Springer, Berlin Heidelberg New York (2006)

9. Egan, A., Ning, C.Z., Moloney, J.V., Indik, R. A., et al.: Dynamic instabilities in Master Oscillator Power Amplifier semiconductor lasers. IEEE J. Quantum Electron. **34**, 166–170 (1998)
10. Gehrig, E., Hess, O., Walenstein, R.: Modeling of the performance of high power diode amplifier systems with an optothermal microscopic spatio-temporal theory. IEEE J. Quantum Electron. **35**, 320–331 (2004)
11. Jakušev, A.: Development, analysis and applications of the technology for parallelization of numerical algorithms for solution of PDE and systems of PDEs. Doctoral dissertation. Vilnius Gediminas Technical University, Technika, 1348, Vilnius (2008)
12. Kumar, V., Grama, A., Gupta, A., Karypis, G.: Introduction to Parallel Computing: Design and Analysis of Algorithms. Benjamin/Cummings, Redwood City, CA (1994)
13. Lang, R.J., Dzurko, K.M., Hardy, A., Demars, S., Schoenfelder, A., Welch, D.F.: Theory of grating-confined broad-area lasers. IEEE J. Quantum Electron. **34**, 2196–2210 (1998)
14. Langtangen, H.P.: Computational Partial Differential Equations. Numerical Methods and Diffpack Programming. Springer, Berlin (2002)
15. Lichtner, M., Radziunas, M., Recke, L.: Well posedness, smooth dependence and center manifold reduction for a semilinear hyperbolic system from laser dynamics. Math. Meth. Appl. Sci. **30**, 931–960 (2007)
16. Lim, J.J., Benson, T.M., Larkins, E.C.: Design of wide-emitter single-mode laser diodes. IEEE J. Quantum Electron. **41**, 506–516 (2005)
17. Maiwald, M., Schwertfeger, S., Güther, R., Sumpf, B., Paschke, K., Dzionk, C., Erbert, G., Tränkle, G.: 600 mW optical output power at 488 nm by use of a high-power hybrid laser diode system and a periodically poled MgO:LiNbO$_3$. Optics Letters **31**(6), 802–804 (2006)
18. Marciante, J.R., Agrawal, G.P.: Nonlinear mechanisms of filamentation in braod-area semiconductor lasers. IEEE J. Quantum Electron. **32**, 590–596 (1996)
19. Ning, C.Z., Indik, R.A., Moloney, J.V.: Effective Bloch equations for semiconductor lasers and amplifiers. IEEE J. Quantum Electron. **33**, 1543–1550 (1997)
20. Pessa, M., Näppi, J., Savolainen, P., Toivonen, M., Murison, R., Ovchinnikov, A., Asonen, H.: State-of-the-art aluminum-free 980-nm laser diodes. J. Lightwave Technol. **14**(10), 2356–2361 (1996)
21. Radziunas, M., Wünsche, H.-J.: Multisection lasers: longitudinal modes and their dynamics. In: J. Piprek (ed) Optoelectronic Devices-Advanced Simulation and Analysis, pp. 121–150. Spinger Verlag, New York (2004)
22. Radziunas, M.: Numerical bifurcation analysis of the traveling wave model of multisection semiconductor lasers. Physica D **213**, 98–112 (2006)
23. Schultz, W., Poprawe, R.: Manufacturing with nover high-power diode lasers. IEEE J. Select. Topics Quantum Electron. **6**(4), 696–705 (2000)
24. Spreemann, M.: Nichtlineare Effekte in Halbleiterlasern mit monolithisch integriertem trapezförmigem optischen Verstärker. Diploma thesis, Department of Physics, HU-Berlin, (2007)
25. Walpole, J.N.: Tutorial review: Semiconductor amplifiers and lasers with tapered gain regions. Opt. Quantum Electron. **28**, 623–645 (1996)
26. Wang, H.H.: A parallel method for tridiagonal equations. ACM Transactions on Mathematical Software **7**(2), 170–183 (1981)

Parallel Algorithm for Cell Dynamics Simulation of Soft Nano-Structured Matter

Xiaohu Guo, Marco Pinna, and Andrei V. Zvelindovsky

Abstract Cell dynamics simulation is a very promising approach to model dynamic processes in block copolymer systems at the mesoscale level. A parallel algorithm for large-scale simulation is described in detail. Several performance tuning methods based on SGI Altix are introduced. With the efficient strategy of domain decomposition and the fast method of neighboring points location, we greatly reduce the calculating and communicating cost and successfully perform simulations of large-scale systems with up to 512^3 grids. The algorithm is implemented on 32 processors with the speedup of 28.4 and the efficiency of 88.9%.

1 Introduction

Use of soft materials is one of the recent directions in nano-technology. Self-organization in soft matter serves as a primary mechanism of structure formation [9]. A new challenge is the preparation of nanosize structure for the miniaturization of device and electronic components. We are interested in structures formed by an important soft matter – block copolymers (BCP). BCP consist of several chemically different blocks, and they can self-assemble into structures on the nano-scale. These can be very simple structures as lamellae, hexagonally ordered cylinders and spheres, and more complex structures such as gyroid. Block copolymer systems have been studied for several decades in a vast number of experimental and theoretical works. For this purpose, applied external fields like shear flow or electric field can be used [9]. The research is driven by the desire to tailor a certain morphology. These experiments are very difficult, and computer modeling guidance can help to understand physical propriety of block copolymers. In general, the computer

Xiaohu Guo · Marco Pinna · Andrei V. Zvelindovsky
School of Computing, Engineering and Physical Sciences, University of Central Lancashire
Preston, Lancashire, PR1 2HE, United Kingdom
e-mail: XGuo@uclan.ac.uk · MPinna@uclan.ac.uk · AVZvelindovsky@uclan.ac.uk

modeling of pattern formation is a very difficult task for the experimental times and spatial resolution required. In the past decades, many different computer modeling methods have been developed. Some of them are very accurate but very slow, others are very fast and less accurate, but not less useful [9]. The latter ones are mesoscopic simulation methods that can describe the behavior of block copolymers on a large scale. One of these methods is the cell dynamics simulation (CDS). The cell dynamics simulation is widely used to describe the mesoscopic structure formation of diblock copolymer systems [5, 4]. The cell dynamics simulation is reasonably fast and can be performed in relatively large boxes. However, experimental size systems and experimental times cannot be achieved even with this method on modern single processor computers. To link simulation results to experiments, it is necessary to use very large simulation boxes. The only way to achieve this goal is to create a computer program that can run on many processor in parallel. Here we present a parallel algorithm for CDS and its implementation.

The chapter is organized as follows: In Sect. 2, the cell dynamics simulation algorithm is presented. In Sect. 3, the parallel algorithm is presented, before drawing the conclusions in Sect. 4.

2 The Cell Dynamics Simulation

The cell dynamic simulation method is described extensively in the literature [5, 4]. We only repeat the main equations of cell dynamics simulation here for a diblock copolymer melt. In the cell dynamics simulation, an order parameter $\psi(\mathbf{r},t)$ is determined at time t in the cell $\mathbf{r}$ of a discrete lattice (see Fig. 1), where $\mathbf{r} = (r_x,r_y,r_z)$.

For AB diblock copolymer, we use the difference between local and global volume fractions $\psi = \phi_A - \phi_B + (1-2f)$ where ϕ_A and ϕ_B are the local volume fractions

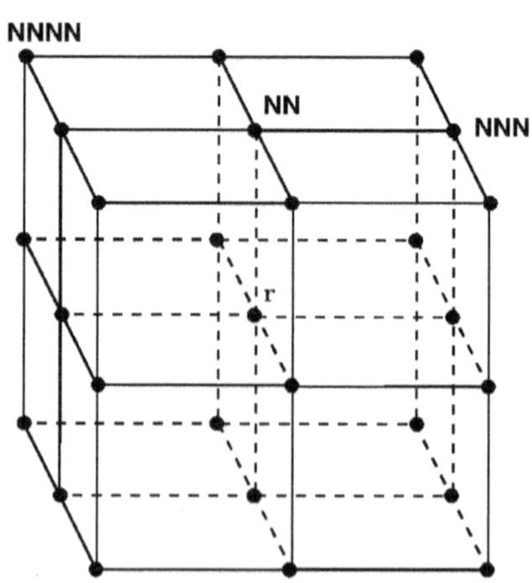

Fig. 1 A stencil for Laplacian, where **NN** denotes nearest neighbors, **NNN** next-nearest neighbors, **NNNN** next-next-nearest neighbors.

of A and B monomers respectively, and f is the volume fraction of A monomers in the diblock, $f = N_A/(N_A + N_B)$.

The time evolution of the order parameter is given by a Cahn–Hilliard–Cook (CHC) equation [3]:

$$\frac{\partial \psi}{\partial t} = M\nabla^2 \left(\frac{\delta F[\psi]}{\delta \psi} \right) + \eta \xi(\mathbf{r},t), \tag{1}$$

where M is a phenomenological mobility constant. We set M=1, which correspondingly sets the timescale for the diffusive processes (the dimensionless time is tM/a_0^2, where the lattice cell size a_0 is set to 1). The last term in (1) is a noise with amplitude η and $\xi(\mathbf{r},t)$ being a Gaussian random noise [6]. The free energy functional (divided by kT) is

$$F[\psi(\mathbf{r})] = \int d\mathbf{r}[H(\psi) + \frac{D}{2}|\nabla\psi|^2] + \frac{B}{2}\int d\mathbf{r}\int d\mathbf{r}' G(\mathbf{r}-\mathbf{r}')\psi(\mathbf{r})\psi(\mathbf{r}'),$$

where

$$H(\psi) = [-\frac{\tau}{2} + \frac{A}{2}(1-2f)^2]\psi^2 + \frac{v}{3}(1-2f)\psi^3 + \frac{u}{4}\psi^4 \tag{2}$$

with τ being a temperature-like parameter and A, B, v, u, D being phenomenological constants. The Laplace equation Green function $G(\mathbf{r}-\mathbf{r}')$ satisfies $\nabla^2 G(\mathbf{r}-\mathbf{r}') = -\delta(\mathbf{r}-\mathbf{r}')$.

The numerical scheme for cell dynamics in (1) becomes [5]

$$\psi(\mathbf{r},t+1) = \psi(\mathbf{r},t) - \{\langle\langle\Gamma(\mathbf{r},t)\rangle\rangle - \Gamma(\mathbf{r},t) + B\psi(\mathbf{r},t) - \eta\xi(\mathbf{r},t)\}, \tag{3}$$

where

$$\Gamma(\mathbf{n},t) = g(\psi(\mathbf{n},t)) - \psi(\mathbf{n},t) + D[\langle\langle\psi(\mathbf{n},t)\rangle\rangle - \psi(\mathbf{n},t)]$$

and the so-called map function $g(\psi(\mathbf{r},t))$ is defined by

$$g(\psi) = [1 + \tau - A(1-2f)^2]\psi - v(1-2f)\psi^2 - u\psi^3. \tag{4}$$

The formula

$$\langle\langle\psi(\mathbf{r},t)\rangle\rangle = \frac{6}{80}\sum_{k\in\{NN\}}\psi(\mathbf{r_k},t) + \frac{3}{80}\sum_{k\in\{NNN\}}\psi(\mathbf{r_k},t) + \frac{1}{80}\sum_{k\in\{NNNN\}}\psi(\mathbf{r_k},t) \tag{5}$$

is used to calculate the isotropized discrete Laplacian $\langle\langle X\rangle\rangle - X$; the discrete lattice is shown in Fig. 1.

We use periodic boundary condition in all three directions. The initial condition is a random distribution of $\psi(\mathbf{r},0) \in (-1,1)$.

3 Parallel Algorithm of CDS Method

The sequential CDS algorithm (implemented in Fortran 77) was parallelized both to overcome the long computing time and lessen memory requirement on individual processors. As a result, a significant speedup is obtained compared with the sequential algorithm. The method adopted in the parallelization is the spatial decomposition method.

3.1 The Spatial Decomposition Method

The computing domain is the cubic, which is discretized by three-dimensional grids (see Fig. 2). A typical simulation example of sphere morphology of a diblock copolymer melt is shown in Fig. 2. The snapshot is taken at 5000 time steps in the simulation box of 128^3 grids points. The parameters used are A=1.5, B=0.01, τ=0.20, f=0.40, u=0.38, v=2.3, which are the same as the ones used in [5]. The grids are divided into three-dimensional sub-grids and these sub-grids associated with different processors. There is communication between different processors. In order to get a better communication performance, the processes on different processors are mapped to a three-dimensional topological pattern by using Message Passing Interface (MPI) Cartesian topology functions [8]. The topological structure of the processes is the same as the computing domain, which also reflects the logical communication pattern of the processes. Due to the relative mathematical simplicity of CDS algorithm itself, there is no need for global data exchange between different processors.

The communication structure in CDS method consists of exchanging boundary data between neighboring sub-grids. In order to reduce communication costs, additional memory is allocated at the boundaries of each of the sub-grids. These extra grid points are called *ghost points* and are used to store the boundary data

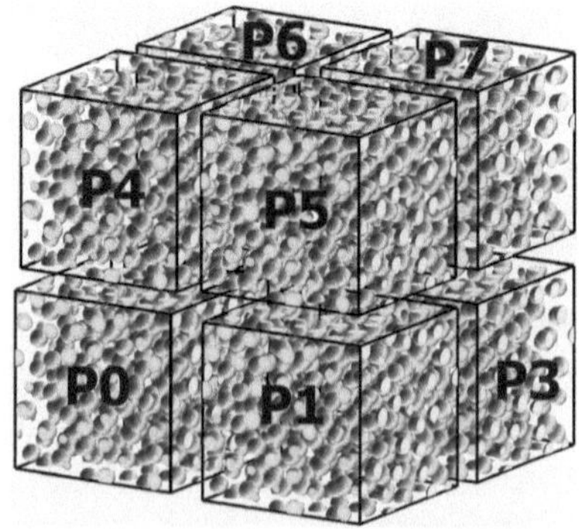

Fig. 2 A cubic computing domain on a three-dimensional grid (x y z). The grid is subdivided into 8 subgrids that are associated with 8 processors (P0 to P7).

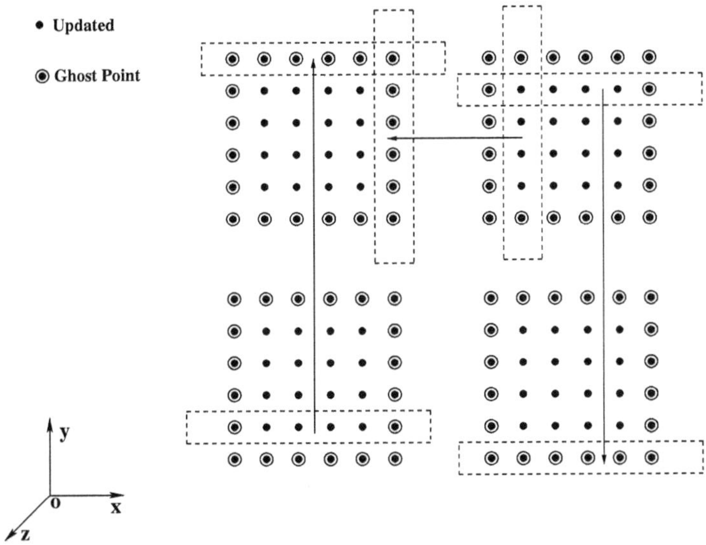

Fig. 3 The domain decomposition of 8^3 grid as viewed in x–y plane. The grid is subdivided into 8 subgrids (only 4 are seen in the x–y plane). The ghost points are added to each of subgrids. The communications between processors in x–y plane are indicated by the arrows.

communicated by the neighboring processors. We arrange the communication in blocks to transmit all the needed points in a few messages, as shown in Fig. 3.

The CDS simulation's parallel pseudo algorithm can be described in the following:

```
Step 1:
     MPI Initialization
     Define Global Grids in X, Y, Z Directions
     Automatic Factorization of Total Processors
     P = NPX * NPY * NPZ
     Construct the toplogical communication structure
     MPI_COMM_X, MPI_COMM_Y, MPI_COMM_Z
     Define MPI Datatypes for boundary communication
Step 2:
     IF(BreakPoint .eq. .TRUE.) THEN
       Read Init File from t=Time_Begin
     ELSE
       Init the value of ψ(r,0) on different processors
       Time_Begin=0
     ENDIF
     DO t=Time_Begin+1, Time_End
        Exchange Boundaries of ψ(r,t−1)
        DO r in 3D Subgrids
```

```
        Compute map function g(r,t) using (4)
      ENDDO
      Exchange Boundaries of g(r,t)
      DO r in 3D Subgrids
        Compute the density functions ψ(r,t)
        using (3)
      ENDDO
   ENDDO
Step 3:
   Parallel Output
```

In step 1, MPI environment is initialized, with various processors numbered, and the grids are automatically divided into sub-grids. Processes topology functions are used to construct communication pattern [8]. In step 2, our program can start from any time and a parallel random generator is used to set the initial value $\psi(\mathbf{r},0)$. Each time step performs two times boundary communication. In step 3, MPI parallel Input/Output functions [8] are used to output results.

3.2 *Parallel Platform and Performance Tuning*

Our code is developed on an SGI Altix 3700 computer that has 56 Intel Itanium-2 CPUs (1.3GHz 3MB L3 Cache) and 80GB addressable memory with SLES 10 and SGI Propack 5 installed. Network version is ccNUMA3, Parallel Library is SGI MPT, which is equal to MPICH 1.2.7, and the Intel C and Fortran Compilers 10.1 are used. For the reference, a typical CDS requires at least 10^5 time steps. For a 512^3 simulation system, which can accomodate a few domains of a realistic experimental soft matter system, such run would take approximately 105 hours on a single processor.

Several optimizational methods are used. In Fig. 4, we present the efficiency of the algorithm depending on the way the processors are distributed along the different axes of the grid (for 8 processors) [2]. We observe that the performance is rather sensitive to the processor distribution. The highest performance is not necessarily achieved for cubic sub-grids (in our illustration, the distribution (x,y,z)=(1,1,8) gives the best performance). This difference is due to the interplay of two factors: the way elements of a three-dimensional array are accessed in a given language (in our implementation, Fortran 77) and the exchange of ghost points data along sub-grids boundaries.

SGI MPI implementation offers a number of significant features that make it the preferred implementation to use on SGI hardware. By default, the SGI implementation of MPI uses buffering in most cases. Short messages (64 or fewer bytes) are always buffered. Longer messages are also buffered, although under certain circumstances buffering can be avoided. For the sake of performance, it is sometimes desirable to avoid buffering. One of the most significant optimizations for bandwidth sensitive applications in the MPI library is single copy optimization [7] avoiding

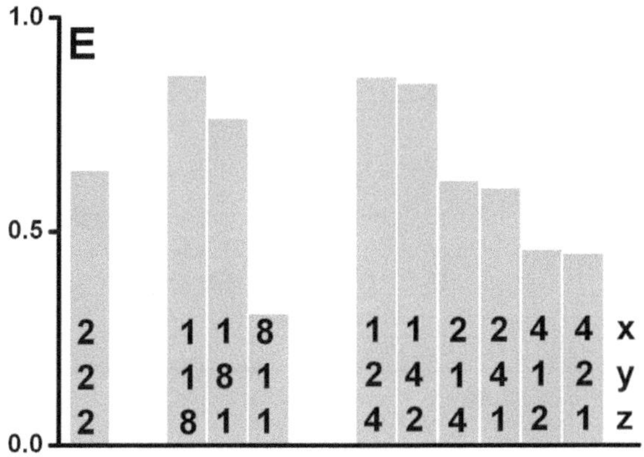

Fig. 4 Efficiency $E = E(256^3, 8)$ for different distributions of 8 processors along x-, y-, z-directions.

the use of shared memory buffers. There are some limitations on using single copy optimization:

- The MPI data type on the send side must be a contiguous type;
- The sender and receiver MPI processes must reside on the same host;
- The sender data must be globally accessible by the receiver.

By default, memory is allocated to a process on the node that the process is executing on. If a process moves from node to another during its lifetime, a higher percentage of memory references will go to remote nodes. Remote accesses often have higher access times. SGI use Dplace to bind a related set of processes to specific CPUs or nodes to prevent process migration.

In order to improve the performance, we enabled the single copy optimization and used Dplace for memory placement on the SGI Altix 3700.

3.3 Performance Analysis and Results

We validate our parallel algorithm by performing simulations on grids that were varied from 128^3 to 512^3.

We analyze the speedup $S(m,P)$ and the efficiency $E(m,P)$, which are defined by [1]

$$S(m,P) = \frac{T(m,1)}{T(m,P)}, \quad E(m,P) = \frac{T(m,1)}{PT(m,P)}, \tag{6}$$

where m is the total scale of the problem, and P is the number of processors. $T(m,1)$ and $T(m,P)$ represent the computing time for 1 and P processors, respectively. The number of processors P was varied from 1 to 32.

The distribution of processors along all three dimensions is illustrated in Table 1.

Table 1 Distribution of N processors along x-, y-, z-directions, $N = x \times y \times z$.

N	1	2	4	8	12	16	20	24	28	32
x	1	1	1	1	1	1	1	1	1	1
y	1	1	1	2	3	2	4	4	4	4
z	1	2	4	4	4	8	5	6	7	8

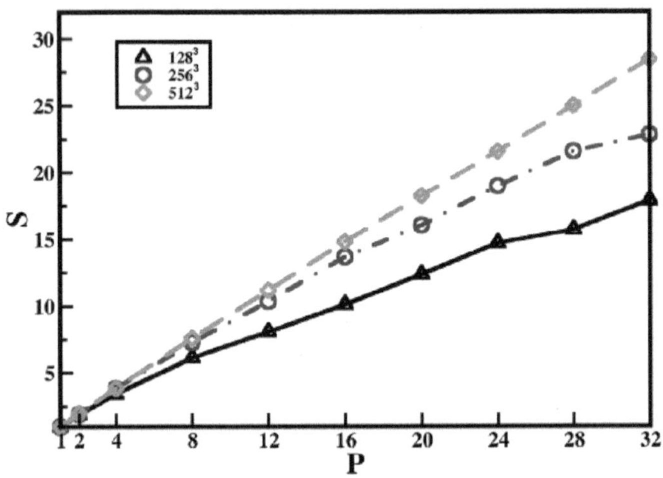

Fig. 5 The speedup $S = S(m, P)$ with single copy optimization.

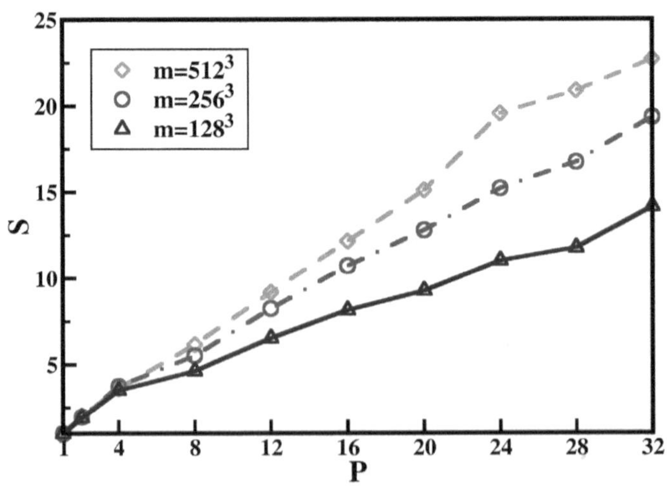

Fig. 6 The speedup $S = S(m, P)$ without single copy optimization.

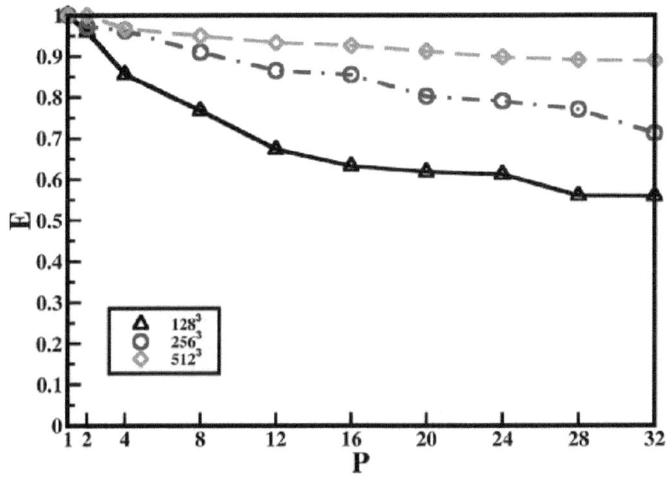

Fig. 7 The effiency $E = E(m,P)$ with single copy optimization.

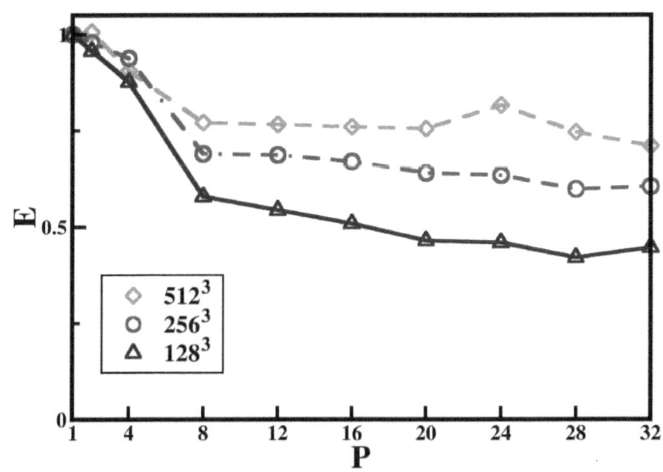

Fig. 8 The effiency $E = E(m,P)$ without single copy optimization.

Figures 5–8 show the algorithm speedup and efficiency with and without using single copy optimization. From these we can see that use of single copy optimization eliminates a sharp drop in efficiency in the region of 1 to 8 processors (compare Fig. 8 and Fig. 7). With 32 processors, only 71% efficiency is achieved for the largest grids scale. With single copy optimization, the efficiency is increased to 88.9%.

From Figs. 5 and 7 we can see that, for the largest system, the speedup varies from 1.0 to 28.4 with the number of processors increasing from 1 to 32, while the efficiency is decreasing from 100.0% to 88.9%. The algorithm maintains a higher efficiency with the number of processors increasing. Due to the block method (see

Fig. 3), the communications have less influence for the larger grids compared with the smaller ones. Therefore, the ratio of communication time to computation time is decreasing with the grids scale increasing. As it is seen from Figs. 5 and 7, the parallel code performance is better for larger grids.

4 Conclusions

We presented a parallel cell dynamics simulation algorithm in details. The code is suited to simulate a large box that is comparable with real experimental systems size. The parallelization is done using SGI MPT with Intel compiler. The program was tested with various grid scales from 128^3 to 512^3 with up to 32 processors for sphere, cylinder, and lamellar diblock polymer structures. Several performance tuning methods based on SGI Altix are introduced. The program with single copy method shows high performance and good scalability, reducing the simulation time for 512^3 grid from days to several hours.

Acknowledgments The work is supported by Accelrys Ltd. via EPSRC CASE research studentship. All simulations were performed on SGI Altix 3700 supercomputer at UCLan High Performance Computing Facilities.

References

1. Dongarra, J., Foster, I., Fox, G., Gropp, W., Kennedy, K., Torczon, L., White, A. (eds.): Sourcebook of Parallel Computing. Elsevier Science, San Francisco (2003)
2. Guo, X., Pinna, M., Zvelindovsky, A.V.: Parallel algorithm for cell dynamics simulation of block copolymers. Macromolecular Theory and Simulations **16**(9), 779–784 (2007)
3. Oono, Y., Puri, S.: Study of phase-separation dynamics by use of cell dynamical systems. I. Modeling. Physical Review A **38**(1), 434–453 (1988)
4. Pinna, M., Zvelindovsky, A.V.: Kinetic pathways of gyroid-to-cylinder transitions in diblock copolymers under external fields: cell dynamics simulation. Soft Matter **4**(2), 316–327 (2008)
5. Pinna, M., Zvelindovsky, A.V., Todd, S., Goldbeck-Wood, G.: Cubic phases of block copolymers under shear and electric fields by cell dynamics simulation. I. Spherical phase. The Journal of Chemical Physics **125**(15), 154905–(1–10) (2006)
6. Ren, S.R., Hamley, I.W.: Cell dynamics simulations of microphase separation in block copolymers. Macromolecules **34**(1), 116–126 (2001)
7. SGI: Message Passing Toolkit (MPT) User's Guide. URL http://docs.sgi.com
8. Snir, M., Gropp, W.: MPI: The Complete Reference. MIT Press, Cambridge, Massachusetts (1998)
9. Zvelindovsky, A.V. (ed.): Nanostructured Soft Matter: Experiment, Theory, Simulation and Perspectives. Springer, Dordrecht (2007)

Docking and Molecular Dynamics Simulation of Complexes of High and Low Reactive Substrates with Peroxidases

Žilvinas Dapkūnas and Juozas Kulys

Abstract The activity of enzymes depends on many factors, i.e., the free energy of reaction, substrate docking in the active center, proton tunneling, and other factors. In our study, we investigate docking of luminol (*LUM*) and 4-(1-imidazolyl) phenol (*IMP*), which show different reactivity in peroxidase-catalyzed reaction. As peroxidases, *Arthromyces ramosus* (*ARP*) and horseradish (*HRP*) were used. For this study, simulation of substrate docking in active site of enzymes was performed. Enzyme–substrate complexes structural stability was examinated using molecular dynamics simulations. The calculations revealed that *LUM* exhibits lower affinity to *HRP* compounds I and II (*HRP I/II*) than to *ARP* compounds I and II (*ARP I/II*). In the active center of *ARP I/II*, *LUM* forms hydrogen bonds with Fe=O. This hydrogen bond was not observed in *HRP I/II* active center. In contrast with *LUM*, *IMP* binds to both peroxidases efficiently and forms hydrogen bonds with Fe=O. Molecular dynamics studies revealed that enzyme complexes with *LUM* and *IMP* structurally are stable. Thus, arrangement diversities can determine the different substrates reactivity.

1 Introduction

The formation of enzyme–substrate complex is crucial to biocatalytic process. The substrate binding and chemical conversion proceeds in a relatively small area of the enzyme molecule known as the active center. This small pocket in the globule of enzyme contains amino acid residues responsible for the catalytic action and substrate specificity. Thus, understanding of biomolecular interactions of active site residues and substrate is the key in solving fundamental questions of biocatalysis.

Žilvinas Dapkūnas · Juozas Kulys
Vilnius Gediminas Technical University, Department of Chemistry and Bioengineering
Saulėtekio Avenue 11, LT-10223 Vilnius, Lithuania
e-mail: Zilvinas.Dapkunas@fm.vgtu.lt · Juozas.Kulys@fm.vgtu.lt

Fungal *Coprinus cinereus* peroxidase (*rCiP*) has enzymatic properties similar to those of horseradish (*HRP*) and *Arthromyces ramosus* (*ARP*) peroxidases. Additionally, *rCiP* crystal structure is highly similar to *ARP* [16]. It is well established that oxidation of substrates by *rCiP*, *ARP*, and *HRP* occurs with two one-electron transfer reactions through the enzyme intermediates, i.e., compound I and compound II formation [1]. Moreover, they share similar active site structure: essential *His* and *Arg* catalytic residues of the distal cavity and a proximal *His* bound to heme iron [11]. Experimental data show that distal *His* acts as an acid/base catalyst [20] and distal *Arg* participates in substrate oxidation and ligand binding [9].

Several studies suggest that proton transfer may be important for peroxidase catalyzed substrate oxidation. It was shown that the oxidation of ferulic acid by plant peroxidase is accompanied by proton transfer to active site *His* [12]. Recently, it was shown that slow proton transfer rate could be the main factor that determines low N-aryl hydroxamic acids and N-aryl-N-hydroxy urethanes reactivity in *rCiP* catalyzed process [14]. The calculations performed by *Derat* and *Shaik* also suggest that proton-coupled electron transfer is important for *HRP* catalyzed substrates oxidation [6].

In our study, we investigated luminol (*LUM*) (Fig. 1), which is known to exhibit different reactivity toward *rCiP* and *HRP*. Kinetic analysis shows that *LUM* reactivity is about 17 times less for *HRP* compared with *rCiP* [15]. In our study, we test whether arrangement in active site of enzyme could explain different *LUM* reactivity. Additionally, we investigated highly reactive *rCiP* and *HRP* substrate 4-(1-imidazolyl) phenol (*IMP*) (Fig. 1) and compared with *LUM* results. Tools we used for this study were robust automated docking method, which predicts the bound conformations of flexible substrate molecule to enzyme, and fast engine, which performs molecular dynamics simulations and energy minimization.

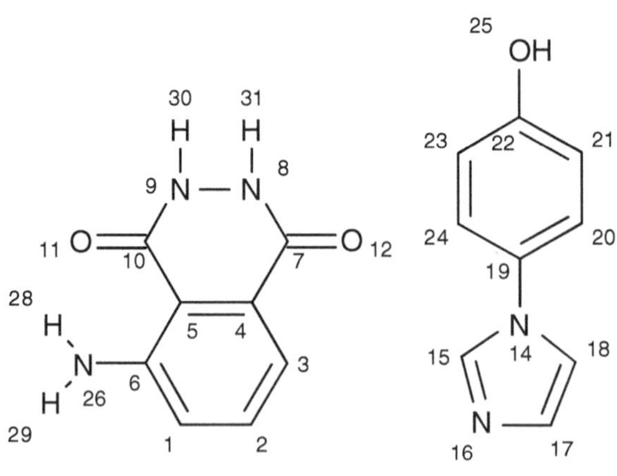

1 **2**

Fig. 1 Structures of investigated compounds: **1**, luminol; **2**, 4-(1-imidazolyl) phenol.

2 Experimental

2.1 Ab Initio Molecule Geometry Calculations

Ab initio calculations of electronics structures of substrates and partial atomic charges were performed using the Gaussian 98 W package [8]. The optimization of substrates geometry was performed using HF/3-21G basis set. Further, optimized geometries were used to calculate partial atomic charges according to HF/6-31 basis set. All calculations were carried out on single processor PC Pentium 4, 3GHz, 1GB RAM, and calculation time lasted up to 30 min.

2.2 Substrates Docking in Active Site of Enzyme

The simulations of substrates docking in active site of enzymes were performed with AutoDock 3.0 [18, 19, 10]. AutoDock uses a modified genetic algorithm, called Lamarkian algorithm (LGA), to search for the optimal conformation of a given substrate in relation to a target enzyme structure. LGA is a hybrid of evolutionary algorithm with local search method, which is based on Solis and Wets [22]. The fitness of the substrate conformation is determined by the total interaction energy with the enzyme. The total energy, or the free energy of the binding, is expressed as

$$\Delta G = \Delta G_{vdw} + \Delta G_{hbond} + \Delta G_{el} + \Delta G_{tor} + \Delta G_{sol}, \tag{1}$$

$$\Delta G_{vdw} = \sum (A_{ij} r_{ij}^{-12} - B_{ij} r_{ij}^{-6}), \tag{2}$$

$$\Delta G_{hbond} = \sum (E(t)(A_{ij} r_{ij}^{-12} - D_{ij} r_{ij}^{-10})), \tag{3}$$

$$\Delta G_{el} = \sum (q_i q_j / \varepsilon(r_{ij}) r_{ij}), \tag{4}$$

where the terms ΔG_{vdw}, ΔG_{hbond}, ΔG_{el}, ΔG_{tor}, ΔG_{sol} are for dispersion/repulsion, hydrogen bonding, electrostatic, global rotations/translations, and desolvation effects, respectively. Terms i and j denote atoms of ligand and protein, and coefficients A, B, D are Lennard–Jones (LJ) potentials. $E(t)$ is a directional weight for hydrogen bonding, which depends on the hydrogen bond angle t. q and $\varepsilon(r_{ij})$ are charge and dielectric constant, respectively.

For the docking simulation, the X-ray crystallographic structures of *Arthromyces ramosus* peroxidase (*ARP*) (PDB-ID: 1ARP) and horseradish peroxidase (*HRP*) (PDB-ID: 1HCH) were chosen. All water molecules in data files were removed, except the oxygen atom of one structural water molecule that was left in the active site. In order to model catalytically the active state of *ARP* of compound I and compound II (*ARP I/II*), the distance of Fe=O bond was set to 1.77 Å, i.e., the average Fe=O distance of compounds I and II (*HRP I/II*) of horseradish peroxidase [16].

The energy grid maps of atomics interaction were calculated with 0.15 grid spacing and 120 grid points forming 18 cubic box centered at the active site of peroxidase. The docking was accomplished using the Lamarkian genetics algorithm. The number of individuals in populations was set to 50. The maximum number of evaluations of fitness function was 2500000, maximum number of generations 27000, and number of performed runs 100. All docking simulations were performed on single node (3.2GHz, 1GB RAM) of "Vilkas" PC cluster (http://vilkas.vgtu.lt). Calculation time lasted up to 2 hours and result files were about 910KB in size.

2.3 Molecular Dynamics of Substrate–Enzyme Complexes

Modeled geometries of *LUM* (*IMP*) and peroxidase complexes containing the lowest substrates docking energies were supplied for molecular dynamics (MD) simulations. MD was performed with GROMACS 3.2.1 package [4, 17] using GROMOS-96 43a1 force field [23]. Parameters for *ARP I/II* and *HRP I/II* modeling were used as described in [24]. The topologies for *LUM* and *IMP* were generated with PRODRG2 server [21].

The substrate–enzyme complexes were dissolved with *SPC* (Simple Point Charge) type [3] water solvent. Total negative charge of modeled systems was neutralized with sodium cations. Thus, overall modeled systems contained up to 15000 atoms. The energy of such systems was minimized using the steepest descent method with no constraints. Minimized structures were supplied to 50 ps position-restrained dynamics, where lengths of all bond in the modeled systems was constrained with *LINCS* algorithm [13]. This algorithm resets chemical bonds to their correct lengths after an unconstrained update. Berendsen temperature and pressure coupling scheme was used [2]. The effect of this scheme is that a deviation of the system temperature and pressure from initial is slowly corrected.

During the position-restrained dynamics calculations, the temperature was 300 K and the pressure was 1 bar. To treat non-bonded electrostatics, Particle-Mesh Ewald scheme [5, 7] was used, which was proposed by *Tom Darden* to improve the performance of the Ewald summation. Twin range cut-off scheme was used for non-bonded Lennard–Jones (LJ) treatment. The long-range cut-off was set 1.0 and 1.0 nm for electrostatics and LJ. System atoms were supplied with velocities generated with Maxwellian distribution. The 0.5 ns duration molecular dynamics simulations with no constraints were performed on the structures obtained after position-restrained dynamics with the same options described above about position-restrained dynamics.

MD calculations were performed at VGTU cluster "Vilkas". Position restrained and with no constraints MD were performed on 2 nodes Intel Pentium 4 3.2GHz, 1GB RAM, Gigabit Ethernet. Number of processors was chosen according to time required to perform 10 ps duration MD. For single and two nodes, the time was about 120 min and 45 min, respectively. Thus, performing calculation in parallel we got results about 2 times faster. Using two processors, 50 ps position restrained MD lasted about 240 min, and 0.5 ns unrestrained MD lasted for 33 hours. After unrestrained MD simulations, obtained final result files were up to 700MB in size.

3 Results and Discussion

3.1 Substrate Docking Modeling

The docking modeling of *LUM* and *IMP* with *ARP I/II* and *HRP I/II* was performed in order to elucidate substrates arrangement in active site of peroxidases. During docking calculations, multiple conformational clusters adopted in active site of enzymes. Further, only numerous conformational clusters with lowest docking Gibbs free energy (ΔG) were analyzed (Table 1).

Results of calculations showed that *LUM* docked in the active center of *ARP I/II* and *HRP I/II* with different affinity. For *ARP I/II-LUM* complex, the docking free energy was about 4 $kJ\ mol^{-1}$ higher than for *HRP I/II* (Table 1). Furthermore, measured distances indicates possible hydrogen bond formation between *LUM* N-H and Fe=O. *LUM* docks near *His 56* residue, which can serve as proton acceptor. The calculated distances are 2.26 Å and 2.73 Å for H30 and H31, respectively (Fig. 2c). Calculations show also that *LUM* was located close to *Arg 52*.

Analysis of *LUM* and *HRP I/II* complexes suggests that *LUM* does not form hydrogen bond with Fe=O (Fig. 2d). The distances of H30 and H31 atoms from Fe=O group were 5.07 Å and 7.35 Å, respectively. In *LUM* and *HRP I/II* complexes,

Table 1 Substrates docking Gibbs free energies and distance measurement in active sites of *ARP I/II* and *HRP I/II*.

Enzyme	Substrate	Cluster No.	#Portion, %	Docking ΔG, kJ mol^{-1}	Distances from active site residues, Å Fe=O	*HIS*	*ARG*
ARP I/II	**LUM*	1	15	−30.5	H30 − 2.26 H31 − 1.75	H30 − 2.26 H31 − 2.73	H30 − 2.38 H31 − 4.22
		2	35	−30.3	H30 − 4.41 H31 − 2.27	H30 − 3.83 H31 − 2.69	H30 − 6.15 H31 − 5.10
	IMP	1	29	−31.4	1.73	3.15	4.63
		2	33	−31.2	1.73	3.12	4.65
HRP I/II	**LUM*	1	61	−26.0	H28 − 4.52 H29 − 5.97	H28 − 2.50 H29 − 3.12	H28 − 3.03 H29 − 4.69
		2	30	−25.2	H28 − 5.03 H29 − 4.81	H28 − 4.56 H29 − 4.51	H28 − 1.92 H29 − 2.61
	IMP	1	31	−30.3	1.82	3.02	2.30
		2	30	−30.0	1.84	3.02	2.31

Portion of conformations in cluster.
* Atom numbers are depicted in Fig. 1.

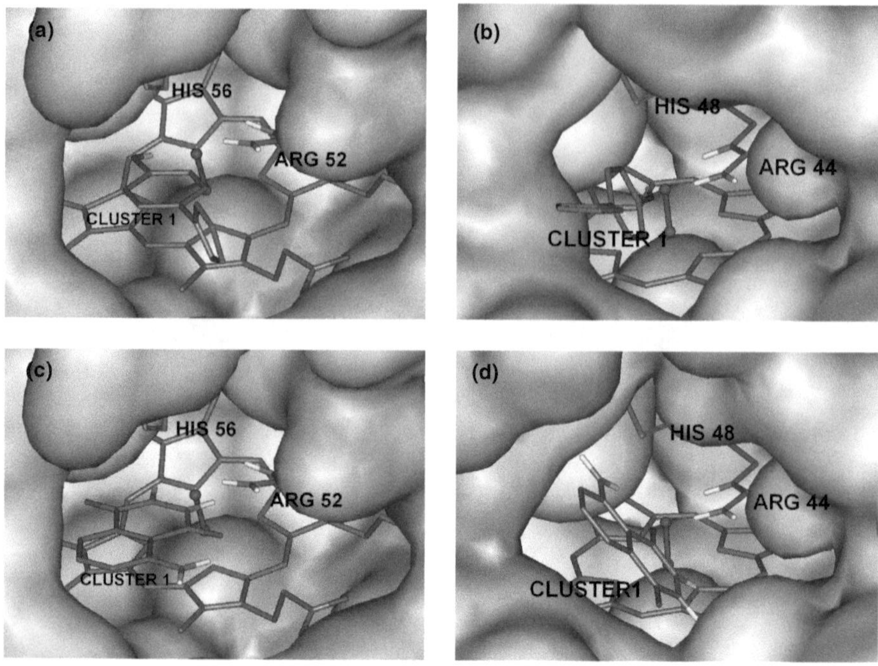

Fig. 2 Peroxidase–substrate modeled structures. *ARP I/II* with *IMP* (**a**), *LUM* (**c**), and *HRP I/II* with *IMP* (**b**) and *LUM* (**d**).

the -NH_2 group is buried in the active center of enzyme. Hydrogen atoms (H28 and H29) were oriented to *His 48* residue and distances were 2.50 Å and 3.12 Å, respectively. Distances from *Arg 44* were 3.03 Å and 4.69 Å for H28 and H29, respectively.

Analysis of *IMP* docking results showed that for both enzymes, docking ΔG were similar (Table 1). *IMP* arrangement in active site of enzymes showed that -OH group was placed deep in the active center. The hydrogen atom of *IMP* -OH group was at distance 1.73 Å from Fe=O in the active site of *ARP I/II*. This allows forming hydrogen bond with Fe=O. *IMP* was located near *His 56* residue at distance 3.15 Å (Fig. 2a).

In *IMP* and *HRP I/II* complex, hydrogen atom of –OH group was 1.82 Å away from Fe=O and 3.02 Å from *His 56* (Fig. 2b). Thus, *IMP* can form hydrogen bond with Fe=O in active site of *HRP I/II*.

3.2 Molecular Dynamics Simulation

In order to test stability of modeled complexes, the molecular dynamics simulations were performed during 0.5 ns. To estimate structural change of the complexes,

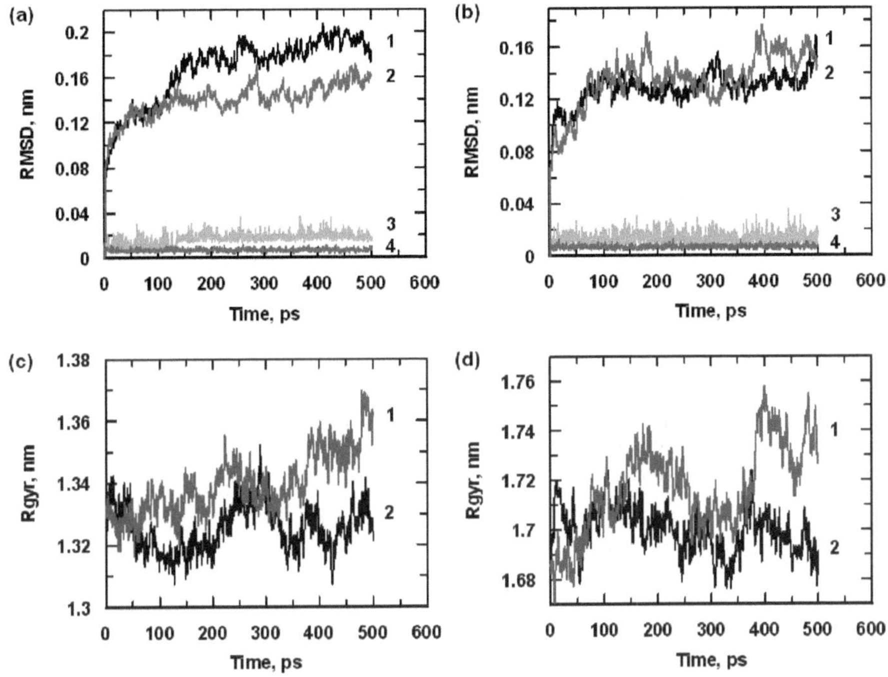

Fig. 3 Structural stability of peroxidase–substrates complexes during 0.5 ns MD. (**a**) Calculated RMSD of *ARP I/II* complexes with *IMP* and *LUM*: curves 1, 2 for enzyme in complexes with *IMP* and *LUM*, and 3, 4 for substrates *IMP* and *LUM*, respectively. (**b**) RMSD of *HRP I/II* complexes with *IMP* and *LUM*: curves 1, 2 for enzyme in complexes with *IMP* and *LUM*, and 3, 4 for *IMP* and *LUM* respectively. (**c**) Calculated Rgyr of enzyme: curves 1 and 2 for *ARP I/II–LUM* and *ARP I/II–IMP*, respectively. (**d**) Rgyr of enzyme: curves 1 and 2 for *HRP I/II–LUM* and *HRP I/II–IMP*, respectively.

RMSD and Rgyr was analyzed. The RSMD value gives information about protein overall structure stability. For ligands atoms, it gives information about position stability in the active center of enzyme. Parameter Rgyr indicates protein molecule compactness, i.e., how much the protein spreads out from calculated center of mass. This parameter increases as the protein unfolds and thus indicates the loss of native structure.

According to the RMSD parameter, both substrates complexes with *ARP I/II* and *HRP I/II* are structurally stable (Fig. 3a and 3b). During all simulation time, *IMP* and *LUM* stayed in the active centers. Rgyr dynamics shows that *ARP I/II* and *HRP I/II* structures in complexes with *LUM* and *IMP* remains unaltered during overall simulation (Fig. 3c and 3d). The MD simulations showed that the structures of *ARP I/II* and *HRP I/II* and *LUM* and *IMP* complexes were similar to those obtained by docking simulations (data not shown).

4 Conclusions

Modeled structures of *IMP* and *ARP I/II* (*HRP I/II*) demonstrate hydrogen bond formation between active site Fe=O group and *IMP*–OH group. Therefore, *IMP* forms in the active centers productive complex in which proton transfer from substrate is favorable. *LUM* exhibits lower affinity to *HRP I/II* than to *ARP I/II*. In *ARP I/II*, active site *LUM* forms hydrogen bonds with Fe=O although in *HRP I/II* such hydrogen bonds are not observed. This might influence different reactivity of *LUM* of these peroxidases. Molecular dynamics simulations show that complexes of *LUM* and *IMP* with peroxidases are stable.

Acknowledgments The research was supported by Lithuanian State Science and Studies Foundation, project BaltNano (project P-07004). The authors thank Dr. Arturas Ziemys for providing *ARP* compound I/II structure files and for help in computation methods and Dr. Vadimas Starikovicius for consulting and providing help in parallel computations.

References

1. Andersen, M.B., Hsuanyu, Y., Welinder, K.G., Schneider, P., Dunford, H.B.: Spectral and kinetic properties of oxidized intermediates of Coprinus cinereus peroxidase. Acta Chemica Scandinavica **45**(10), 1080–1086 (1991)
2. Berendsen, H.J.C., Postma, J.P.M., van Gunsteren, W.F., DiNola, A., Haak, J.R.: Molecular dynamics with coupling to an external bath. Journal of Chemical Physics **81**(8), 3684–3690 (1984)
3. Berendsen, H.J.C., Postma, J.P.M., Van Gunsteren, W.F., Hermans, J.: Interaction models for water in relation to protein hydration. In: B. Pullman (ed.) Intermolecular Forces, pp. 331–342. D Reidel Publishing Company, Dordrecht (1981)
4. Berendsen, H.J.C., Van der Spoel, D., Van Drunen, R.: GROMACS: A message-passing parallel molecular dynamics implementation. Computer Physics Communications **91**(1-3), 43–56 (1995)
5. Darden, T., York, D., Pedersen, L.: Particle mesh Ewald: An Nlog(N) method for Ewald sums in large systems. Journal of Chemical Physics **98**(12), 10,089–10,092 (1993)
6. Derat, E., Shaik, S.: Two-state reactivity, electromerism, tautomerism, and "surprise" isomers in the formation of Compound II of the enzyme horseradish peroxidase from the principal species, Compound I. Journal of the American Chemical Society **128**(25), 8185–8198 (2006)
7. Essmann, U., Perera, L., Berkowitz, M.L., Darden, T., Lee, H., Pedersen, L.G.: A smooth particle mesh ewald potential. Journal of Chemical Physics **103**(19), 8577–8593 (1995)
8. Frisch, M.J., Trucks, G.W., Schlegel, H.B., Scuseria, G.E., Robb, M.A., Cheeseman, J.R., Zakrzewski, V.G., Montgomery, J.A., Jr., Stratmann, R.E., Burant, J.C., Dapprich, S., Millam, J.M., Daniels, A.D., Kudin, K.N., Strain, M.C., Farkas, O., Tomasi, J., Barone, V., Cossi, M., Cammi, R., Mennucci, B., Pomelli, C., Adamo, C., Clifford, S., Ochterski, J., Petersson, G.A., Ayala, P.Y., Cui, Q., Morokuma, K., Salvador, P., Dannenberg, J.J., Malick, D.K., Rabuck, A.D., Raghavachari, K., Foresman, J.B., Cioslowski, J., Ortiz, J.V., Baboul, A.G., Stefanov, B.B., Liu, G., Liashenko, A., Piskorz, P., Komaromi, I., Gomperts, R., Martin, R.L., Fox, D.J., Keith, T., Al-Laham, M.A., Peng, C.Y., Nanayakkara, A., Challacombe, M., Gill, P.M.W., Johnson, B., Chen, W., Wong, M.W., Andres, J.L., Gonzalez, C., Head-Gordon, M., Replogle, E.S., Pople, J.A.: Gaussian 98. Gaussian, Inc., Pittsburgh PA (2001)
9. Gajhede, M.: Plant peroxidases: substrate complexes with mechanistic implications. Biochemical Society Transactions **29**(2), 91–99 (2001)

10. Goodsell, D.S., Olson, A.J.: Automated docking of substrates to proteins by simulated annealing. Proteins: Structure, Function, and Genetics **8**(3), 195–202 (1990)
11. Henriksen, A., Schuller, D.J., Meno, K., Welinder, K.G., Smith, A.T., Gajhede, M.: Structural interactions between horseradish peroxidase C and the substrate benzhydroxamic acid determined by x-ray crystallography. Biochemistry **37**(22), 8054–8060 (1998)
12. Henriksen, A., Smith, A.T., Gajhede, M.: The structures of the horseradish peroxidase c-ferulic acid complex and the ternary complex with cyanide suggest how peroxidases oxidize small phenolic substrates. Journal of Biological Chemistry **274**(49), 35,005–35,011 (1999)
13. Hess, B., Bekker, H., Berendsen, H.J.C., Fraaije, J.G.E.M.: LINCS: A linear constraint solver for molecular simulations. Journal of Computational Chemistry **18**(12), 1463–1472 (1997)
14. Kulys, J., Ziemys, A.: A role of proton transfer in peroxidase-catalyzed process elucidated by substrates docking calculations. BMC Structural Biology **1**, 3 (2001)
15. Kulys, Y.Y., Kazlauskaite, D., Vidziunaite, R.A., Razumas, V.I.: Principles of oxidation of luminol catalyzed by varous peroxidases. Biokhimiya **56**, 78–84 (1991)
16. Kunishima, N., Fukuyama, K., Matsubara, H., Hatanaka, H., Shibano, Y., J.T.Amachi: Crystal structure of the fungal peroxidase from Arthromyces ramosus at 1.9 Å resolution: structural comparisons with the lignin and cytochrome c peroxidases. Journal of Molecular Biology **235**(1), 331–344 (1994)
17. Lindahl, E., Hess, B., Van der Spoel, D.: GROMACS 3.0: A package for molecular simulation and trajectory analysis. J Mol Mod **7**, 306–317 (2001)
18. Morris, G.M., Goodsell, D.S., Halliday, R.S., Huey, R., Hart, W.E., Belew, R.K., Olson, A.J.: Automated docking using a Lamarckian genetic algorithm and an empirical binding free energy function. Journal of Computational Chemistry **19**(14), 1639–1662 (1998)
19. Morris, G.M., Goodsell, D.S., Huey, R., Olson, A.J.: Distributed automated docking of flexible ligands to proteins: Parallel applications of AutoDock 2.4. Journal of Computer-Aided Molecular Design **10**(4), 293–304 (1996)
20. Poulos, T.L., Kraut, J.: The stereochemistry of peroxidase catalysis. Journal of Biological Chemistry **255**(17), 8199–8205 (1980)
21. Schüttelkopf, W., van Aalten, D.M.F.: PRODRG: a tool for high-throughput crystallography of protein-ligand complexes. Acta Crystallographica Section D **60**(8), 1355–1363 (2004)
22. Solis, F.J., Wets, R.J.B.: Minimization by random search techniques. Mathematics of Operations Research **6**(1), 19–30 (1981)
23. Van Gunsteren, W.F., Billeter, S.R., Eising, A.A., Hunenberger, P.H., Kruger, P., Mark, A.E., Scott, W.R.P., Tironi, I.G.: Biomolecular Simulation: The GROMOS96 Manual and User Guide. Hochschulverlag AG an der ETH Zurich: Zurich, Switzerland (1996)
24. Ziemys, A., Kulys, J.: An experimental and theoretical study of Coprinus cinereus peroxidase-catalyzed biodegradation of isoelectronic to dioxin recalcitrants. Journal of Molecular Catalysis B: Enzymatic **44**(1), 20–26 (2006)

Index

A
annular jet, 224
antisymetrization operator, 215

B
beamforming, 159
Bio model, 171
black box optimization, 104
block copolymers, 253
block preconditioners, 39
box relaxation, 173

C
cell dynamics simulation, 254
Cholesky decomposition, 88
coefficients of fractional parentage, 213
computational aeroacoustics, 195
condition estimation, 11

D
data analysis, 161
direct numerical simulation, 224
domain decomposition, 48, 187, 227, 246

E
efficiency of parallelization, 72

F
fictitious region method, 183
filtering process, 182
finite difference scheme, 31
finite volume method, 184, 209
fractional time step algorithm, 183

G
geometric multigrid, 170
global optimization, 69, 70, 93, 103
grid computing, 51

H
Hartree-Fock algorithm, 59
HECToR system, 115, 118
HPC, 197
HPC services, 115
HPCx system, 115, 116

I
idempotent matrix eigenvectors, 217
interior point methods, 85, 149
invariants of the solution, 30

K
Krylov subspace methods, 39

L
Large Eddy Simulation, 194
lasers, 237
linear algebra, 89
Lipschitz optimization, 93
load balancing, 50, 52
local optimization, 104

M
mathematical model, 51, 196, 208, 224, 227
mathematical model of multisection laser, 240
mathematical programming, 146
matrix computations, 3
Minkowski distances, 72
mixed mode programming, 135
  funnelled version, 136
  master-only version, 136
  multiple version, 137
modelling languages, 147
molecular dynamics, 125
molecular dynamics simulation, 266, 268
multi-physics, 38

multicommodity network flow, 147
multicore computer, 188
multidimensional scaling, 72
multidimensional scaling with city-block distances, 75
multigrid method, 172
multisection lasers, 238
multisplitting method, 49
multithreading, 26

N
Navier-Stokes-Brinkmann system, 182
nonlinear finite difference scheme, 243
nonlinear optics, 30
nuclear shell model, 213

O
optimization of grillage-type foundations, 104
orthogonalization, 214

P
Padé approximation, 226
parallel access I/O, 127
parallel algorithm, 33, 185, 198, 211, 227, 229, 246
parallel algorithms for reduced matrix equations, 9
parallel branch and bound, 94
parallel eigensolvers, 58, 60
parallel explicit enumeration, 78
parallel genetic algorithm, 80
parallel multigrid, 176
parallel packages, 126
parallelization, 40
parallelization tools, 25
ParSol library, 27
ParSol tool, 248
periodic matrix equations, 19
portability of the algorithm, 126
predictor–corrector algorithm, 210
pressure correction method, 183
pulsar, 158

R
RECSY, 3
reduced matrix equations, 6
Runge-Kutta algorithm, 226

S
scalability analysis, 247
ScaLAPACK, 57
SCASY, 3
SKA, 157
spatial decomposition, 256
speedup, 72
stabilized difference scheme, 171
staggered grids, 242
substrate docking modeling, 265, 267
SuFiS tool, 182
support vector machine, 86
swirl, 227
Sylvester-type matrix equations, 3

T
template programming, 28
two phase, 223
two-phase flows, 223

V
vortical structure, 229

FSC
www.fsc.org
MIX
Papier | Fördert gute Waldnutzung
FSC® C083411

Kolibri 360 GmbH
Ferdinand-Jühlke-Straße 7
99095 Erfurt, Deutschland
produktsicherheit@kolibri360.de